The Islamic Design Module in Latin America

The Islamic Design Module in Latin America

Proportionality and the Techniques of Neo-Mudéjar *Architecture*

JOHN F. MOFFITT

McFarland & Company, Inc., Publishers
Jefferson, North Carolina, and London

Library of Congress Cataloguing-in-Publication Data

Moffitt, John F. (John Francis), 1940–
The Islamic design module in Latin American : proportionality and the techniques of *neo–Mudéjar* architecture / John F. Moffitt.
p. cm.
Includes bibliographical references and index.

ISBN 0-7864-1806-0 (softcover : 50# alkaline paper) ∞

1. Architecture — Latin America. 2. Architecture, Colonial — Latin America — Mudéjar influences. 3. Architecture — Composition, proportion, etc. 4. Building — Technique. I. Title.
NA702.2.M64 2004 720'.98'0903 — dc22 2004002928

British Library cataloguing data are available

Manufactured in the United States of America

McFarland & Company, Inc., Publishers
Box 611, Jefferson, North Carolina 28640
www.mcfarlandpub.com

In memoriam:
François Bucher and Santiago Sebastián

This work is dedicated to Mike Laurance,
with thanks for showing me how to do it,
and to Scott Ehlers, who helped me to find the essential schematum.
Además se dedica lo siguiente a dos detectives de la arquitectura española:
Pedro Lavado Paradinas y Juan Antonio Ramírez Domínguez

"Le bon Dieu est dans le détail."
— Gustave Flaubert

"Der liebe Gott steckt im Detail."
— Aby Warburg

Table of Contents

Introduction

Serving as an introductory exemplar, this investigation takes as its point of departure the façade of a Jesuit church in Tepotzotlán (Estado de México). (Fig. 1) Completed in 1762, this is treated here as a representative Hispanic "*fachada retablo*" (altarpiece façade) incorporating compositional principles that, although current throughout Latin America, were largely foreign to contemporary European architectural design. The larger issue raised here is the often remarked upon, distinctive nature of most *Spanish* buildings— and a number of representative examples of these characteristic structures are similarly examined here. Grounded in historical (and physical) data, and seeking to resolve an unanswered question —*how* did they actually do it?— what follows is a computer-assisted detective story, with this showing for the first time a simple workshop technique which consistently accounts for a distinctively Hispanic design procedure.

Generations of art historians have repeatedly commented upon the anomalous character of Spanish (and Hispanic) architecture. As they observe, when compared with contemporary European examples having similar functions, clearly the Spanish (or Latin-American) building looks "different." Nonetheless, to date nobody has come up with a systematic, meaning wholly practical, explanation for the Hispanic architectural phenomenon. The answer must be found in a broader context, a historical fact unique to Spain: it was the only European, trans-Atlantic colonial power to have once been occupied by *Islamic* overlords. In fact, that occupation endured for centuries, from 711 to 1492, at which point Spain took physical possession of what would become Latin America. Since the focus of this investigation is the "how" of the actual execution of those Hispanic buildings, one logically turns to an Islamic design model for an explanation of their visual deviations from coeval European architectural norms.

The specific technical issue treated here is *architectural proportionality* (which is explained in further detail in the Appendix, and in figs. 74–84). Typically, Hispanic façades are profusely decorated, at times to the point of *horror vacui*. Nonetheless, and has been noted by all students of architectural design, the overall effect is one of an aesthetically pleasing, proportionate distribution of all the parts. Unfortunately, to date no scholar has attempted to explain the actual workings of a simple mechanical process consistently generating all the intricately repeated and

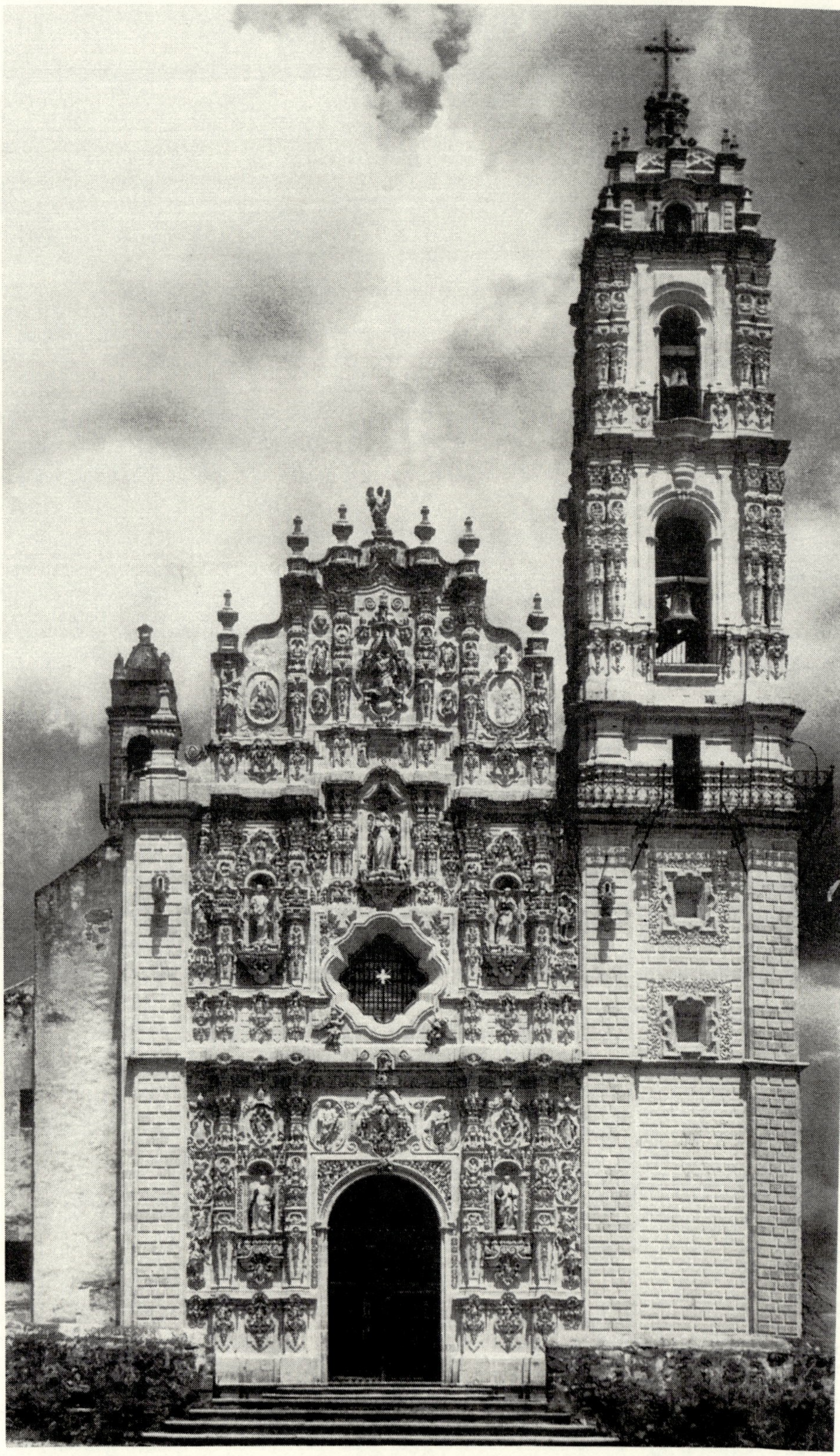

proportionately distributed design elements so distinctive of Hispanic architectural creativity. Our task is to recreate the routine operations of once-common (but now forgotten) instrumental practices that had dictated a certain kind of design procedure. That clearly long-standing, native tradition of mundane workshop operations proves to be the factor that decisively distinguishes Hispanic architecture from its contemporary European counterparts.

The current Spanish term for this traditional Hispanic type of patterned architectural design is *neo-mudéjar*. Hence, the design techniques recreated throughout Spanish America can be specifically called "neo–Islamic" methods, for that means the same thing as *neo-mudéjar*. First pioneered by Islamic artisans in Andalusia, these now-forgotten *mudéjar* workshop techniques were still being employed centuries later in Spanish America by journeyman designers and simple artisans. As is also shown here, even the canonical format of the Islamic mosque was consciously employed for the first open-air chapel in New Spain, San José de los Naturales in Mexico City, which was itself taken as a structural model and so it was replicated at other sites.

Close formal analyses of the façades of various Latin American churches (beginning with the example at Tepotzotlán) reveal them to reiterate certain formats and procedures worked out by in Andalusia by Islamic builders in the ninth and tenth centuries, particularly for portals to be applied to the Great Mosque at Córdoba. Common to both examples, although they are widely separated both in time (tenth versus eighteenth century) and place (Andalusia versus America), is a unifying, modular and proportionate, system determining abstract ratios. As found in many Latin American churches, the imposed schematic — only expressed in ratios, with no numbers being initially applied — has been called the "*trazado regulador*," or design by pre-determined regulating-lines and template-modules. Whether applied in Islamic Andalusia or in Spanish America, detailed analysis now reveals that the regulating format most commonly applied derived from manipulations of the 3:4:5 geometrical figure called the "Pythagorean triangle." The function of such inscribed geometric figures, including their specific on-site design applications, is illustrated here with 84 visual examples (including measured drawings of Egyptian, Greek, Roman and medieval European prototypes).

This study takes a new approach to a perennial problem of architectural design. With the aid of a computer and the appropriate graphic software, these "inscribed" (or invisible) *trazados reguladores* are now applied to a number of Spanish and Latin-American buildings (both their elevations and their plans). So doing, one reveals the actual operations of "inscribed" intersecting networks generating a recognizably unique kind of Hispanic architectural design adhering to a simple, modular-proportional system. Originating in Islamic Spain long before, along with the architectural grid-plan generating it, the triangulating system was based on certain, wholly pragmatic and long-established, anonymous artisan techniques. Transferred to the New World, where the demand for churches was unquestionably pressing, the result

***Opposite:* Fig. 1. Iglesia de San Francisco Javier, Tepotzotlán, Mexico. Façade erected 1760–64, and probably designed by Lorenzo Rodríguez.**

was a large body of systemic and consistently harmonious, "architecture without architects."

Completely lacking in "architectural theory," the traditional process of transferring a simple but proportionate line-drawing on to a rising wall—"design by string"—is reconstructed in the way it probably had been seen put into practice in eighteenth-century Mexico. Vitruvius' well-known architectural treatise (ca. 25 BCE) is also re-examined in the specific light of these traditional artisan procedures. Although scholars today heavily stress the "theoretical" aspects in Vitruvius, the more logical conclusion drawn here is that the Roman's statements regarding proportionate design were usually later read by builders in a much more practical, or "hands on," way. Four practical builder's manuals, one from Seville (ca. 1619), with another published in Madrid (1639), and two others composed in Mexico (ca. 1640 and ca. 1800), provide verbal confirmation for what was, initially, a wholly graphic hypothesis.

Such is the evidence, both visual and verbal, attesting to a hitherto unsuspected Islamic physical presence to be encountered throughout Latin America. Although the graphic argument has been largely developed by wielding a post-modernist artifact, a computer, it is appropriate that the written argument interpreting the visual phenomena reverts to a mode of discourse appropriate to the period when these design schemes evolved. According to the medieval *ars dictaminis* (or art of rhetorical discourse), the implicit (or "inscribed") structure of a judicial dissertation should be divided into five parts. First comes the introduction, an *exordium* or *proemium*. Next follows the main body of the argument. This comprises the second element, the *narratio*, that is, a narrative exposition of the facts in the matter. The third essential component is the *argumentatio* or *probatio*, that is, a recitation of the physical evidence. This is complemented by a fourth element, the *refutatio*, a running refutation of opposing opinions, and as demonstrated by the various *evidentia* sequentially presented in the *argumentatio* or *probatio*. The final, fifth, component is the closing argument, the *peroratio* or *epilogus*, providing resolution to the foregoing: *Quod est demonstrandum*. The same procedures also apply to the conventionalized format structuring the modern detective story.

All the graphic material generated by me to be presented here—with that serving as the essential "documentation" for a forgotten design procedure (here described in full for the first time)—was made possible through consultation with (and considerable mentoring from) a highly skilled, graphic designer-consultant, Mike Laurance (to whom professional inquiries may be addressed at: www.zianet.com/laurance). All funding for this nearly two-decade-long investigation—including the expenses of much foreign travel, as well as purchase of computer hardware and the most advanced graphic software—was generously provided by the H.M.S. Phake-Potter Literary Foundation (http://www.xlibris.com/HMSPhakePotter.html). I am also most appreciative of written comments criticizing a much earlier draft from George Kubler (Yale University) and, especially, my conversations with the late François Bucher (Florida State University); François was largely responsible for directing me to practical (versus theoretical) explanations of the phenomena examined here. I also recall with great pleasure many lively discussions about Latin-American architecture with my late friend (and co-author), Dr. Santiago Sebastián López (Universidad Literaria de Valencia).

I really must acknowledge the great stimulus initially given to the evolving "geometric" character of this investigation provided by my former graduate student, Mr. Scott Ehlers (whose specific, and truly essential, contributions are described in my chapter 4, note 3). A resounding "*Gracías, amigo*!" goes to Dr. Eugene Cunnar, my colleague in the English Department at New Mexico State University; Gene was so kind as to read my manuscript from the perspective of a lay person (i.e., not an art historian), so assuring me that one was able to make sense of the formal discussions. I also wish to acknowledge the help of Janine J. Henri (University Library, U-T Austin) in providing me with invaluable bibliographical data, and Mardith Schuetz-Miller kindly shared with me some specialized information contained in an as-yet unpublished book manuscript. I am also most appreciative of comments made on my manuscript by Professor Juan Antonio Ramírez (Universidad Autónama de Madrid), whose many monographs on Hispanic art and architecture have inspired in me (to be quite honest) great envy. Special thanks go to Dr. Pedro Lavado Paradinas (Madrid, Ministerio de Cultura), an acknowledged expert in *mudéjar* design, for sending photocopies of essential documentation allowing me to coordinate my graphic thesis with the extant historical records. ¡*Vale*!

JFM
Las Cruces, New Mexico
April 2004

1

Hispanic versus European Architectural Principles

To be discussed here are some acknowledged (also some others not yet recognized) persistent peculiarities of architectural design in Spanish America.[1] These so-called "*invariantes castizos*" (native constants) span a wide range of historical periods, those which might be called in European terms Mannerism, Baroque and Rococo.[2] The particular question addressed here is: Exactly how were these stylistically anomalous, Hispanic buildings actually designed? More specifically, what were the actual workshop procedures, and how did these routine craft operations directly influence the initial conception of a given architectural design? In effect, I wish to recover the "secret" of the Hispanic builders, specifically the employment by simple artisans of a standard set of proportionate figures, with these being used to transfer small preliminary sketches to "true size" façades.[3]

Therefore, this investigation is framed by a larger issue, one central to the entire range of art history, *proportionality*.[4] However, equally important is a medieval precedent set in Spain itself, and that one is—rather than Christian—instead *Islamic*. In his opening sentence, Bernard Bevan succinctly stated the case in his standard *History of Spanish Architecture* (1939): "Spanish architecture is the product of two civilizations, the Christian and the Islamic, two mentalities that were diametrically opposed," but historically melded.[5]

Sinclair Gauldie, a practicing architect in Scotland, provides a general context for the technical issue examined in this investigation:

> Architects have always been susceptible—for purely aesthetic reasons—to the fascination of mathematical ratios. This interest is reinforced by the fact that proportional systems have practical as well as aesthetic uses. They may, for instance, help to systematize the whole building operation, and indeed they were very often necessary in the past for the setting out of large and complicated buildings. Before calibrated tape-lines, foot-rules, and theodolites were invented, dimensions and angles had to be set out by some such device as knotting a cord [and a sketch shows a knotted cord formed into a triangle with the ratios of 3:4:5], and in these circumstances there was great practical convenience in being able to set up all the major sizes of the building as fixed proportions of one part of it.... It will be a good deal easier to do this if the wall (and indeed the whole building) has already

> been designed on a grid with proportions which relate naturally to each other. Consequently, a special fascination attaches from the designer's point of view to relationships which yield the greatest number of similar shapes similarly related.[6]

To get at the problem posed here, specifically the nature of forgotten workshop techniques accounting for the distinctive formal characteristics of Hispanic architecture, one should begin with the physical analysis of a concrete example. Placed under the patronage of Saint Francis Xavier, the town of Tepotzotlán (Estado de México) was originally founded around 1585 by Jesuit missionaries as a school for Indian children. Construction of the present Church of San Francisco Javier only began eighty-five years later, in May 1670. Taking twelve years to complete, the edifice was officially dedicated in September 1682. Of all this Jesuit complex, only the decorative church-façade — designed however much later by an unknown architect, and erected between 1760 and 1762 — will concern us now. (Fig. 1) For as much formal as well as contextual reasons, it is believed that the otherwise unknown designer most likely was of Andalusian origins, a native-born Spaniard.[7] However that may be, the Tepotzotlán portal-façade will be discussed here as a prototypical expression of those "enduring ethnic traits" (*invariantes castizos*) characterizing Hispanic architectural design. Treated as a formal exemplar, it also conveniently serves as the epitome of many such design schemes with intricate ornamental networks belonging to ecclesiastical structures erected throughout Latin America.

The Hispanic portal-façade of the New World is strikingly different from its "Baroque" counterparts in Catholic Europe, and the strictly formal characteristics of Baroque architecture in Europe have been well defined since 1915, with the publication of Heinrich Wölfflin's *Kunstgeschichliche Grundbegriffe.*[8] Wölfflin based his definition of "European Baroque" mainly upon Italian monuments and those that closely imitated them. Spanish architecture of the late sixteenth and early seventeenth centuries *is* different from the Italian model, but so is English, French, or Czech architecture of the same period. Nonetheless, the Hispanic monuments are peculiar in their own way. This is only natural: the cultural influences belonging to Spanish history are different from the conditions affecting the development of English, French, or Czech architecture.

In representative European counterparts (or non–Hispanic examples) hewing to the Italian models— kindred in time, but quite distinct in formal conception — the distinguishing architectural characteristic is a massive plasticity suggesting "weightiness." This metaphorical factor typically is expressed visually by sturdy, often free-standing, columns conforming to the canonic classical orders. These supporting members clearly indicate the vertical, continuous distribution of weights, descending from the top of the structure down to its very substantial base. In non–Hispanic portal-façades, typically there is additionally a clear sense of an architectural focus and of an accelerating visual climax. Complementing the dramatic, top-to-bottom, figurative weight-distribution, the design elements also initiate a regularized, lateral-inwards, movement. Commencing at the ends of the façade, the viewer's eye movement comes to a logical, emphatic and orchestrated, perspectival conclusion at eye-level in the lower center of the building. Such scenographic effects unilaterally direct the interest of the passerby into the welcoming portal of the church.

Quite to the contrary, those Late Baroque design factors common to mainstream European ecclesiastical architecture are largely absent in representative examples of the divergent compositional principles expressed in any number of similarly functioning and nearly contemporary Hispanic structures. Well over fifty years ago, Fernando Chueca Goitia had explained the formal characteristics of a certain architectural corpus he called "Trans-Hispanic" (*transhispánico*).[9] This term represents a conglomerate of older styles of strictly Spanish peninsular origins transported to, and then "re-duplified" in the New World after 1525. In particular, the Trans-Hispanic portal-façade pertains to a phenomenon nearly unique to Spanish architectural history. This is what Chueca labels the "*fachada-retablo*" (altarpiece-façade), a device initially developed in Spain during the medieval periods, and later called an "*imafronte*" in Mexico.

It frequently happens that the external portal-motif will reiterate the internal motif of a huge, floor-to-ceiling, towering wall-screen, the high altar or *retablo mayor*, rising majestically above and behind the altar-table.[10] Chueca had defined the Spanish retablo as "a polyptych described by architectonic elements." The same definition applies to a typical Hispanic *fachada-retablo*. Initially, he says, "its squared compartments served to enclose different images, either painted or sculpted, and these are separated by thin moldings and juxtaposed in files and in stories." Subsequently, as he explains, "this tendency reached its climax in America, where one is not dealing with altarpieces as such, but instead with a series of columns or pilasters, where each element is placed so closely together as to disallow space between them for statuary." At this stage, "the *retablo* becomes an abstracted decoration, based upon monotonously reiterated architectural sub-units in which all representational significance has been lost."

Chueca has also made a very important historical observation, namely that "the system has its clearcut antecedents [specifically] in those façades belonging to Hispano-Islamic architecture which had been typically treated as petrified tapestries [*tapices pétreos*]."[11] In short, regardless of the motifs actually employed, the fundamental *design principles* characterizing Hispanic architecture are, rather than "European," "Arabic." And, as I would argue, said Hispanic *design principles* arose for the most part from what may be called Islamic *construction techniques*, the exact nature of which has been, unfortunately, largely forgotten. It is my task to try to reconstruct these simple workshop techniques, the kind directly provoking a distinctive kind of architectural design.

The historical, Christian medieval (or post–Islamic) precedents in Spain may be quickly reviewed. An apposite example of an earlier, architecturally-scaled *retablo mayor* (high altar) was designed around 1496 by Diego de Siloé for the Cartuja de Miraflores, near Burgos. (Fig. 2) Although filled with three-dimensional components—a crucifixion group, a monstrance in a niche, donor figures, and so forth—the overall effect is one of "planarity," as though the figurative elements had been pressed between two plexiglass sheets. Another effect is that of *horror vacui*; all the profusely packed elements—either figurative or architectural—are similar in size

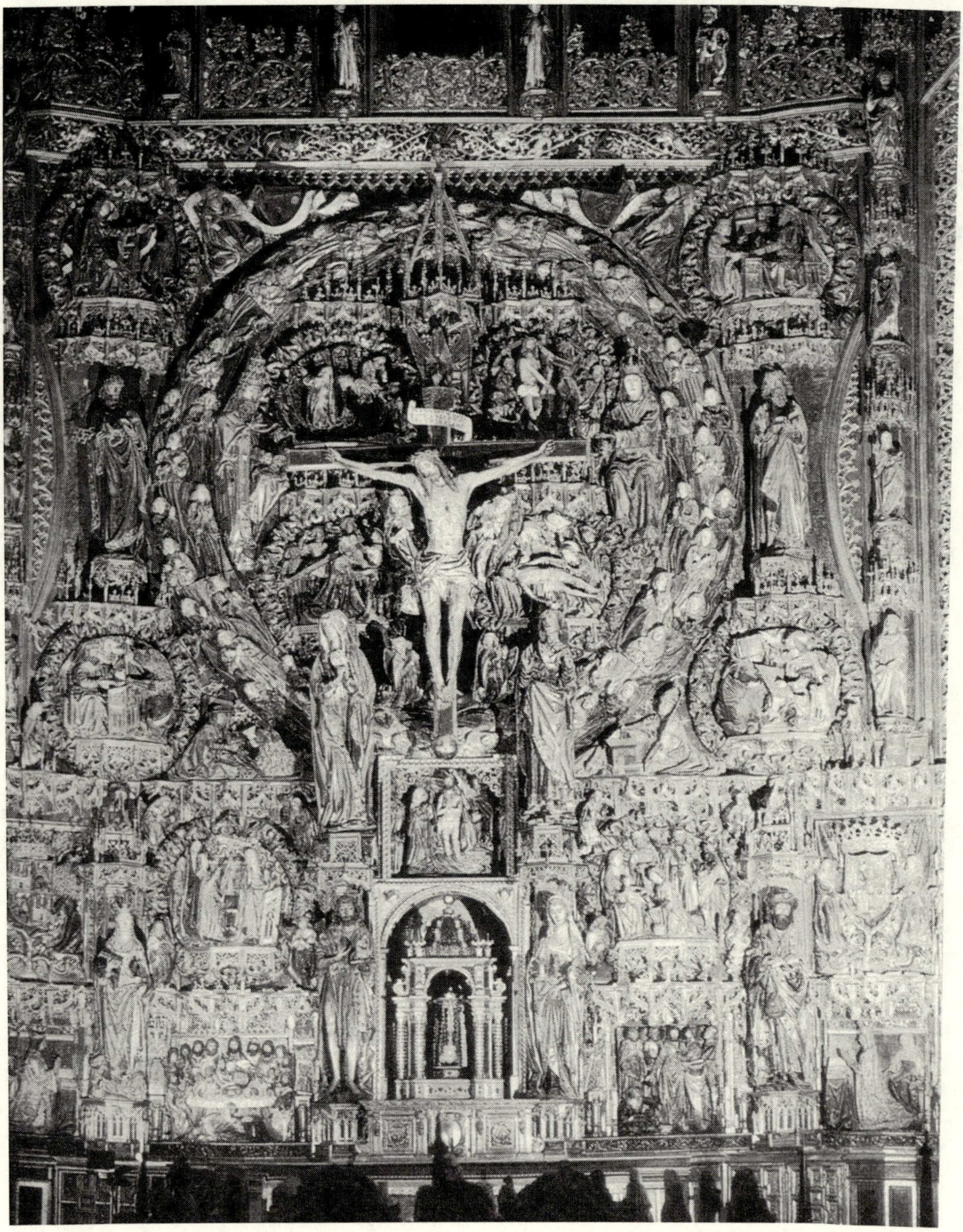

Fig. 2. Gil de Siloé, High Altar (***retablo mayor***) of the Carthusian Monastery of Miraflores, Burgos, ca. 1496–9.

and squeezed into proportionately spaced compartments. The overall organizing system seems "geometric," one employing glittering circles, squares, rectangles, and triangles. Another characteristic example of the Spanish external *fachada-retablo* scheme, prefiguring the altar-wall situated within, is represented by a portal-façade designed before 1525 for the church of San Esteban in Salamanca.[12] These examples (and many others could, of course, be cited) are exemplary of a transitionary, Late Medieval to High Renaissance, architectural style in Spain. Now commonly called "Plateresque" (like chased silver), this kind of architectural expression had commonly (but generally only superficially) employed classical motifs, "*a lo romano*."[13]

Another, often cited, exemplar of Spanish Plateresque design principles is the façade of the University of Salamanca, completed in 1529. (Fig. 3) Here we are dealing with a secular building embellished with miniaturized classical motifs and propelled by a highly complicated and erudite humanistic iconographic program.[14] Besides the (often tentative) introduction of various (largely decoratively applied) "classical" motifs—and so designated: "*a lo romano*," the predominant characteristic of Plateresque design is a richly embellished ("petrified tapestry") approach to certain selected, rigidly enframed, zones. The applied decoration remains just that, "*decoración*." As seen at Salamanca and elsewhere in Spain, a rectangular field of thickly textured, sculptural veneer adheres to a massively flattened, and otherwise quite bare, masonry wall. The final result is "pseudo-architectural": a scheme that is at once additive, prismatic, cellular and laminated. Characterized by a persistent sense of *horror vacui*, this is ornamentation that denies outright both structural mass and unilateral climax.

As also described by Chueca, the impression overall is decidedly "anti-classical." He explains that, in Spanish architecture during the Renaissance, "either the wall is left audaciously bare or it is profusely decorated, either one or the other, with peremptory contrastive results." The functional consequence was that "Plateresque entablatures turned into mere decorative bands due to the inflation of the frieze and an improbable atrophy of architraves and cornices," even though "these are, nevertheless, the very elements which, at least according to classical logic, should have been treated as the most active and emphasized elements." The immediate result is that, "columns are scarcely utilized; replaced instead by jointed socles, they become little more than vertical friezes." As it appears to Chueca, the underlying "intention stems from an attempt to define narrow pathways, and to create boxy, vertical and horizontal, compartments, within which the caprices of applied grotesques are played out." The larger end result is that "the underlying logic behind any employment of the classical architectural elements is subverted."[15]

By whatever term we choose to call them, the planarized and weightless—hence, "pseudo-architectural"—designs of Plateresque façades do anticipate, by some two and a half centuries, many subsequent formal variations in the New World. Including Tepotzotlán (fig. 1), those later reiterations of Plateresque design principles may be conveniently called "Neo-Plateresque."[16] The metaphorical sense of "weightlessness," common to both Plateresque and Neo-Platesque structures, is readily apparent at Tepotzotlán, and the actual mechanics of the anti-gravitational effect are easily identified. As is readily ascertained, none of the emphatically stressed and heavily

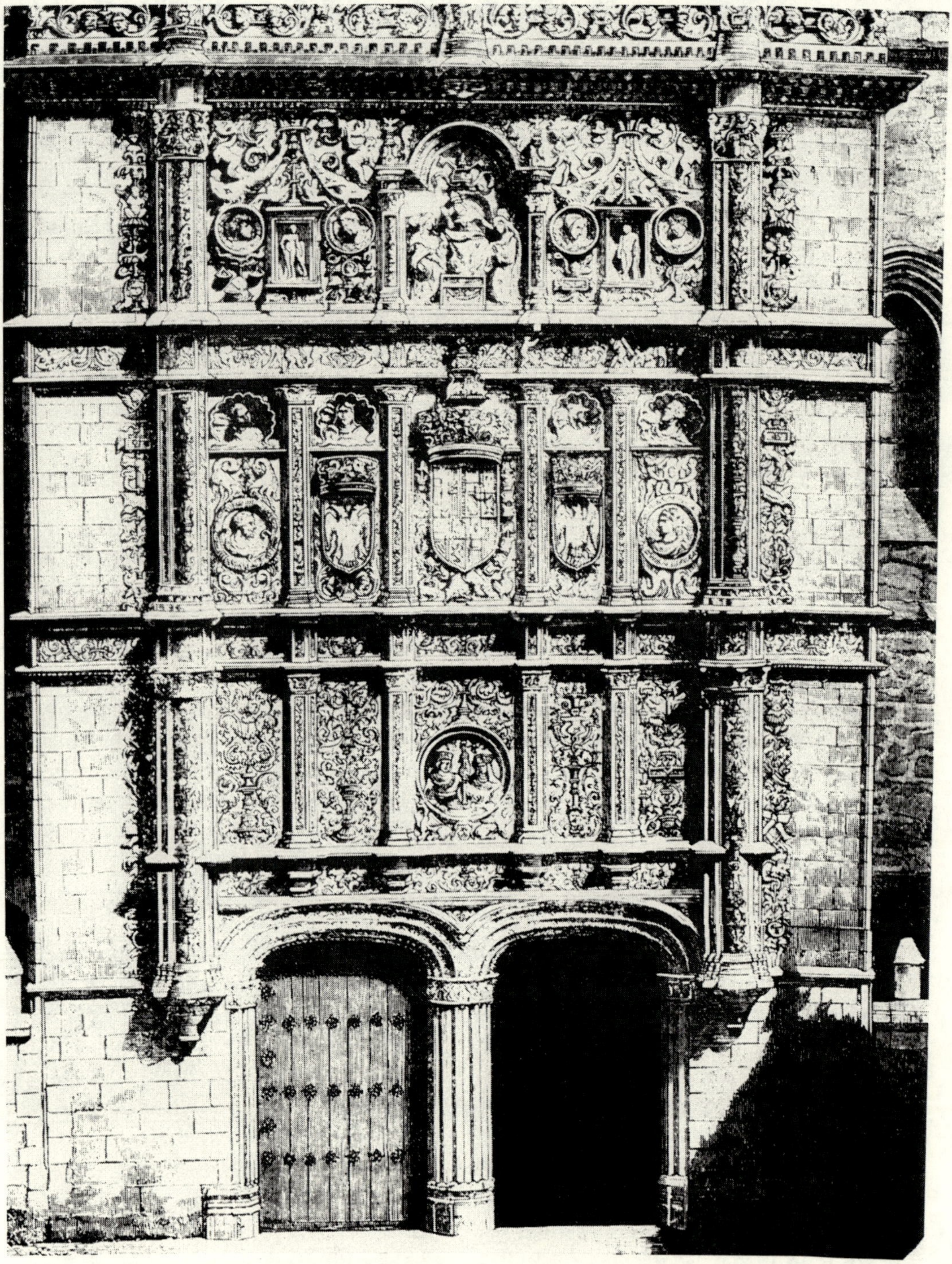

Fig. 3. Unknown Plateresque designer, the portal-façade of the University of Salamanca, Spain. Designed ca. 1525. (Anonymous engraving, ca. 1865.)

overlaid, vertical members ever actually reaches the ground. Instead, appearing to hover as "dissolved socles" a few feet above ground level, they merely seem to recede into the wall. The perceived result is that the elaborately and repetitiously carved, "petrified-tapestry" façades have neither base nor pedestal. Carpet-like, seemingly floating or "suspended," their putative means of support is to "hang" from the top of a building; the consistent result is what Chueca calls architectural "*decoración suspendida.*"

Seemingly perversely (at least according to a classicistic rationale), it is to the most elevated points in the building that the greatest surface complexity and visual emphasis typically adheres. The perceived result is an architecture apparently turned upside down; this kind of architectural design figuratively overturns the naturalistic-gravitational orientations of classical norms, the kind encountered in mainstream European architecture since the Renaissance. That notwithstanding (and somewhat paradoxically), classical principles are figuratively alluded to on Hispanic façades by a conspicuous display of classicistic motifs of the "*opus romanorum*" type. Nonetheless, in the strictly visual (even functional) sense, this is complete *anti*-classicism. Curiously, this "anti-classicism" appears to have been founded (at least in part) upon the indispensable textual source for all subsequent classical functions and principles in European architecture, Vitruvius.

Joseph Armstrong Baird has usefully discussed the unique stylistic features of Mexican architecture distinguishing it from its European counterparts during the Baroque period. He observes that Mexican architecture definitely "was not concentrated on the dramatic plane and space effects of Baroque Europe." To the contrary, its "special character was founded on an aesthetic preoccupation with ornamental and ornamental-sculptural, rather than architectural, developments, and on a general rejection of the over-all planning and spatial conception of the Baroque period in western Europe." Even more to the point, essentially,

> the architects were ornamental designers. And it is here that the notable character of Mexican architecture can be seen. The strong preference in the viceregal period was for bilateral symmetry of design, with an enrichment of certain parts.... What one might call the sense of organized pattern in viceregal Mexico was reinforced by Islamic background in Spain.... The Mudéjar, or Islamic, influence was [revived] in terms of geometric, non-figurative ornament of rigidly controlled patterns.... [Consequently,] there was generally more interest in an animated surface — what one might call the "ornamented wall" — than in the structural problems of architecture proper.... Here again, although individual features had a climactic organization, there was a general disinterest in the [European] Baroque in terms of complex space and structural forms. The involvements of seventeenth-century Rome [or] the melting curvilinear rhythms of eighteenth-century Bavaria and Germany, were almost entirely lacking in Mexico.[17]

Viewed in the global sense, the explanation for the heterodox characteristics of Hispanic architecture is historical, hence logical, even simple. It is the "Mudéjar, or Islamic, influence," and such as that is consistently expressed in "geometric, non-figurative ornament of rigidly controlled patterns." In short, even though borders were permeable to cultural and artistic exchange, and Islamic influence was present throughout the Mediterranean basin from the seventh century on, the only country in western Europe (with the exception of Sicily, itself a Spanish possession since 1282) that had been continually occupied by Moslem overlords was Spain. In fact, Islamic dominion over the Spaniards endured from AD 711 until 1492, when the Kingdom of Granada was finally taken over by the Christian *Reconquista*. The effects of nearly eight centuries of Islamic dominition over most of the Iberian peninsula left their indelible imprint upon Spanish architecture, also upon the Hispanic architecture belonging to its New World colonies.

2

The Grid and Its Role in Islamic Design-by-Geometry

Whatever its putative Greco-Roman theoretical basis in texts printed during the Renaissance (and from which I shall quote), as a strictly practical exercise, Spanish colonial architectural design generally betrays a consistently applied organizational principle that had its historical roots in the Iberian peninsula. The practical basis—shown here to derive from a specific workshop technique—produced a kind of design format looking decidedly "foreign" to the rest of Baroque architecture in Catholic Europe.

Its apparently alien character derives from the fact that the distinctively Hispanic compositional pattern had been specifically derived from the rich architectural heritage of Islamic *Al-Andalus* (meaning "Land of the Vandals," after a Germanic tribe once occupying the area), and Andalusia was later to become the birthplace of the majority of the doughty *conquistadores* of Central and South America. Christians residing in Islamic Spain were usually allowed to follow their own faith, even to build churches. The architectural result was a hybrid known nowhere else in medieval Europe, "*mudéjar*" art, a unique fusion of Islamic design techniques adapted to Christian liturgical and domestic needs.[1] Such was the product of skilled Muslim artisans later allowed to remain in the reconquered Christian territories, and the name of these *mudéjares* comes from the Arabic, *mudayan*, "one permitted to remain."[2] Treated as an architectural phenomenon, the distinctive *mudéjar* style was the longest lasting in Europe; the period of its florescence is contemporaneous with the other, coeval European styles now called Romanesque, Gothic and Renaissance. Much later, for it endures in Latin America well into the eighteenth century, the artistic phenomenon resulting from the trans-Atlantic migrations of Spanish conquistadors to the New World is now generally known as "*arte mudéjar en América.*"[3] In effect, even Christian Spaniards had been decisively "arabized" by their long cohabitation with the adherents of the Islamic faith. So, and not surprisingly, they brought their Moorish proclivities to America.

Implicitly following the art-historical, formalist-analytical model of Heinrich Wölfflin, in 1948 Oskar Hagen published a book in which he usefully sought to define the enduring "Elements of Spanish Art," meaning those which distinguish it visually from other, contemporary European arts.[4] In the strictly formal sense, he called attention to persistent compositional traits of "a simple frontal view," "the inflexible angular outline," "a solemn symmetrical composition," even "excessive symmetry," with this trait being manifested in "a special manner of ornamentation, a profuse interlace, controlled by rigid borders of geometrical shape." In fine, "whatever the medium, [the design] extends as an all-over pattern upon every available plane."

This *Grundbegriffe* was dubbed by Hagen the overriding "Principle of Planarity in Spanish Art." Hagen was convinced that "*el planismo*" represents something like a primordial value in Spanish art, and he even found it at work in the *Dama de Elche*, a bust supposedly executed by an Iberian sculptor around 450 BC.[5] Logically however, one should instead date emergence of this essential "Principle of Planarity in Spanish Art" only *after* AD 711, meaning during the extended period of Moslem occupation.

What we may instead call the enduring "Elements of Islamic Art," meaning those which distinguish it visually from the European arts—except for *Spanish* art and architecture—are well known. As summed up by Dalu Jones, the specific formal traits are:

> increasingly intricate surface decoration by use of reflecting and shining materials and glazes, the repetition of designs, the deliberate contrasting of textures and the manipulation of planes. However, despite this tendency towards effects of decorative richness and intricacy, there is always an ultimate sense of sobriety.... Islamic art is an art of repose, intellectual rather than emotional, where tensions are resolved. It is an conceptual art.... The principles are those of repetition and the continuous permutation of motifs and designs.... Many devices typical of Islamic architectural decoration—for example, the *muqarnas* [stalactite-like pendentives]—are explained by a desire to dissolve the barriers between those elements of the buildings that are structural (load-bearing) and those that are ornamental (non–load bearing).... The tendency is for surfaces to be fluid: decoration helps to make the transition, imperceptibly, from one plane to another.... In Islamic decoration—whether applied to objects or buildings—each motif is merely part of an overall patterning of surface, achieved by the counter-position and super-imposition of a variety of different designs and materials, each of which retains its identity within the whole composition.... Thus façades and wall surfaces are mostly decorated with symmetrical repetitions of units which, in turn, are made up of smaller repeated units.

Finally, Jones gives us a brief explanation of exactly how it was all done—with a *grid*:

> Primary grids indicate the principle elements of the [Islamic] decorative scheme [in which] each overall surface is subdivided. Patterns are mostly contained in rectangular or square panels, which are themselves strongly framed by horizontal and vertical bands. The framing bands of course produce a primary grid of their own, which reflects and underlines the contours of the building [and] links façades with portals, secondary openings with the main portals and windows, or different patterned or textured surfaces. Their function is to hold these elements together visually, as if with a belt....
>
> Secondary grids control the patterning within each of the elements of the primary grid. Here are found the counter-position of designs and the contrasts of textures, materials, colors and patterns which produce the decoration. Within

these secondary grids are also found the qualities of repetitions which unify each element with the other, and each surface with the rest of the building. These common elements within a decorative scheme harmonize and bring together the different designs and patterns that are found near each other, but are visually different or contrasting. Primary and secondary grids in Islamic decorative surfaces demand to be "read."[6]

As every student of Islamic architecture now recognizes, in traditional building practice a preliminary drawing certainly was used, and, according to Ronald Lewcock, "the design is executed upon a grid that represents the *dire*, or cubit, while the grid of the smaller squares establishes the size of the brick module for the builder."[7] This scheme is truly ancient; as Erwin Panofsky explained,

> The Egyptians effected subdivision of the stone or wall surface by means of a finely meshed network [or grid] of equal squares.... The network used by the Egyptian artists [and architects] precedes the design and predetermines the final product....The result was a foregone conclusion, once the figure's [or building's] absolute size was determined. Conversely, the absolute size is, of course, determined by a single square of the network, thus making it possible to reconstruct the whole figure from the merest fragment of such a network.... As we know, the Egyptian artist-theoretician first constructed a network of equal squares and he inserted into this network the outlines of his figure — unconcerned as to whether each line of the network coincided with one of the [structurally] significant junctures.... The idea [was] constructing on the basis of an absolute module.[8]

Inasmuch as all the important architectural plans were drawn upon papyrus, they have not been preserved, that is, with the single exception of the so-called "Turin papyrus,," which shows the somewhat tattered plan for the elevation (façade) of the tomb of Ramses IV (reigned ca. 1167–61 BCE). (Fig. 4) In this unique case, but which may be taken to represent a standard practice, the architectural features are drawn in black ink whereas the imposed grid pattern was itself drawn in red. Spiro Kostof has described the actual working process in some detail:

> Drawings were made with reed pens on such surfaces as papyrus and leather. These materials, being expensive as well as fragile, were used for the more important drawings, such as the master schemes maintained in institutional archive. Only rare fragments have survived.... The instruments of design were the ruler, the square, and the triangle. Rulers were made of wood, and divided into cubits; the cubit, the major unit of Egyptian architecture, was in turn subdivided into 7 palms, and each palm into 4 digits. These were measurements of human derivation. The royal cubit stood for the length of the forearm from the elbow to the tip of the middle finger....
>
> Plans and elevation drawings, often used in conjunction, were governed by a central axis line and the rule of bilateral symmetry. They were overlaid by squared grids. A famous instance can be seen in one of the very few papyrus drawings in existence, now in Turin [fig. 4], showing the front and side elevations for a shrine. The line [delineating the architectural features] is itself [drawn] in black, the grid in red. Similar grids were known to have been applied to walls and blocks of stone, to control the carving of figures in relief and in the round. The practice may have corresponded to the modern conception of *mise aux carreaux*, the squared grid....

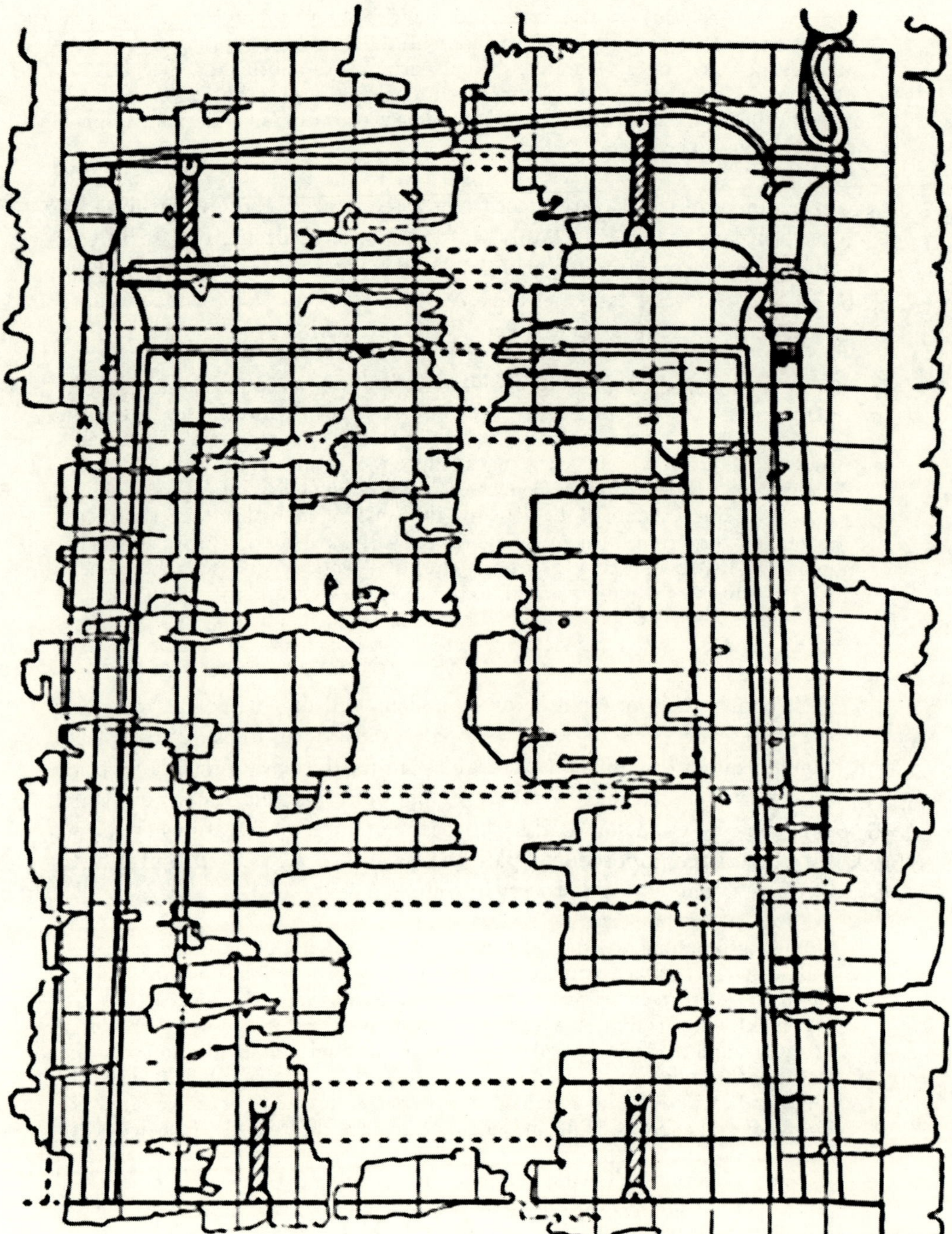

Fig. 4. Unknown Egyptian architect's squared-grid, working drawing of the elevation (façade) of the tomb of Ramses IV (reigned ca. 1167–61 BCE). (Smith, *Egyptian Architecture*, 1938.)

> The theory of architectural design was seemingly empirical, and only minimally indebted to mathematics. According to one scholar, A. Badawy, the architect employed simultaneously a modular system and a geometric system. The module would be derived from a major dimension in the building under design, [for instance] the width of the inner room in a temple, which might be, say, 10 cubits. Multiples and fractions of the module would then determine all the other dimension of the building, as well as the placement of columns and piers.... The geometric system depended on a few simple figures, mainly the square and a specified number of triangles. Among them was the so-called "Osiris," or sacred, triangle, where the height to the base is in the relation of 4 to 3 [and with a diagonal of 5 units]....
>
> The next step after the completion of the design would have been to transfer the building onto the site. This was done by reference to something called the "plan-set" (Badawy's translation); the original word was related etymologically to the verb "to plan, to project," and the noun for "cord." The procedure seems to have involved staking the ground and stretching a cord to delineate the outline of the building and the formative [central] axis. The length of the building was predetermined and laid out accordingly; the width was then established proportionally to it, in accordance with some recognized rule "Its length is exact," a text from Dendera states, "its width [is then determined] according to the formula; its norm is [realized] in excellent work."
>
> The initial phase of this laying out of the plan-net was a solemn ceremony presided over by the king and overseen by the goddess Seshat. The [architectural] details would then be filled in under the direction of the architect, and in reference to working drawings. The main instrument for the entire procedure was a cord knotted at twelve equal intervals. With it, one could fix a right angle, or even a catenary cord. We have a statue of [the architect] Senmut holding such a cord, neatly rolled up. It is also the standard attribute of land surveyors, who measured fields and plots.... This is as much as can be gleaned about the profession of architecture ancient Egypt from the limited evidence at our disposal.[9]

As Kostof just explained, from the outset, a basic module was the 3:4:5 triangle. As Siegfried Giedion further observed, "No architecture is more imbued with proportion than that of Egypt, which stood at the beginning.... The ceremony of 'Stretching the Cord' (knotted at intervals of 3, 4, and 5 to form a right-angled triangle) was performed by the Pharaoh himself at the erection of important sanctuaries."[10] The knotted cord was also used by the Egyptian "arpedonapts," or ritual surveyors, for strictly practical purposes; as Matila Ghyka states, one can find "any right angle by using a continuous rope divided by knots into $3 + 4 + 5 = 12$ equal parts. This triangle is also the only right-angled one having its sides in an arithmetical progression. The Pythagoreans called a right angle the 'angle of equity,' and [pre–Islamic] Acheminid and Sassanid Persian architects made use of this triangle to establish the profile of their elliptic domes."[11] (Specific architectural applications are given, and illustrated, in the Appendix, and in figs. 74–84.)

As is well known, it was from the Egyptians that the idea of proportionate composition was passed on the Greeks, and from them on to the Romans. The Romans, of course, also occupied the Iberian peninsula, where they left behind them a considerable number of their proportionately designed edifices. These must be reckoned to have exerted some influence on the Islamic architecture later erected in Al-Andalus. Concerning the design principles determining the composition of classical, Greek to

Roman, architecture, we have a unique written record: the treatise called *De architectura, libri decum* that had been composed by Marcus Vitruvius Pollio around 25 BCE.[12] Moreover, Vitruvius's writings were known, also consulted, throughout medieval Europe.[13]

As defined by Vitruvius (in Book I, chapter 2, section 1), "architecture [*architectura*] consists of order [*ordinatione*], which in Greek is called *taxis*, and of arrangement [*dispositione*], which the Greeks name *diathesis*, and of proportions [*eurythmia*] and symmetry [*symmetria*], also decor [*decore*] and distribution [*distributione*], which in Greek is called *oeconomia*." Overall, what we will be looking for in Hispanic architecture is the post–Vitruvian evolution of specific applications of the key terms *eurythmia* and *symmetria*, and such as these standards were employed within the larger concerns of *ordinatione* and *dispositione*.

Specifically, mastery of mathematics (*geometria*), states Vitruvius in an extremely practical manner (in I, 1, 4), "affords many advantages to architecture, mainly by teaching us proportion through the use of the measuring compasses [*primum ex euthryramis circini tradit usum*] since these are tools that make it easy [for us] to lay out site-plans for buildings [*e quo ... in areis expendiuntur descriptiones normarumque et librationum et linearum directiones*]." In this case, as he states, for all kinds of preliminary workshop planning there are to be inevitably applied "the set-level, the compasses, and the plumblines"—*norma, librationum, circini, lineas*— because these simple tools facilitate both design and execution.

In Vitruvius own words (I. 1. i), geometric "theory [*scientia*] is the ability to demonstrate and rationalize [that is, *a posteriori*] the manual productions of dexterity [*res fabricatas*, which are themselves derived *a priori*] within the intellectual principles of proportion [*rationis proportione*]." Although admittedly prone afterwards to such "rationalizations," proportion really originates from — meaning, was initially (or *a priori*) "demonstrated"— in extremely simple workshop practice. Its procedures were, in short, tool-derived: "*per instrumentum*," hence, the rationalization (*scientia*) only came long after the fact of a long-standing, pre-*scientia* practice.

In short, proportions are derived by Vitruvius very practically, namely, from (I. 2. ii) "the selection of module-units [*modulorum*], arising from the members of the work itself, and, starting from the individual parts of the members, one constructs the whole work to correspond accordingly to those modules—*ex partibusque separatis ad universae figurae*." On a much more practical — and literally instrumental level, Vitruvius advises (I. 2. ii) that, at all stages of the operation, "layout and design demand competency in the manipulation of compasses and rulers; by these [instrumental] means, working plans— laid out upon flat surfaces— are obtained." To repeat the same statement in Latin: "*Ichnographica* [the initial design worked out on a flat sheet of paper] *est circini regulaeque modice continens usus, e quo capiuntur formarum in soliis arearum descriptiones.*"

A review of the basic points laid down by Vitruvius reveals that a simplified linear drawing is the initial step. This example of "*ichnographia*" is a proportionate ground-plan that naturally and inevitably arises from the manipulations of compasses and straight-edges. The next stage described by Vitruvius (I. 2. iv) is called the "*Ortho-*

graphia," explained as the representation of "the elevation, which is an image of a vertical façade [*erecta frontis imago*]; it will be lightly sketched in to show the outlines of the work for future stages [*operis futuri figura*]." Those "future stages" are those actions later done at the actual working site, where the initial *frontis imago* is be applied, and then a few, minor adjustments will arise due to local exigencies.

As we now know, Egyptian architects had also drawn their plans using a ruler and a square, and also by employing triangular templates. One drawing on papyrus, a rare survivor (fig. 4), in fact shows both the plan and the elevation of temple; the architectural features are inscribed in black ink and the grid was drawn in red.[14] From depictions in bas-relief, also traces of red lines found upon a temple mound, it is now confirmed that knotted ropes were used to lay out the ground plan at the building site. Some of these builder's cords, with twelve knots, have also been unearthed; folded into the 3: 4: 5 triangle, these were essential to establish the right angles fixing the corners of a gigantic net of ropes which recreated on the ground the grid underlying the designer's working drawing. The artisan's corded grid-plan probably postdates the fisherman's net, but most likely the latter inspired the former. Besides knotted cords and measuring rods, the surveyors also used a portable sighting device, a *merkhet*, usually the split center-rib of a palm leaf. Besides squared grids, templates, ground nets, sighting devices and the ritualized 3:4:5 triangle module, the ancient Egyptians also gave their basic unit, the cubit, to the Islamic builders.

Unfortunately, as far as we are aware, no documentation — that is, no architectural pattern books, or theoretical treatises— have survived from the Islamic world; if they had a Vitruvius, we have no record of his architectural writings. The exception to the rule of documentary silence is Spanish, the writings (and illustrations) of Diego López de Arenas and Andrés de San Miguel (as will also be cited below: figs. 64–70). Nonetheless, as shown in some old Islamic miniature paintings, the architectural draftsmen really had, as Lewcock explains, created on paper "buildings drawn in pure elevation and plan, showing every detail and decorative pattern." Research by modern art historians has further shown that the grid was subdivided in squares, with each unit measuring between 42 and 62 mm (depending upon local standards for brick sizes), and these cubit-units "represented the structural modules of the plan." Moreover, on-site analysis of a number of "buildings from the 9th century onwards in terms of such grids has found that their design conforms exactly to them, even in detail." Overall, we now know that the Islamic designers' task was to create a "simple method of establishing a commensurate systems of proportions throughout a building, [one] deriving its ratios from the perfect square." The Arabic name for builders or sub-contractors is *al-banna*; in Spain, centuries later, construction workers are still called *albañiles*.

Fortunately, one or two of the preliminary drawings actually used in designing earlier Islamic buildings have survived. (Fig. 5) Our example is a sixteenth-century reconstruction of the ground-plan of part of the north-west *iwan* (porch) of the Friday Mosque in Isfahan, Iran, which was completed in 1088. Unfortunately, what have *not* survived are the Islamic designers' actual working-drawings, or preliminary sketches. Nonetheless, and following their rigorous empirical analysis of extant buildings, it has now been determined by various scholars that the appearance of those

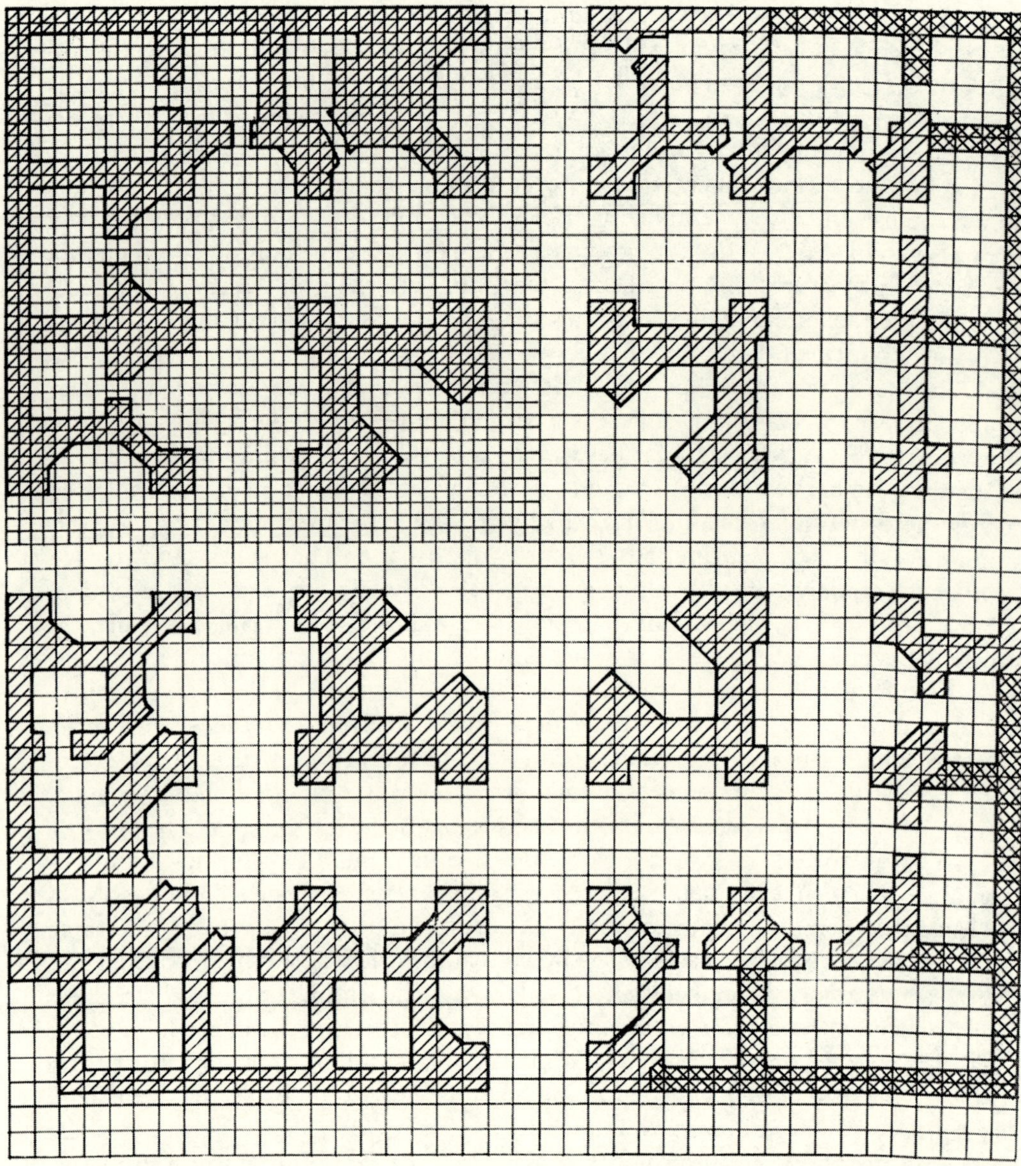

Fig. 5. Gridded ground-plan of a detail of the north-west *iwan* (recessed porch) belonging to the Friday Mosque, Isfahan, Iran. Plan ca. 1550; re-drawn after the original drawn in 1088. (Michell, *Architecture of the Islamic World*, 1978.)

preliminary sketches would have been, in effect, a series of wholly geometric figures drawn over grids, wherein each mini-square would have represented the basic cubit-unit. Specifically, the geometric figures so imposed upon the ubiquitous squared grid-underlay would have been compass-drawn circles, with these then being subdivided into triangular, rotating motifs.

The square, itself having a center-point (as determined by the intersection of lines drawn from its four corners), perhaps has its own symbolic valence. As explained by James Dickie,

> At the heart of Islamic Law stand the Five Pillars (the Arabic means, literally, "corners"). This does not mean that the edifice of faith is pentagonal, for the Pillars are disposed in the pattern of a quincunx, with the First Pillar, the *Shahada*, located at the certer, to which the remaining four are peripheral: *salat* (prayer), *siyam*, (fasting), *zakat* (payment of the poor-tax), and *hajj* (pilgrimage). Textually present in the Koran, the Five Pillars are accordingly an obligation, with the partial exception of the *hajj*, performance of which is contingent upon one's means. The Pillars that concern architecture are the Second [*salat*-prayer] and the Fifth [*hajj*, the spiritual journey to Mecca, upon which all mosques, anywhere, are physically aligned].[15]

Fortunately, a British architect, Dr. A. G. Walls, has carried out an exhaustive analysis of the "regulating-patterns" determining the design of a single Islamic religious structure erected in Jerusalem around 1482, a *madrasa* (theological school), and the results of his investigation were published in 1990 as a monograph, extensively illustrated with dozens of his drawings and diagrams.[16] His overall conclusions can be quickly summed up: "the plan of the [entire] Madrasa was controlled by the fundamental and primary geometries centered on point S," with that centric-point be centered upon the exact center of the *sahn* (patio-courtyard).

This point about "architectural-design-by-geometry" is best illustrated graphically by one of Walls' many drawings. (Fig. 6) In effect, "Point S" represents a specific compass-point, from which a large circle had been drawn (apparently over a grid, such as that shown in fig. 5). This *compass-point* in turn determines both the vertical axis of the open courtyard and the center lines for two covered, basilica-like halls attached, at right angles, to the *sahn*. This *vertical axis* is also symbolic, since it aligns itself upon "point M," the *mihrab*, a prayer-niche precisely situated to point to Mecca. The outer edge of this circle also established "point 5," from which another circle was drawn, that determining the outer edge — and proper width — of the adjacent vestibule, a rectangular entrance-way. As a kind of "cloning-through-geometry," the procedure was repeated throughout the madrasa in order to establish *a self-generating system of proportionate relationships* between all components making up the entire architectural complex.

Particularly interesting in the context of my own investigation is the fact that, as shown in Walls' diagram, the circle generating the nucleus of the entire design has also produced a non-inscribed (or invisible), six-branched, "star-pattern," with that figure in turn generating the six tangent points yielding straight lines describing a hexagon (at points 1-2-3-4-5-6). Further bisection, horizontally and vertically, of the square inscribed within this sexpartite figure then fixed the anchor points for minor architectural extensions into the courtyard. The inscribed star-pattern itself yields 12 micro-motifs, a dozen triangles, each with ratios (as ascertained by using a graduated ruler) of, respectively, 3 (base), 4 (height), and 5 (diagonal). The standard name for a triangular figure with these 3:4:5 proportions is the "Pythagorean triangle," the same geometric figure appearing in the ancient Egyptian architectural ceremony called "Stretching the Cord."

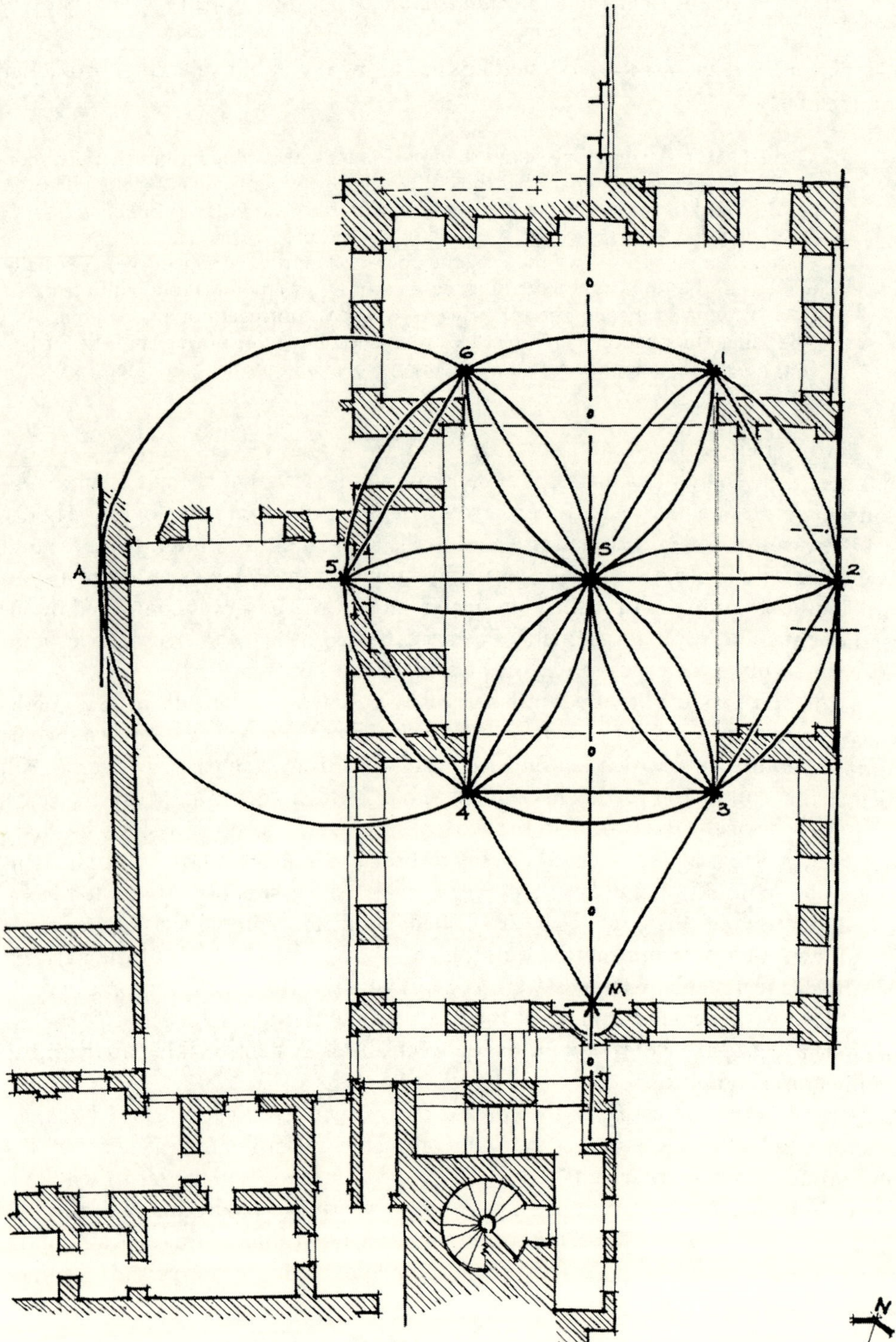

Fig. 6. A. G. Walls, "Anchoring the Fundamental Geometry Predetermining Islamic Architectural Design" (Walls, *Geometry and Architecture in Islamic Jerusalem*, 1990.)

The larger context for this kind of "architectural design by geometry" is religious: Islam, to the contrary of Christianity, was iconoclastic. By forbidding representation of the human figure, Islamic theology made Moslem art, especially all manner of its architectural decoration designed for installation in mosques, turn to elaborate geometric patterns and linear designs, "pure abstraction" in short. The severely limited, allowable range of Islamic iconographical parameters were compressed into three basic categories: 1) the "epigraphical" (highly stylized Koranic inscriptions); 2) highly abstracted floral patterns, the so-called "arabesque" (or *ataurique* in Spanish; from *al-turiq*, "leafy design"); 3) finally, and the most important: geometry-derived, band-like series of interwoven, linear patterns. The latter were evidently grid-derived, just as we know was the case with the overall architectural design (see fig. 5).

3

A Graphic Analysis of the Geometric Procedures Generating the "Arabic Interlace"

The unique formal result of geometry-derived Islamic design procedures, the micro-cosmic "Arabic interlace," was functionally analyzed in great detail long ago by an industrious French draftsman, Jean Bourgoin. For his illustrated, large-folio album, *Les Eléments de l'art arabe: le trait des entrelacs* (1879), he made 190 pen-tracings of diverse Islamic decorative ensembles. In the upper left quadrant of each of his plates, he also added dotted-line patterns; these reveal the underlying (or non-inscribed) geometric design module which had predetermined the layout of each component making up any given decorative ensemble. (Fig. 7) Tracing over one of his many dotted-line drawings,[1] I have emphasized (by using a pencil and ruler to draw parallel lines, with these stacked above the shorter base-line) the omnipresence, also the exact operations, of a governing triangular motif. The series of striated triangles which I have added to Bourgoin's tracing represent the Islamic designer's essential tool, a *template* (as I will later show); this template is what the Hispanic designers later called a "*cartabón*."

In this particular example, the repeated *triangle-module* has a base of 3 units, a height of 5, and a diagonal 6 units long. Put otherwise, the figure is one half of an equilateral triangle; as such, it covers one-quarter of the area of a square (equilateral figure) having four, 6-unit sides. As it turns out, this is the standard — or most frequently encountered — Islamic (versus Hispanic) design module. Simply derived from an equilateral matrix, the geometric design is capable of infinite variation, so accounting for its common employment. As shown imposed Bourgoin's drawing (fig. 7), the added 3:5:6 triangle-templates can serve to establish the proportionate, vertical spacings between each star-pattern, also the intersections of their radii extending from their centers. The location for the axis of each star-pattern determined the placement of the leg of an adjustable compass, and this commonplace instrument then fixed (as circumference) their proportionate widths and extensions. Once flipped over, and then doubled in size, the same 3:5:6 triangle-template then serves to place the vertical center of the immediately adjacent row of star-figures.

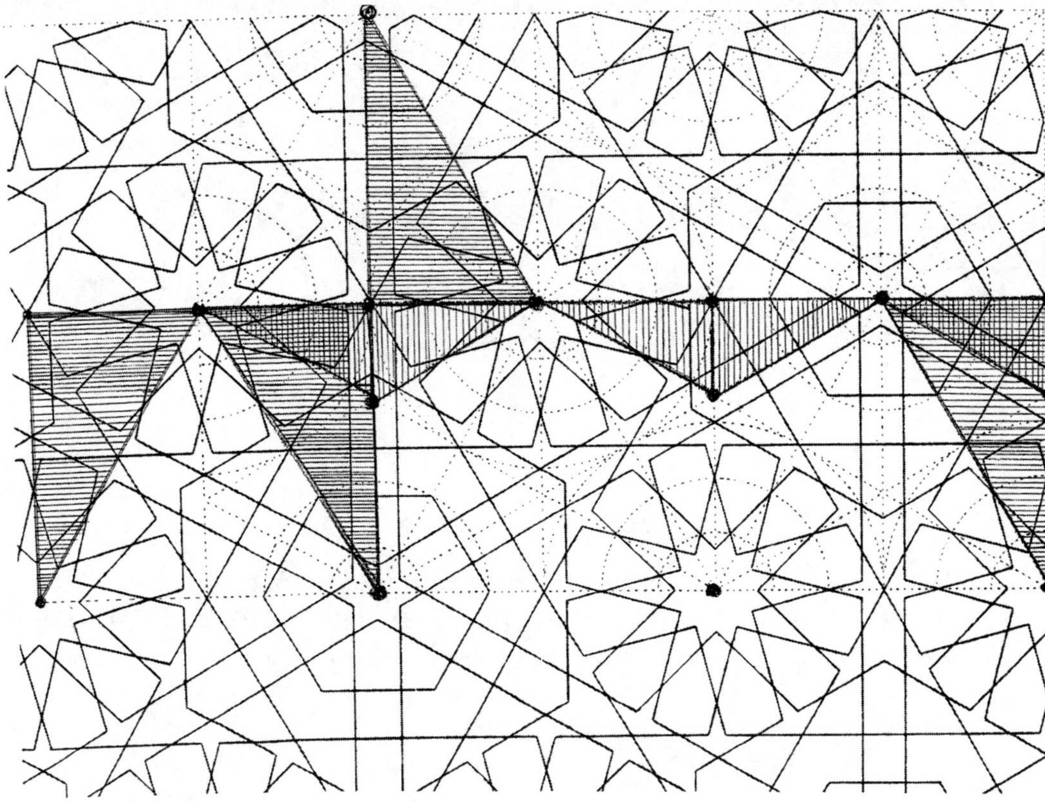

Fig. 7. J. Bourgoin, "Underlying (or Non-Inscribed) Geometric Design-Modules Determining the Generation of Arabic Star-Patterns," *Les Eléments de l'art arabe: le trait des entrelacs* (1879), Plate 113 (with Bourgoin's dotted-line under-drawing here enhanced with tracings of non-inscribed, 3:5:6 triangle-modules, with these explained as having been used as "templates").

As we see, given physical substance as a template, the ubiquitous triangular-module may be, besides expanded or shrunk, also turned on its side or folded over its axis, and each procedure reveals further, *proportionately spaced loci,* with many of these specifically becoming *compass-points.* Overall, the primary function of Arabic "design-by-triangulation" is to create vertically aligned, parallel lines of vertical zones or ascending rows. Once so established, these vertical courses may be filled with different motifs—but the center-points of each macro-motif will be in proportionate relationship to every other mini-motif inserted into the entire composition, whether running from top to bottom, or from left to right. Thus, each micro-cosmic unit becomes harmoniously adjusted to the macro-cosmic design.

It is now easy to show the final results, those producing a harmonious design ensemble, and specifically by using the repeated, equilateral-triangle module. (Fig. 8) In another of Bourguin's many drawings—this one analyzing a typically Islamic "star-pattern" motif[2]—I have "layered" (as is explained below) a computer-generated grid-pattern of non-inscribed, 3:5:6 triangle-modules. This pattern simply rep-

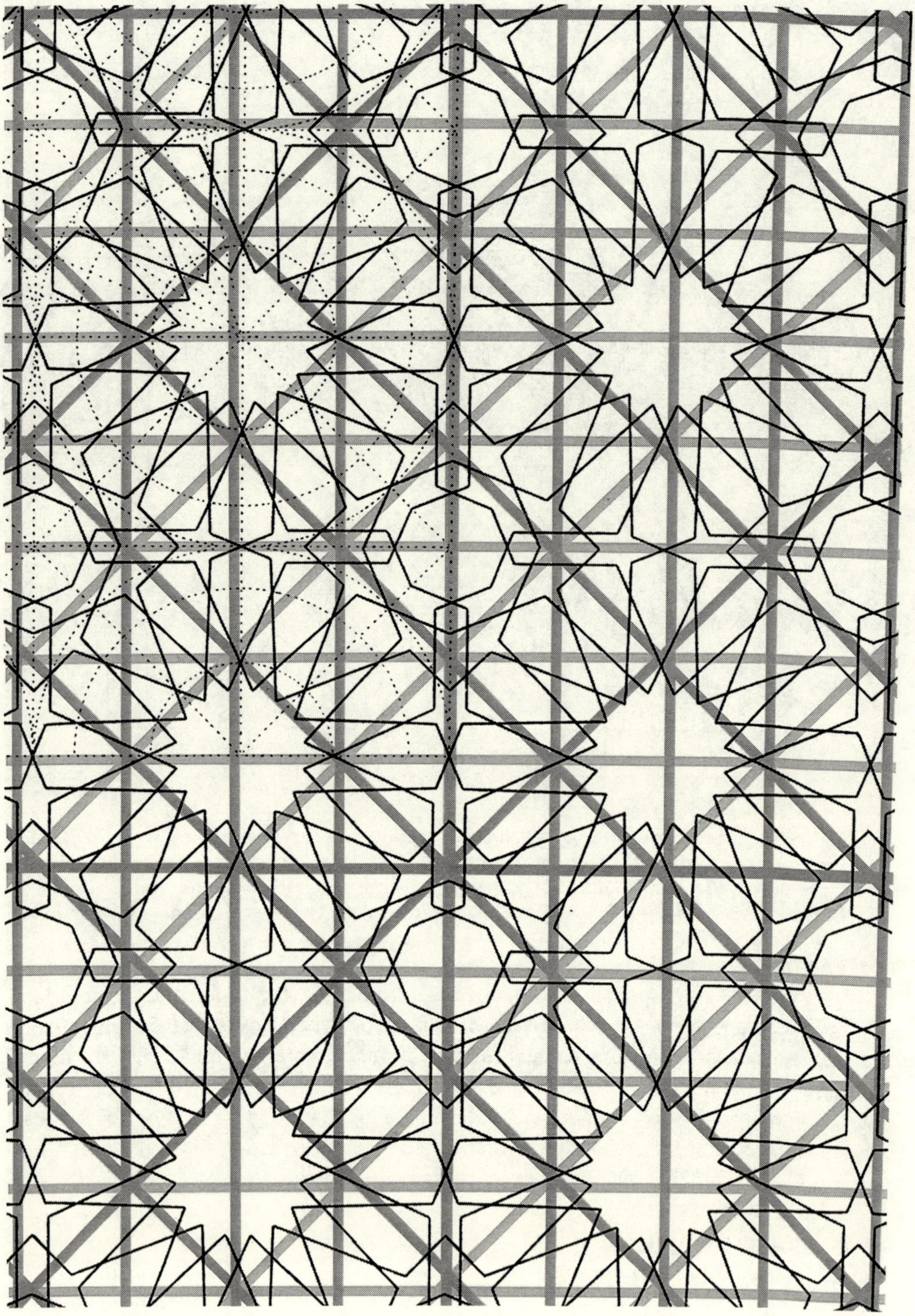

Fig. 8. J. Bourgoin, "Underlying (or Non-Inscribed) Geometric Design-Modules Determining the Generation of Arabic Star-Patterns," *Les Eléments de l'art arabe: le trait des entrelacs* (1879), Plate 57 (computer-generated grid-patterns of non-inscribed, 3:5:6 triangle-modules layered over Bourgoin's tracing).

resents the square grid-pattern known to have been used by Arabic builders (as shown in fig. 5), but with the addition of yet another one, having the same squares, overlaid upon it, and now tilted at a 45-degree angle. What now becomes a doubled "lozenge-unit" grid produces perfectly aligned — or proportionately spaced — rows of, equally, vertical, horizontal, *and* diagonal rows. Although obviously done without the aid of a postmodernist computer, this Islamic mode of "design-by-triangulation" is what will become the distinguishing characteristic of much later Hispanic (versus European) architecture. However, as will be explained later (in figs. 24 ff., as also shown in fig. 58), the distinguishing characteristic of traditional Hispanic (versus strictly Islamic) design-by-triangulation procedures is — rather than the 3:5:6 module — instead a different triangular figure, one having different ratios, those belonging to a 3:4:5 triangular figure.

In a brief, but thought-provoking, conference paper given in 1986, Alberto Donaire (School of Architecture, University of Seville) analytically treated the formal characteristics of "Lace-Design Methods in Late Medieval Spain."[3] The reference is to what is usually called, in English, the "Arabic Interlace" patterns, "*arabesques,*" or "*entrelacs,*" in French; the corresponding Spanish term is "*lacerías,*" from *lazo* (knot). As Donaire states, "geometry gives it unity, and architecture itself used the same geometric patterns." He also observed that "no documents from the Islamic world" verbally described the actual procedures employed in creating these intricate geometric designs, the kind where "the central motif radiates to cover the whole design extension."

In fact, *geometry* is a persistent characteristic of not only Islamic architecture, particularly that built in Andalusia (Moslem Spain), but also of numerous Christian buildings erected during the Renaissance and Baroque eras, either in Spain itself or in Latin America. And since we do know that grid-drawings were used in the preliminary design work for strictly Islamic architecture (see fig. 5), the logical presumption is that they were also used to design Arabic "interlace" patterns appearing as architectural decoration.

As Donaire recognizes, "in a given design, if we link the star-centers generated in the lacework patterns, *a structural grid emerges,*" and he adds that "Moroccan artisans today use squared paper to make their designs." In short, the employment of a *structural grid* is indeed the key to the whole interlace operation. Indeed (as I would further argue), it even provides the key to routine workshop procedures by which one can now define the sources of the distinctive (and often so recognized) stylistic characteristics of Hispanic architecture itself. Returning to his elaborate analysis of the Spanish *lacerías,* Donaire specifically proposes a rather recondite explanation for their origins (and he illustrates his thesis with an elaborate diagram looking a bit like fig. 7, but without intervention of the inscribed "lozenge-pattern" grid), namely:

> A geometric progression with a ratio of the square root of two; the terms indicate *c* main units: ½, square root of ⅔, 1, square root of 2, 2, square root of 2, 4, 4 times the square root of 4. If we put in *p* units (one-fifth of the main unit, *c*), now taking 7/5 as the square root of two, the first of the following series is obtained. The second is derived by adding every two consecutive terms to the first. The third series, doing the same procedure over again with the consecutive terms of

> the second series, reads as follows: 2½, 3½, 5, 7, 10, 14, 20, 28, P [then comes] 6, 8½, 12, 17, 24, 34, 48, 68 [then comes] 14½, 20½, 29, 41, 58, 82, 116, 164.
>
> All the dimensions in [the diagram Donaire drew] belong to this canon, except 13, which is the only one measured in a direction other than the interlace lines; it is the hypotenuse of a triangle whose *catheti* (acute angles) are 5 and 12. The whole or partial dimensions of every possible Andalusian interlace design will be represented by canonic whole numbers or combinations of them. Therefore, the working method could have been as follows: By drawing a structural grid suitable for the panel to be decorated, an obligated lace-work design derives automatically.

Unfortunately, since I got a "D" in algebra in high school (for I am "left-brain" deficient), I can't make heads or tails of any of this numerical mumbo-jumbo. More importantly, I don't know either of any professional craftspeople who could make any more sense out of all this than I can. Nonetheless, what really works best here is the principle of "Occam's razor." This holds (instead) that the simplest — or most mundane — explanation is always preferable. In this case, the professional craftspersons I know tell me that their working procedure boils down to a simple query: "What is the easiest way for me to get this particular job done?" Certainly that does not involve recitation of an arcane litany like "2½, 3½, 5, 7, 10, 14, 20, 28, *P* [then comes] 6, 8½, 12, 17, 24, 34, 48, 68 [and then comes] 14½, 20½, 29, 41, 58, 82, 116, 164," etc. Moreover, since I was once an exhibited painter and a professor of perspective, I understand the strictly *practical* uses of geometry and, especially, the unquestionable usefulness of the grid for transfer of one's preliminary graphic ideas onto a much larger working space (or canvas).

Well, there you have it, the "Occam's razor" explanation for the Arabic interlace. In short, it was most likely initially designed on paper ruled out with a "lozenge-grid pattern." Besides being easy to manipulate, perhaps its most desirable feature is that the procedure obviates any need for those tiresome lists of numbers hypothesized by Donaire. The actual means of execution of the linear design at the site would have been by means of triangular *templates*, wooden patterns cut in different sizes; given their unquestioned utility, they were even illustrated in practical handbooks intended for use by Spanish craftsmen (see figs. 65–67). These were given specific names, according to their acute angles (*catheti*).

As we shall see (in Chapter 16), in Spain centuries later, one of these was called the *cartabón cuadrado* (square template); it had two 45° acute angles, plus the inevitable right-angle (90°). However, the most frequently used template was the "*cartabón de cinco*." The five (*cinco*) in the name "5-unit template" refers to five, evenly spaced units that were ticked off along the diagonal; adjacent to the right-angle, there were three units marked on the shorter leg and four along the intermediary length. The result is, of course, a 3:4:5 triangle, meaning the one commonly called the "Pythagorean Triangle." Next in popularity (or frequency of employment) was the "*cartabón de seis*"; given its six-unit hypotenuse, its proportions were a bit narrower. Nonetheless, it is a near match to the 5-unit template; the angle of one acute angle belonging to a given 6-unit template is 30° (actually 29.92, and its other acute angle is 56.25 degrees), while the angle of any given 5-unit template is 36° (36.87, to be precise, and the other acute angle is 53.13 degrees). A rounded-off mean between these two angles—30° and 36°—is 33°, and those are the lesser acute angles uniquely defining the 3:4:5 figure (or five-unit template) commonly celebrated as the *Pythagorean Triangle.*

Such as it was given in Euclid's *Elementa Geometrica* (I. 47), S. K. Heninger explains the basic theorem as follows:

> The theorem that the square of the hypotenuse of a right triangle is equal to the sum of the squares of the two sides has been ascribed to Pythagoras since the earliest times. Here the theorem is represented spatially by the use of geometrical figures [see fig. 63, left side] depicting the smallest possible whole numbers which will exemplify it [with these being 3 and 4 and 5]. The line AB [3 units] and BC [4 units] form a right angle, and the line AC [with 5 units] completes a triangle, of which it is the hypotenuse. In the modern notation of algebra, the theorem would be stated: $(AB)^2 + (BC)^2 = (AC)^2$ In this diagram [again see fig. 63]: (AB) = 3, and $(AB)^2 = 9$ [and] (BC) = 4, and $(BC)^2 = 16$ [and] (AC) = 5, and $(AC)^2 =$ 25. Finally: 32 + 42 = 52 [and] 9 + 16 = 25. *Perspicue probatum est.*[4]

As it was known later to all literate architects, the authoritative explanation of this scheme was that one given by Vitruvius (*De architectura*, liber IX, Proemio, v–viii).

> Pythagoras showed that a [right angle] set-square [*norma*] can be formed without the aid of an artisan [*sine artificis*]. Thus, the result—which carpenters only reach very laboriously, but scarcely with exactitude—can be demonstrated from the reasoning and methods of his teaching. First, we take three measured rulers [*regulae*]—with one three-feet long, the second four, and with the third one being five feet in length. Then these three rulers are then to be joined together with tips touching each other; so doing, they make a triangular figure [*trigon*]. So joined, they will also form a right-angle [*norma emendata*].
>
> Now, if a square figure is extended from the original length of each one of these rulers, the square on the side of the three-foot-long ruler will have an area of nine square feet, while that of four feet gets sixteen; and the one with a five-foot length now yields twenty-five square feet. And it turns out that this cubed area [which may be either horizontal: a "ground-plan"; or, if vertical, it becomes an "elevation"], which was made up from the two squares, respectively the three and four feet in length [and containing 9 and 16 squares respectively], is equal to that of the square described on the [third] side, which is five feet long [and it now shows 25 squares]. When Pythagoras discovered this fact, he had no doubt but that the Muses had inspired him to the discovery, and it is said that he very gratefully offered sacrifices to them.[5]

4

The Installation of Islamic Architectural Design-by-Geometry in Spain

Just as Chueca had demonstrated in a formally focused manner back in 1947, the conceptual roots of those distinctively planimetric, suspended "tapestry-screen," *neo-mudéjar* façades common to Trans-Hispanic architecture ultimately should be traced back to the canonic Hispano-Islamic portal-façade. This decorative scheme was apparently first formulated in Spain in the Great Mosque of Córdoba, begun in 785. Sometimes called a "Friday mosque" (after the designated day for collective prayer), the Arabic name for such a building is *masjid alhama* (or *al-masjid al-jami*), meaning the principal congregational prayer hall in the city.

Although not in any fundamentally functional sense, Spanish designers were eventually to "Europeanize" the old Moslem frontispiece-format. This effect was canonized during the Renaissance, and it was achieved by superimposing a currently fashionable, "plateresque" vocabulary of miniaturized classical motifs upon what remained basically an Islamic compositional arrangement. Needless to say, the religiously-heterodox design sources of Hispanic church façades had long since been opportunely forgotten by the Renaissance builders.

The story of the architect in medieval Islam, as far as it can be retrieved, is unfortunately slim. Some major trends do, however, stand out.[1] With few exceptions, the "architects" (but not in our modern sense of the word) had little theoretical training. Nonetheless, what can best be described as a wholly pragmatic "applied geometry" was always an essential component of building practice, as just as is proved by the visible appearance of all kinds of Islamic monuments. An Arabic historian, Ibn Khaldun (1332–1406) spoke in his *Maqaddima* of the importance of geometry for architects; their trade inevitably "requires either a general or a specialized knowledge of proportion and measurement, in order to bring the forms [of things] from potentiality into actuality in the proper manner, and for the knowledge of proportions one must have recourse to the geometrician."[2] We additionally know that a geometrician, Maslama bin Abdallah, who worked at the court of Córdoba in the tenth century, was also a practicing architect.

However, just before the Moslems arrived upon Spanish soil, there was already operative in Andalusia a notion of geometrically ordered architecture. For this we have the testimony of the erudite Saint Isidore of Seville (560–636). Following Vitruvius, whom he cited, Isidore succinctly defined *architectura* in his *Etymologiae* (Liber XIX) as requiring *dispositio*, *constructio* and *venustas*, that is, the initial layout, and then follows the actual construction, and finally comes "beauty" in the applied details. The principle task of the *architectus* is stated to be the design of the ground plan: *disponere in fundamentis*. He also listed the essential "Instruments for Building," with these including the *norma* (carpenter's square), *regula* (ruler), *perpendiculum* (plumb-line) and *linea* (stretched cord).[3]

More typically, and just like his contemporary Christian counterpart in medieval Europe, the Islamic designer started his career in one or more of the traditional crafts: masonry, cabinet-making, faience, metalwork, and the like. In an architectural tradition that thrived on decorative effects, this early training in similarly geometrically decorated craft work made perfect sense. Yet for all the splendid ornament that sheathes the palaces, mosques, and tombs of Persia, Egypt, North Africa (*al-Magreb*), also *al-Andalus*, the chief obligation of the architectural designer was structural stability. Compared with their European counterparts, cathedrals for the most part, the rate of completing buildings in medieval Islam, even ones of considerable size, tended to be fast. Hence, in Islam most of the important buildings could be both conceived and executed by their original designer.

Buildings like these could not have been achieved without seasoned, albeit entirely empirical, experience in standard construction techniques. As it worked out, there was no clear line drawn between a foreman-mason and the building designer. The three terms used to designate what we would today call a "builder" were *mi'mar*, *banna* and *muhandis*. Recognized "architects" (and somewhat in the modern sense of the word) were referred to as *mu'allim* or *sahib*, also *Ustad*, all honorary titles corresponding to "master," so signifying proper respect for some special ability. (And this suggests that the modern Spanish word *Usted*, the formal address—versus "*tu*," might have had an Arabic derivation, *Ustad*, and hence not from *Vuestra Merced*, "your honor.")

To see how these concepts were actually applied in construction, we may cite a famous Hispano-Islamic monument. The best preserved of the several portals of the Great Mosque in Córdoba, "la Puerta de Al-Hakam II," is found on the southwest corner of the huge building. (Fig. 9; see also fig. 73, a ground plan of the Mosque, where this door is no. 6.) Designed around 960, this arrangement may be taken to represent the epitome of certain underlying design principles common to many later Moslem buildings—and also to Hispanic *neo-mudéjar* architecture. This formal pattern is manifested in both traditional Islamic portal-façades and in many, much later, Latin American *fachadas-retablo* (among others, see figs. 1, 54). In either case, the effects represented include a complicated, rectangular and planarized, wall-pattern that is symmetrically flanked by bulky but undecorated buttresses.

The result is a stark visual contrast between the vertically flattened, rectilinear entrance ensemble and its huge, blank background with pillar-like enframements. Centered within the tenth-century Córdoban portal ensemble, an arched doorway

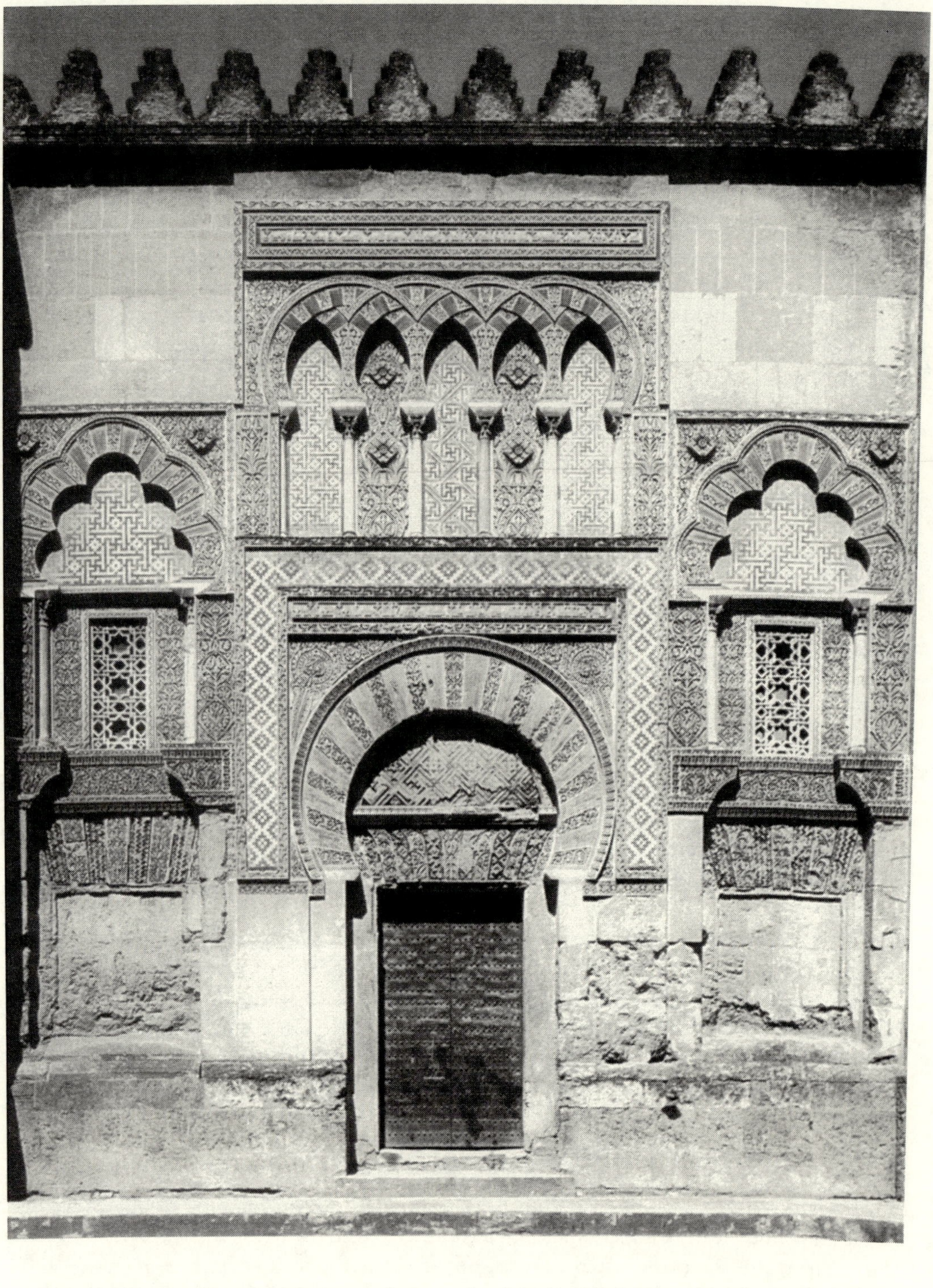

Fig. 9. Unknown Moslem designer, "Puerta de Al-Hakam II," Great Mosque of Córdoba, Spain. Designed ca. 960.

is enframed by a large, pseudo-architectural sub-unit. This anomalous element is the "*alfiz*," a three-sided, rigidly rectilinear moulding serving no apparent structural purpose.[4] Significantly, the *alfiz* device — looking like an upside down, squared base "U" — does not extend down to the ground; instead, it terminates about halfway down the length of the portal. Later, with an applied classical veneer, this same "baseless" effect was replicated in many Neo-Plateresque portals in Latin America.

Placed above the *alfiz* in the Córdoban portal is a blind-arcade system of stacked, interlaced and flatly overlapping, "horseshoe-arches"; this motif visually corresponds to the complicated gable-work surmounting the Mexican church in Tepotzotlán (fig. 1). In both examples, Islamic or Christian, the viewer's eye is emphatically directed upwards, toward a visual climax that is "misplaced," inasmuch as it is situated far above the door or, more broadly put, away from the earth. However, a central portal represents the point of gravitational focus at which — at least according to classical (or European) logic — emphasis should properly have been placed. In the middle of other portal ensembles at Córdoba (for instance, the adjacent "Puerta de San Esteban," first installed in 787, restored in 855, and displaying the earliest known *alfiz*) there is also found the "*tejaroz*," a projecting eave, looking like an atrophied cornice and tentatively supported by two rolled brackets. At Tepotzotlán, for example, the formal equivalents are two broken cornices enframing the centered rose-window and extending from side to side in the portal-façade (see fig. 1). Additionally, in both the tenth-century Moorish portal and in its eighteenth-century Mexican counterpart, one finds a regularized sequence of multiple — seemingly stratified, baseless, and boxed — planarized motifs vertically ascending on either side of the entrance way.

Equally applicable to the situation in Tepotzotlán, eight hundred and more years later, is the careful formal description of the tenth-century Córdoban "Puerta de Al-Hakam II" provided by Fernando Chueca in order to epitomize the Islamic phenomenon. In the particular instance (fig. 9), this prototypical *fachada-retablo hispano-islámica* is characterized by,

> In the first place, strict planarity, that is, by low relief-work, for otherwise a vigorous plasticity would have destroyed the geometric purity of the cubic volumes of the flanking elements. In order that these prismatic forms may impose themselves, it is essential to preserve the smoothness of their faces; therefore, the exaltation of volumes is linked to a pious respect for the plane corresponding to them. In absolutely planarized Hispano-Islamic architectural decoration every component is ordered by a rigid geometrical division, firmly governed by the right-angle....
>
> Enframements or cuadratures are the distinctive and differentiating sign [and], in this case, the extremely important *alfiz*-motif must be emphasized as a characteristic of Spanish architecture.... When Al-Hakam II canonized this kind of portal, the die was cast, and all subsequent Spanish architecture, to a greater or lesser degree, would hew to the [Andalusian-Islamic] model.... Whether round or horseshoe-shaped, it is an impulse to cuadrature which causes the arch to be enframed by an *alfiz*.... Lintel, arch and *alfiz*, all end up forming a stratified series, lacking any tectonic logic due to their redundancy.... In those Córdoban doors there was initially implanted the distinctively Spanish type of weightless, "suspended decoration" [*decoración suspendida*].[5]

In order to recreate a simple design-to-execution workshop-procedure followed by the Islamic designers, I present as evidence a linear schematic of the Portal of Al-Hakam II. (Fig. 10) Over Chueca's original line-drawing of the tangible architectural elements— those "real" physical objects— with the aid of a computer and using the appropriate graphic software, I have superimposed a digitized "layer," a "grid" drawn with repetitive patterns of overlying and evenly spaced, parallel and diagonal, lines (easily done, since my software has a "grid" option).[6] As now laid over Chueca's drawing (previously scanned, like all my other illustrations), the digital-grid now graphically reveals how the placement of all the major vertical and horizontal alignments in the archetypal Islamic portal had been determined by the regularized intersections of parallel diagonal lines with their parallel horizontal counterparts.

First, one needs to find the generative *center* for the entire portal composition,

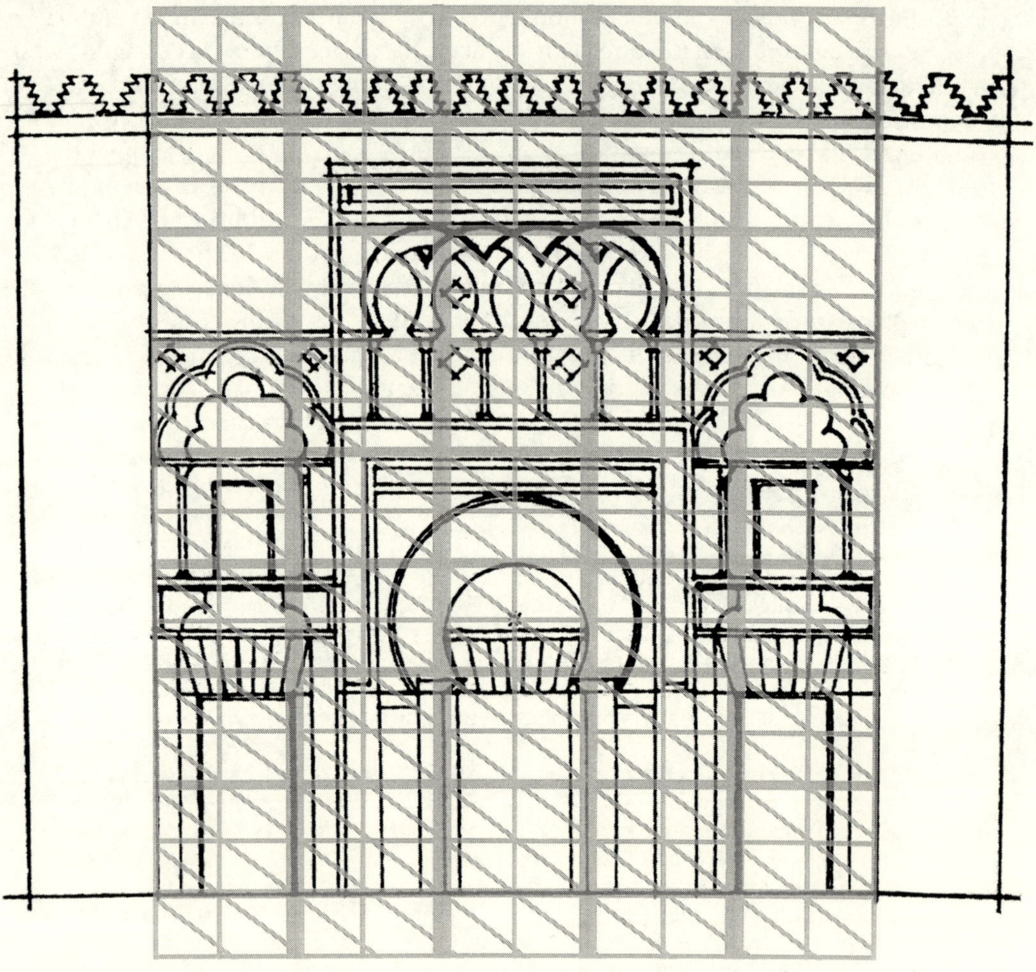

Fig. 10. Inscribed "Regulating Lines" generated by Pythagorean triangles and predetermining the design of the "Puerta de Al-Hakam II." (Computer-generated grid-patterns layered over a drawing reproduced in Chueca, *Invariantes castizos*, 1947.)

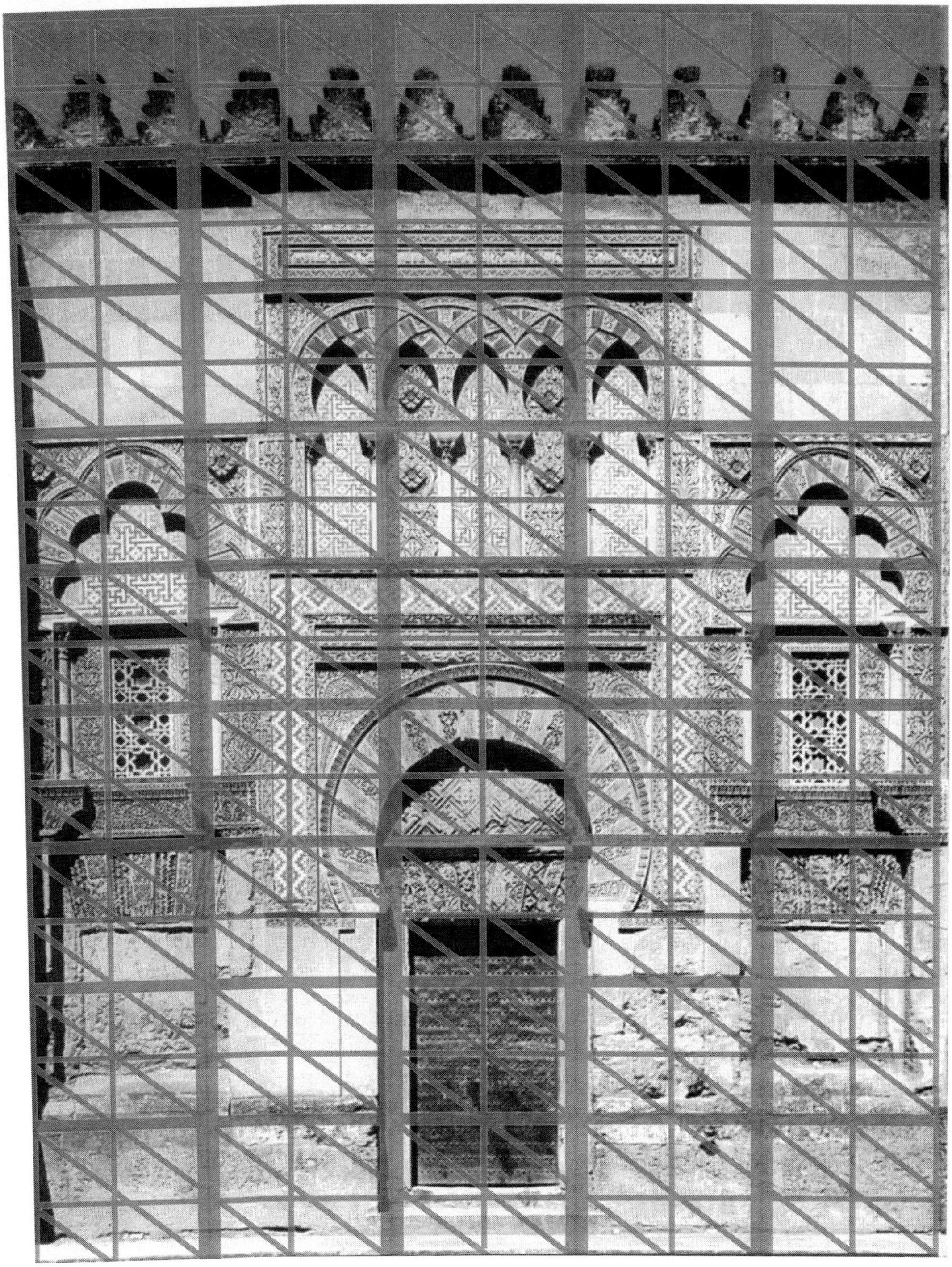

Fig. 11. Inscribed "Regulating Lines" generated by Pythagorean triangles and predetermining the design of the "Puerta de Al-Hakam II." (Computer-generated grid-patterns layered over site-photo: fig. 9.)

if only because the idea of a "center" played important role in Islamic cosmological thought as a symbol of the "unity of being" (*wahdat al-wuyud*).[7] Here, and logically so, it is found in the middle of the horseshoe arch over the door. Here marked with a star, this essential point is where the designer's compass-point was placed to draw the primary horseshoe arch. Through that critical starting point, an initial diagonal line—with a 33° angle—was drawn. The lower end of this diagonal line fixed the bottom, right-hand end of the entire composition. Then, rising through the compass-point, it nicely established the level of a horizontal string-course, and this establishes a base-line for the typically Moorish, interlaced "horseshoe" arches making up a blind arcade.

The same schematic operation defines the rest of the façade. A diagonal rising from the lower right-hand corner of the alfiz-frame running around the blind arcade then fixed the far corner of its cornice. Another parallel diagonal, drawn from the far end of the horizontal base-line for the colonnade, passes through the inscribed compass-point for the arcade, then rises upwards to fix the exact height of crenellations rising above the roof line. Below, another diagonal, rising from the suspended bottom end of the major alfiz framing the door, has fixed the upper-left corner of the same alfiz. Two inscribed, parallel vertical lines (shown in broader bands) rise from the pillars framing the door upwards, alongside two colonettes in the second-story blind arcade; the spaces between the other colonettes are then determined by smaller triangular sequences.

To confirm these findings, as initially based on Chueca's somewhat free-hand sketch (fig. 10), the same grid is now to be applied to the on-site photograph (fig. 9), and the graphic result is Fig. 11. Since this new application—photo versus drawing—yields an outcome matching the findings already deduced (from fig. 9), once again, and throughout the entire composition, one sees that the overall pattern which emerges is both proportionate and that it is, once again, wholly "triangle-based." Hence, throughout the entire composition, a system of design-by-triangulation has produced an infrastructure of perfectly proportionate relationships between all the macro-motifs and all the micro-motifs. More specifically, given the 33° angle, also the inevitable right (90°) angle of the vertical line, then the third angle is also inevitable: 57°. Put otherwise, any triangle so constructed has an overall ratio of three units on the shorter, or upright leg, four on the bottom leg, and five units on its diagonal. What results is a 3:4:5 figure, and it was used here as a *module*. Another conclusion emerges: this is the same 3:4:5 figure as that one once belonging to the ancient Egyptian ceremony of "Stretching the Cord."

5

The "Secret" of the Hispanic Masons: The Inscribed Pythagorean-Triangle Motif

As is now made visually apparent, *a diagonal line intersecting a series of evenly spaced, parallel vertical lines determines the proportionate placement of every significant decorative element* inserted into the "Portal of Al-Hakam II." To verify my findings, anyone following the same graphic procedures can duplicate these results (likewise those later resulting in my figs. 10 ff.; 24 ff.), and independently duplicated results also usefully define those procedures presently characterizing "science" (as opposed to mere speculation or wishful thinking).[1]

Credit must be given to whom it is due; my initial recognition of the fact of a consistently applied Islamic canon of triangle-derived design was due to yet another schematic drawing executed by Chueca, with this one illustrating the geometric pattern predetermining placement of the major elements in the decorative ensemble belonging to the *mihrab* (prayer-niche) in the same mosque. (Fig. 12; located at point 23 in the ground-plan of the Mosque: fig. 73.) As he observed, his drawing demonstrated that it was all due "to the predominance of squared proportionality."[2] According to his findings, here the generating pattern was one of equilateral triangles. That works, that is, until you open the question of a *compositional center*, perhaps even a "spiritual center" (*wahdat al-wuyud*), the kind shown to have been essential by close analysis of Bourgoin's much earlier graphic investigations (figs. 7, 8), a point now confirmed in the strictly architectural sense by Wall's exhaustive research (fig. 6), also by my recent graphic analysis of the Portal of Al-Hakam II (fig. 10).

As Chueca had shown me, obviously, the Islamic designers' exclusive purpose was to generate *predetermined proportionate ratios*, a quest that has fascinated designers since ancient Egypt.[3] However, when I again applied the same computer-based procedure (as used in fig. 10) elsewhere in the Great Mosque of Córdoba, an identical geometric scheme was found to have been employed inside for the decorated portal placed before the "*mihrab*." (Fig. 13) A purely Islamic architectural component, the *mihrab* is a directionally-oriented (towards Mecca), prayer-niche that roughly corresponds in function to a Christian apse. The same geometric ordering patterns emerge here as those we found earlier in the Portal of al-Hakam II (figs. 10, 11) which was designed at about the same time.

However, as I have discovered (initially, somewhat to my surprise), rather than an equilateral (3:3:3) triangle, such as that one shown in Chueca's sketch (fig. 12), the basic motif is instead the so-called *Pythagorean Triangle*, an irregular figure with a proportionate 3:4:5 relationship. As A. G. Walls has shown (fig. 6), this particular geometric figure was actually used in Islamic architectural design to predetermine the proportionate organization of subdivided components in a given building. I will later present documentation of various sorts (see also the examples given in the Appendix, and in figs. 74–84) proving that 3:4:5 design module has always been recognized by builders as being unsurpassed for establishing proportionate relationships, especially in situations where workers have only rudimentary tools and even less grasp of geometrical theory.

My evidence for this conclusion is, somewhat paradoxically, Chueca's sketch (fig. 12), supposedly showing that the Islamic portal design was pre-determined by an equilateral (3:3:3) triangular module. (Fig. 14) The first step was to remove his drawn diagonal lines. The next essential step was to determine the exact placement of the compass-point originally centering the horseshoe arch, since that *center-point*— in this case, the inner axis needed to draw a proper diameter for the arch and its concentric enframements— now logically appears (as in figs. 10, 11) to have served as the central, generating axis-point for the composition of the entire wall. This crucial point was located in the old-fashioned way, by manually wielding a compass (and it is now marked with a star-pattern in my on-site photo: fig. 13).

Once my compass was fixed at that crucial centering point, the one establishing the compositional axis, I was then able to draw the complete circle of the incomplete horseshoe arch. Next, over Chueca's drawing (fig. 12) I drew with a ruler a vertical line through the compass-point, so fixing the pivotal, central vertical axis determining the entire, typically ascending, composition of this example of wholly Islamic "*decoración suspendida*" (with later Christian examples shown in figs. 1, 2, 3). The final step (as applied in fig. 14) was again to overlay the grid-pattern (as used initially in fig. 10), and the result was a verification of the hypothesis that the Islamic designers had actually predetermined their proportions by using a 3:4:5 triangle-module (instead of Chueca's 3:3:3 triangle). The corollary conclusion is that, as reiterated in the most sacred precinct of the Great Mosque of Córdoba, these ratios represent the same proportions as those once generated by the ancient Egyptian "Ceremony of the String/Cord," a solemn ritual personally performed by the Pharaoh when dedicating a great temple to his gods.

Chueca had given me the clear lead, the kind pointing to a underlying (non-inscribed), traditional design-motif explaining the unique qualities of Trans-Hispanic architecture, with that governing module being most likely specifically *triangular*. (Fig. 15) For further verification, once again the same grid was overlaid, this time over my on-site photograph of the same mihrab portal in the Great Mosque of Córdoba (fig. 13), and the same results were forthcoming. As we again see from the imposition of this schematic overlay, a complete rectangle encloses the inner *alfiz*; diagonals rising from either base determine the upper corners. Other, smaller or pro-

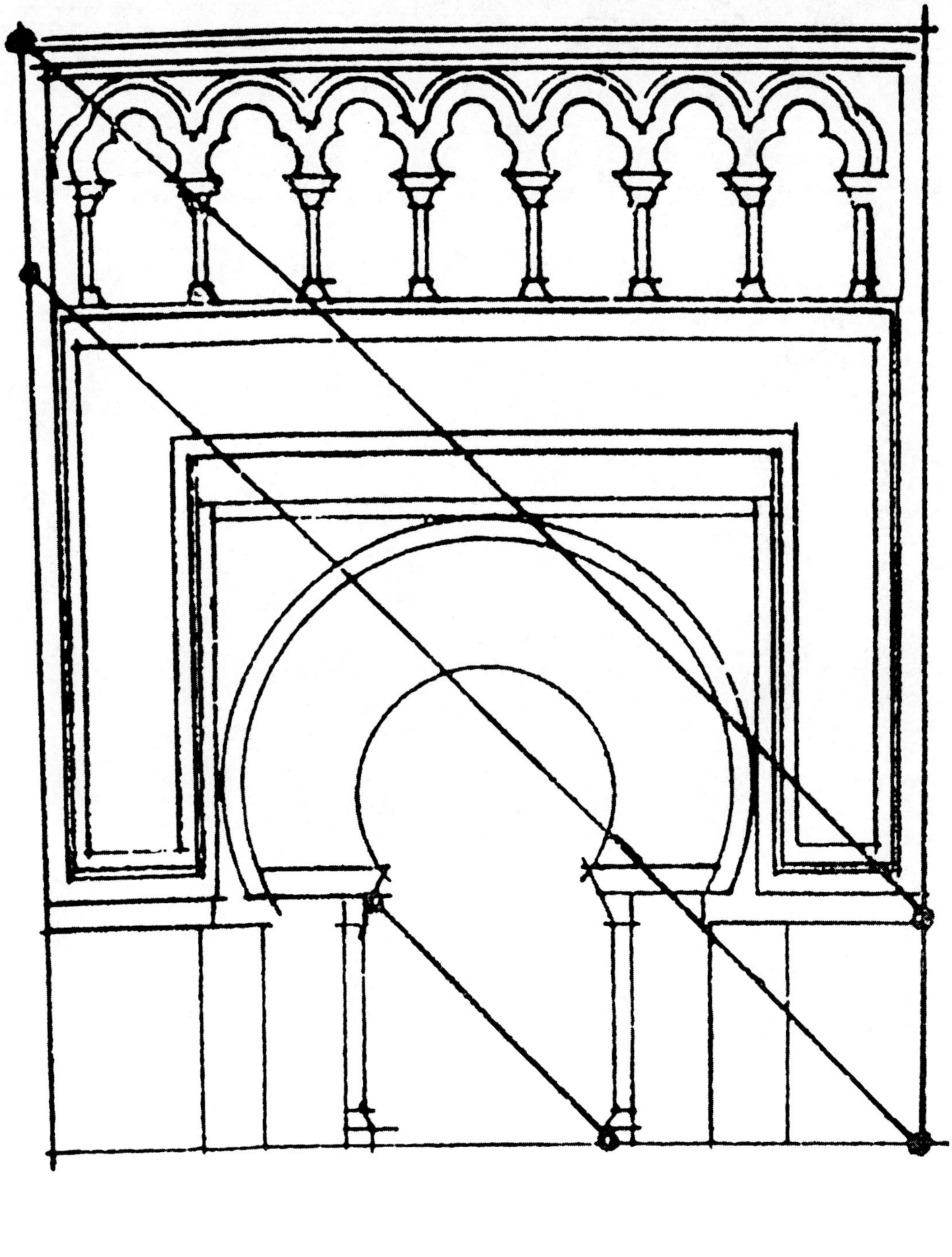

Fig. 12. F. Chueca Goitia, Inscribed "Regulating Lines" generated by equilateral triangles, and so predetermining the design of the portral enframement of the Mihrab (ca. 960), Great Mosque of Córdoba, Spain. (From Chueca, *Invariantes castizos*, 1947.)

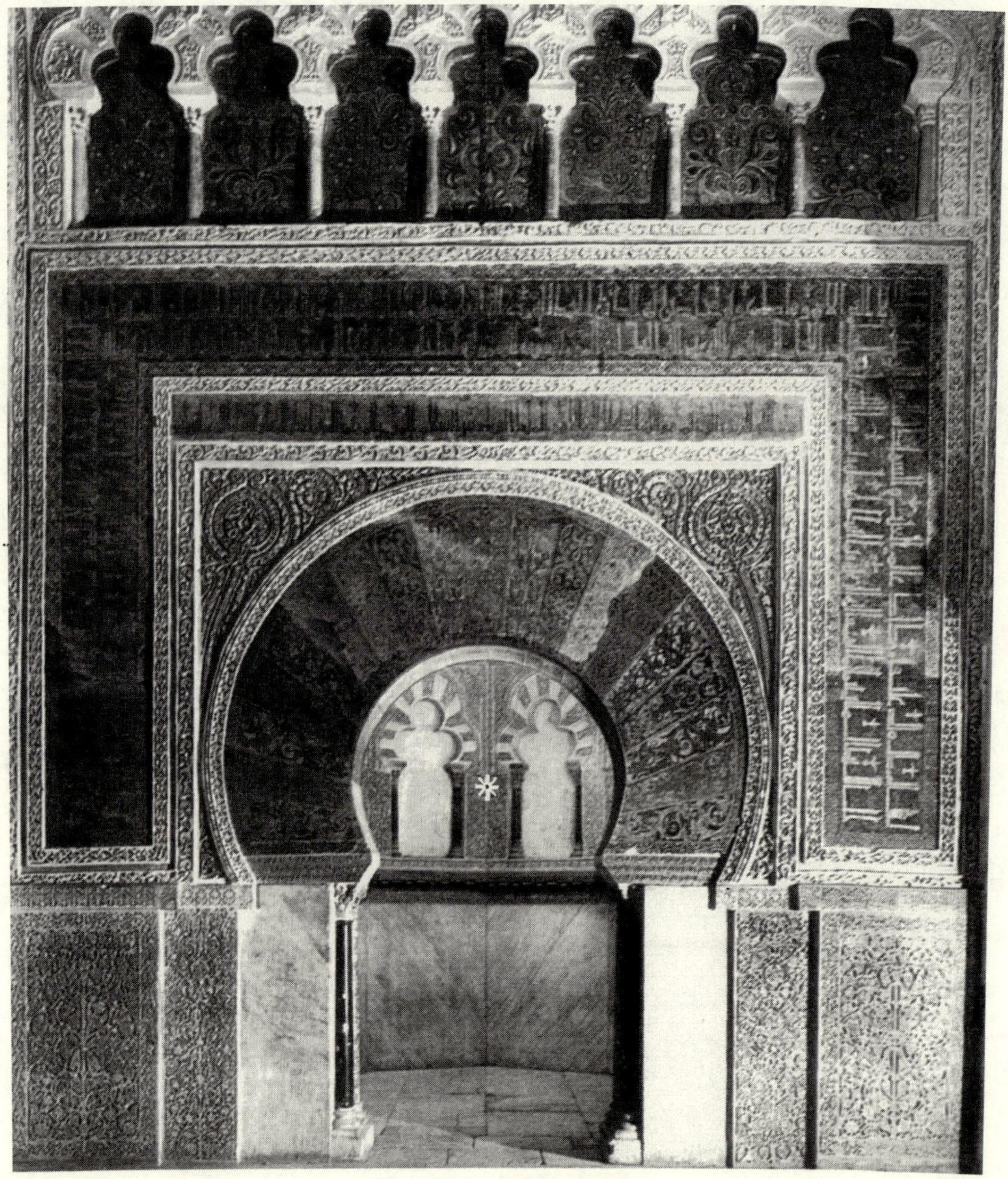

Fig. 13. Entrance to the Mihrab (ca. 960), Great Mosque of Córdoba, Spain. (Star-pattern added to indicate the compass-point originally fixing the diameter of the horseshoe-arch.)

portionately subdivided, squares internally enframe sub-units and generate the other diagonal intersectional points for the vertical and horizontal alignments, including two columns ("*in antis*") placed at the entrance to the mihrab. The axis of this all-encompassing diagonal-patterning — itself ordered by *a 3:4:5 ratio*— was fixed upon the radius point centering the horseshoe arch. The vertical intersections determined

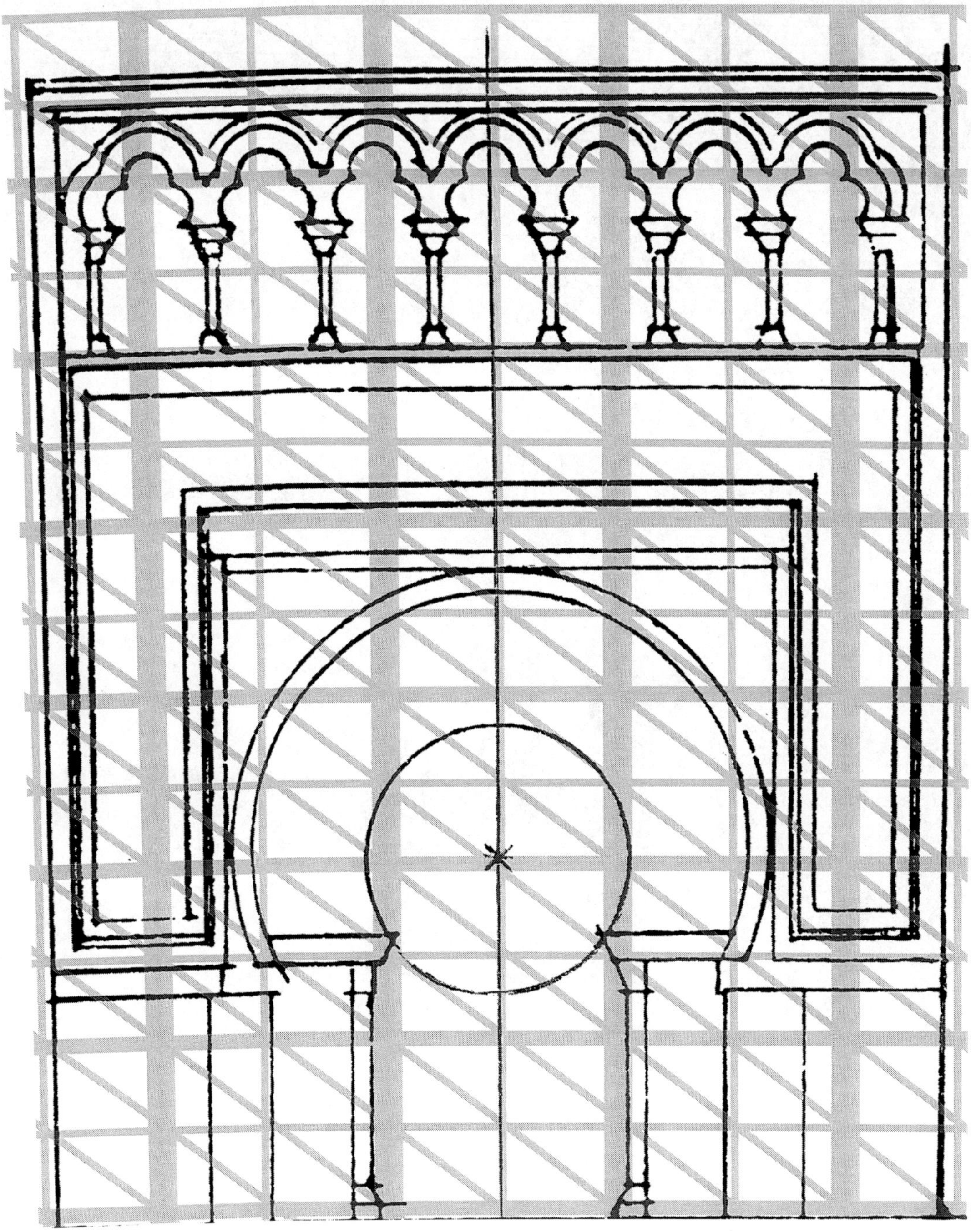

Fig. 14. Inscribed "Regulating Lines" generated by Pythagorean triangles and predetermining the design of the Cordoban *mihrab.* (Computer-generated grid-patterns layered over a drawing—fig. 12—reproduced in Chueca, *Invariantes castizos,* 1947, and now shown here after Chueca's diagonals were removed, and with the horseshoe-arch compass-point and the vertical axis-line restored.)

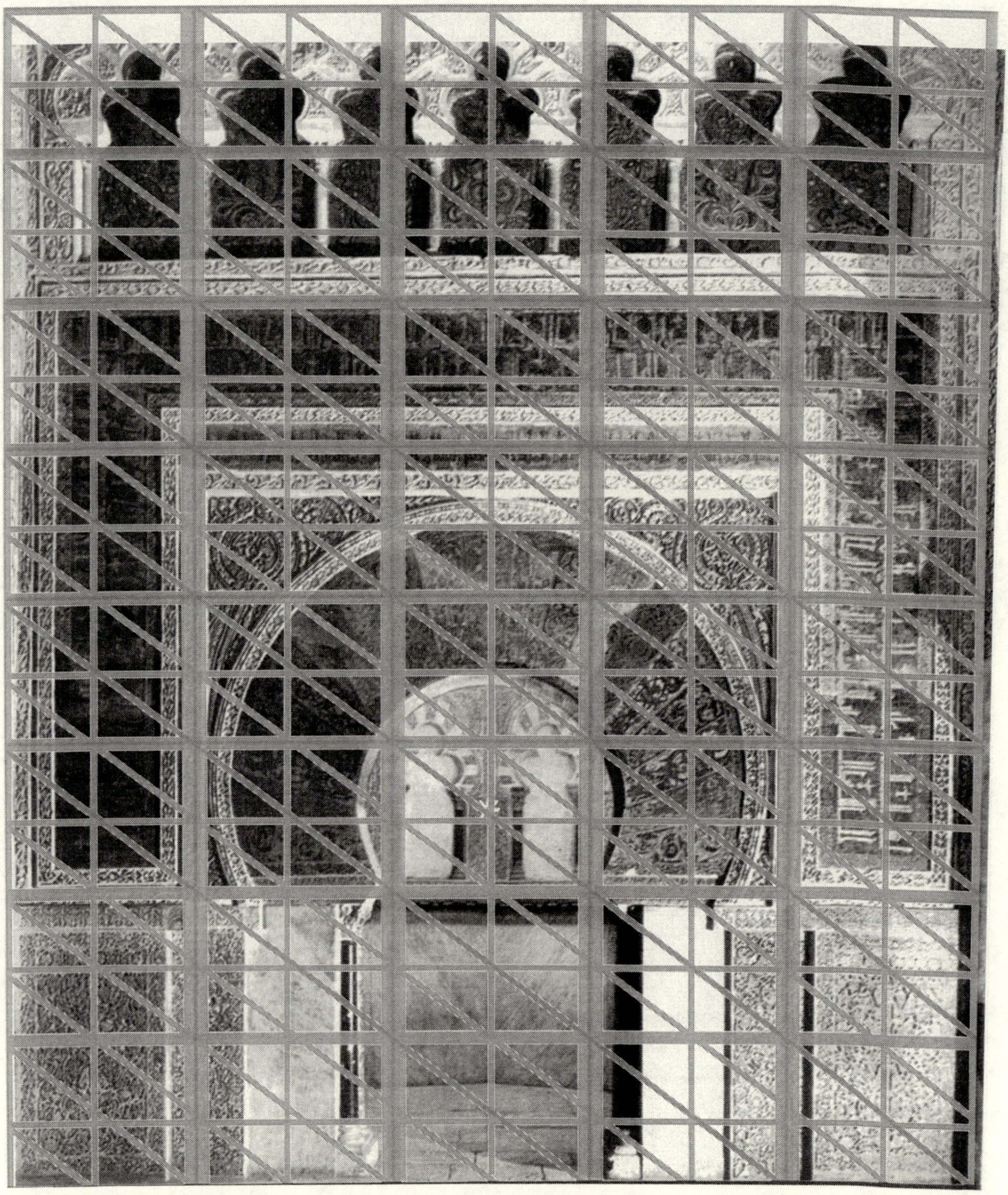

Fig. 15. Inscribed "Regulating Lines" generated by Pythagorean triangles and predetermining the design of the portal enframement of the Cordoban *mihrab*. (Computer-generated grid-patterns layered over photograph.)

by the diagonals set up the widths of all of the vertical elements, including the diameters of the blind arcades in the upper panel as well as the heights of all parts of this crowning, pseudo-architectural feature.

Were the superimposed grid pattern to be further subdivided according to the predetermined pattern, then placed over further, close-up photos, the additional markings would show how the center-points of nearly every sub-motif, many of which are obviously compass-derived, conform to the predetermined pattern of diagonal intersections. Perhaps most significantly, identification of the central radius-point for the entire decorative composition as having been established upon the very center of the horseshoe arch, as drawn with a compass (with the essential axis-point being marked here with an X), points out a specific analogy to the scheme adorning the initial design center-point at Tepotzotlán. There (as shown below in fig. 24), the center-point for the generation of the entire architectural composition is the rose-window (*rosetón*) placed directly above the main portal, and to which it is vertically aligned.

In the larger sense, this pattern of external-to-internal, portal-to-niche, replications sets up a scheme of complementary symbolic functions. Whether in a Moslem mosque or in a Christian church, the larger implications are clear. In either case, a metaphorically "liminal" entrance-way leaves "the world" and directly leads to a "Holy of Holies." Christian Spaniards, who must have picked up the idea from Iberian Mohammedans, were much later to set up an analogous, although more generalized, formal relationship between their *fachadas retablo*, placed outside at the entrance to a church, and the towering *retablos mayores*, situated behind the altar-table in the hemispherical apse situated at the end of the nave inside the church.

For the Christians however, the formal, strictly geometrical, relationships between inner and outer components seem to have never been made so rigorously identical in character as was the case with the Great Mosque at Córdoba. One convenient explanation for the disparity of formal details between one religious structure and the other is "*geometry*"; another is religious orthodoxy. Being inherently "non-figurative," only geometry was admitted as a governing design factor in a mosque. To the contrary, because Christianity is inherently "anthropomorphic," for being based in so many ways upon the pathos-ridden figure of the crucified Christ, geometry could naturally play only a peripheral role in its increasingly anthropomorphic kind of liturgy (and in the anthropomorphic artistic imagery generated by it).

In both cases, either Islamic (Córdoba) or Christian (Tepotzotlán), the underlying, geometrical-proportionate, basis of the organizational pattern — actually a "leit-motiv" — was/is the so-called "Pythagorean triangle." This standard geometrical figure has a right-angle (90°) base of three units or modules; this is joined to a vertical of four units, and the resulting diagonal will comprise five units. The acute angles belonging to the diagonal formed by this 3:4:5 figure are, respectively, 57 and 33 degrees. During the Middle Ages, a builder's system employing the 3:4:5 design module was correctly called "*ad triangulam*"; otherwise, "*ad quadratum*" referred to another favored triangular scheme, that one which was (instead)based upon the equilateral (3:3:3) triangle (see figs. 16. 62).

In short, all kinds of medieval (and, presumably, most Renaissance) proportionate "regulating" line-patterns, the kind of design modules — "inscribed figures" — used by ordinary artisans (versus gentlemen-architects), were derived from triangles (as is further demonstrated with the examples discussed — and illustrated, and in figs. 74–84 — in the Appendix). Much favored by Christian designers in the Middle Ages and the Renaissance, this ubiquitous 3:4:5 geometrical figure additionally links architecture to music, as it generates the next two major "geometric chords" of the Pythagorean scale following the octave, or *diapason* (1:2), and the fifth, or *diapente* (2:3), namely, "the musical fourth," or *diatessaron* (3:4), and then "the musical third," or *sesquialtera, tonus maius* (4:5).[4] Placed back-to-back, and then quadrupled, an ensemble of four Pythagorean triangles yields a rectangular (2:3) figure returning the harmonic cycle to its origins in the musical fifth.

6

The Evidence for "Inscribed Figures" in Architectural Design

An important issue raised here is the very existence of preliminary drawings depending upon "inscribed figures," meaning those geometric patterns that initially generate proportionate relationships, but which are not literally drawn in the finished building (for "inscribed figures," see the discussion — and illustrations, figs. 74–84 — given in the Appendix). Nonetheless, such inscribed figures are essential for establishing the initial layout of the architectural complex, even its final decorative ensembles.

In the case of European architecture, the existence of inscribed figures has generated much scholarly speculation. Rudolf Wittkower provided a list of publications arguing that "harmonic proportion results from a repetition throughout essential parts of a building of the same geometric pattern" — or inscribed, "abstract" figures — developed in the architect's initial drawing. His conclusion regarding such modern research was that, apparently, "a result was here taken as a cause."[1] Nonetheless, industrious research into the archives of the Cathedral of Milan, including close inspection of some original drawings still kept there, has conclusively shown that the preliminary layout for that building was wholly derived from simple repetitions of triangular schemes; in fact, the favored motif was the 3:4:5 figure.[2] Otherwise, there is abundant, although somewhat circumstantial, evidence attesting to the consistent use of inscribed figures as a necessary preliminary stage in architectural design.

Some of the evidence is graphic. A case in point is an engraving published in 1521 inside the magnificently illustrated Italian translation of Vitruvius that was prepared by Cesare Cesariano. (Fig. 16) Besides showing the "inscribed" geometrical system underlying the entire architectural design for the Cathedral of Milan, with a compass being depicted as an essential tool, it also shows (upper left) how a grid pattern, with diagonal intersections, was used to shape the intricately carved base of a column.[3] There many other illustrations of architectural designs with inscribed figures in the *Cesariano Vitruvius*; by my count, there are at least twenty-three examples of diverse kinds of inscribed figures (see also fig. 62). Probably the best known

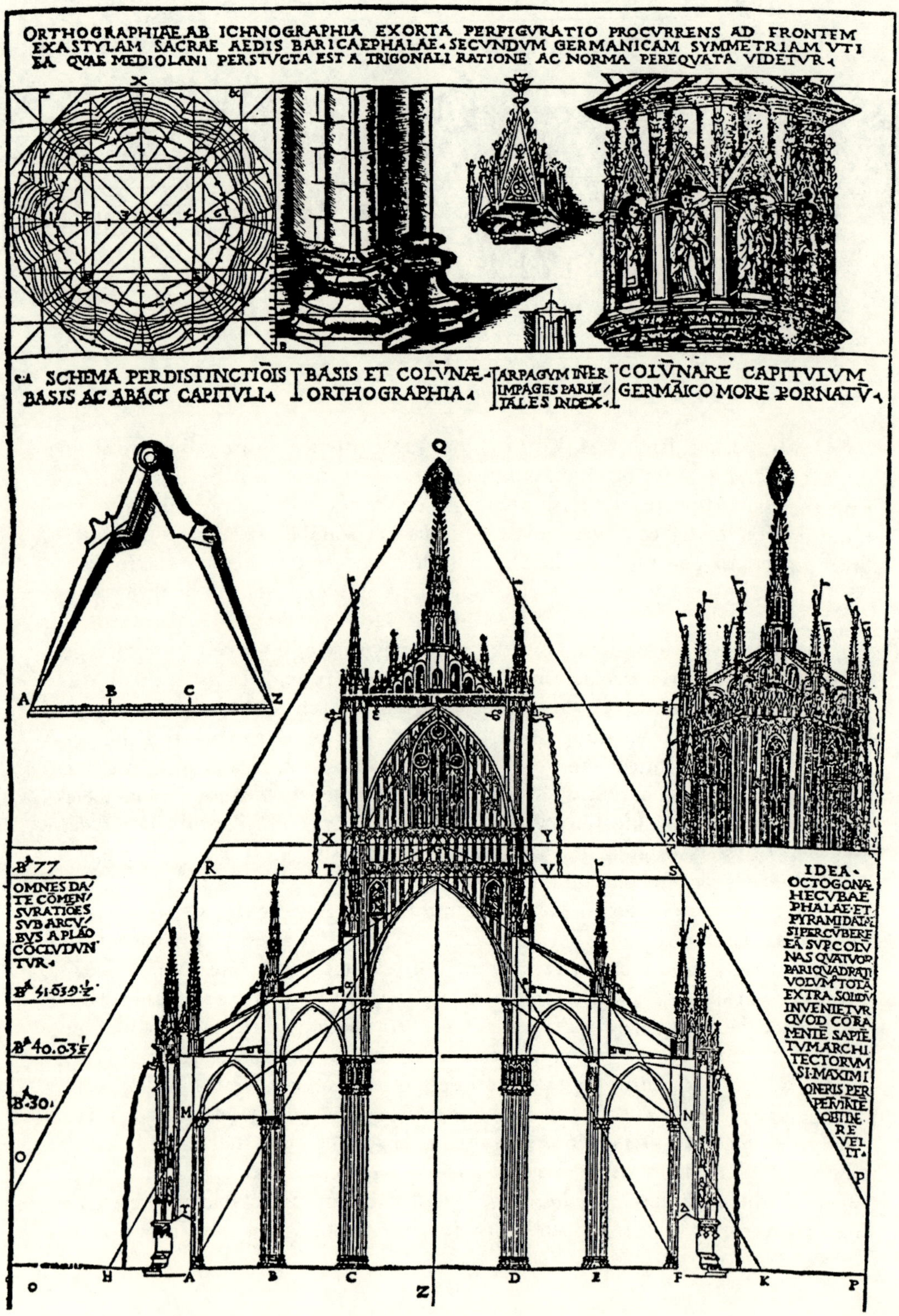
ORTHOGRAPHIAE AB ICHNOGRAPHIA EXORTA PERFIGVRATIO PROCVRRENS AD FRONTEM EXASTYLAM SACRAE AEDIS BARICAEPHALAE. SECVNDVM GERMANICAM SYMMETRIAM VTI EA QVAE MEDIOLANI PERSTVCTA EST A TRIGONALI RATIONE AC NORMA PEREQVATA VIDETVR.
SCHEMA PERDISTINCTIŌIS BASIS AC ABACI CAPITVLI.
BASIS ET COLVNAE ORTHOGRAPHIA.
ARPAGVM INTER IMPAGES PARIETALES INDEX.
COLVNARE CAPITVLVM GERMĀICO MORE ORNATV.
OMNES DATE CŌMENSVRATIŌES SVB ARCVBVS A PLĀO CŌCLVDVNTVR.
IDEA. OCTOGONAE HECVBAEPHALAE ET PYRAMIDATAE SVPERCVBERE EA SVP COLVNAS QVATVOR BARIQVADRATI VOLVM TOTA EXTRA SOLIDV INVENIETVR QVOD CORAMENTE SAPIETVM ARCHITECTORVM SI MAXIMI ONERIS PERPETVITATE OBTINERE VELIT.

are Cesariano's two illustrations of the "Vitruvian figure" (appearing on folios 49 and 50 recto), showing a nude man inscribed upon a squared grid pattern with diagonal intersections. The other inscribed figures are "abstract," intersecting patterns of lines and circles, with most of these appearing in architectural plans, and these make a clear distinction between inscribed ground-plans ("*ichnographia*") and non-inscribed elevations ("*orthographia*").[4]

An even earlier architect's preliminary drawing also shows the inscribed geometrical figures initially fixing the dimensions in the elevation of another Italian church; this sketch appeared in Francesco di Giorgio Martini's *Tratatti di architettura ingegneria e art militare* (composed between 1483 and 1492). (Fig. 17) The architect's written description throws useful archaeological light on what I take to have then been a standard preliminary procedure:

> And in order to demonstrate some other geometric proportions [*geometriche proporzioni*], those commensurate for churches with oblong naves, I first make an equilateral square [*quadrato d'equali lati*]. From one corner to the other angle two diagonal lines [*due linee diagonee*] are drawn. The base of the square is divided into four equal parts. From the [ends of the] base line CD two vertical lines [*due rette linee*] are drawn diagonally; these determinate [placement] of the horizontal line AB. Next a semi-circle is drawn from the far ends of the base line CD; its height passes through the intersection of diagonal line X, where the circle intersects points M and N, from which points horizontal lines [*transverse linee*] are drawn. These provide the proper elevation for the length of the side aisles.
>
> From there a line is dragged passing through the middle of the bigger and smaller squares, and two others pass through the mid-point of the base-line. Passing through the intersections of the vertical and diagonal lines, and extending to the far end of that portion of the semicircle, you find points E and F. This line will serve as the module for the entire church [*sarà modulo a tutto il tempio*]. And two [diagonal] lines are drawn from point Q, and these go to the upper ends of the square at A and B; intersecting points V V, these establish the width and height of the portal. The same height and width are given to the cupola or lantern on the roof [and gives further specifications for the other intersections].... In this way the church becomes rightly commensurate in all its height and breadth, and just as is shown in the inscribed drawing [*come per la figura e disegno si manifesta*].[5]

Francesco provides us with yet another example of an "inscribed figure." (Fig. 18) In this case, it is a *human* figure; this figure is both inscribed upon a squared grid-pattern (*quadratura*) and also inserted within the plan of a basilica. As he explains, "Vitruvius discusses the measurements and proportions of temples and columns, and he states that without symmetry in them no artisan can ever operate well and with reason," and Francesco goes on to describe the proportions of the human body.[6] As it turns out, this apparently fanciful image really was the initial design concept for a church that was actually built in 1485, that being Santa Maria delle Grazie in Calcinaio, near Cortona.[7]

Probably the most elaborate verbal description of an architect's inscribed figure

Opposite: **Fig. 16. Cesare Cesariano, Inscribed triangular motifs and grid-patterns employed for the design of the Cathedral of Milan (begun 1386). (Engraving published in Cesariano, *Vitruvius De Architectura*, 1521).**

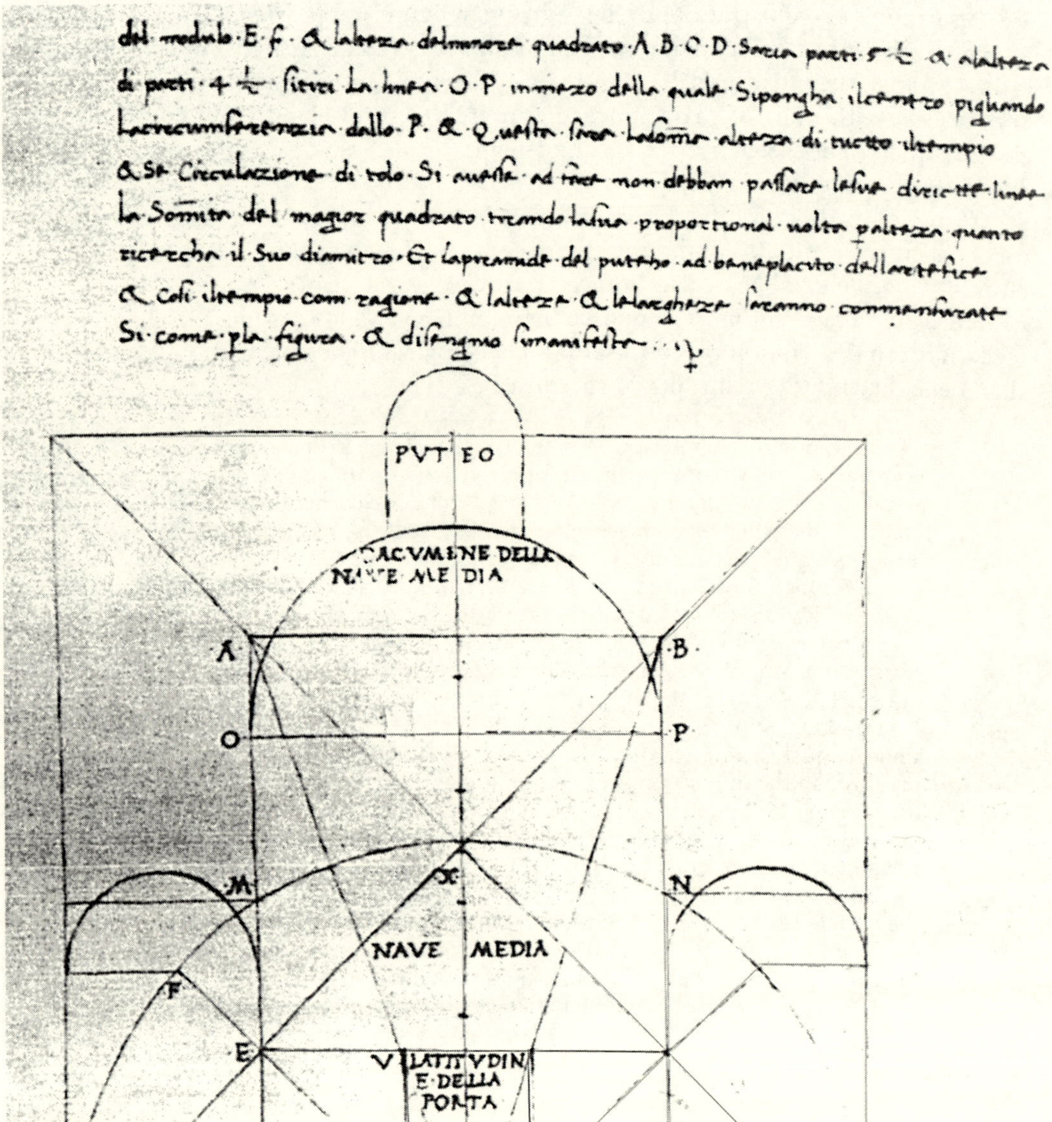

Fig. 17. Francesco di Giorgio Martini, Inscribed triangular and circular motifs determining the initial design for a church (ca. 1490, *Tratatti di architettura ingegneria e art militare*).

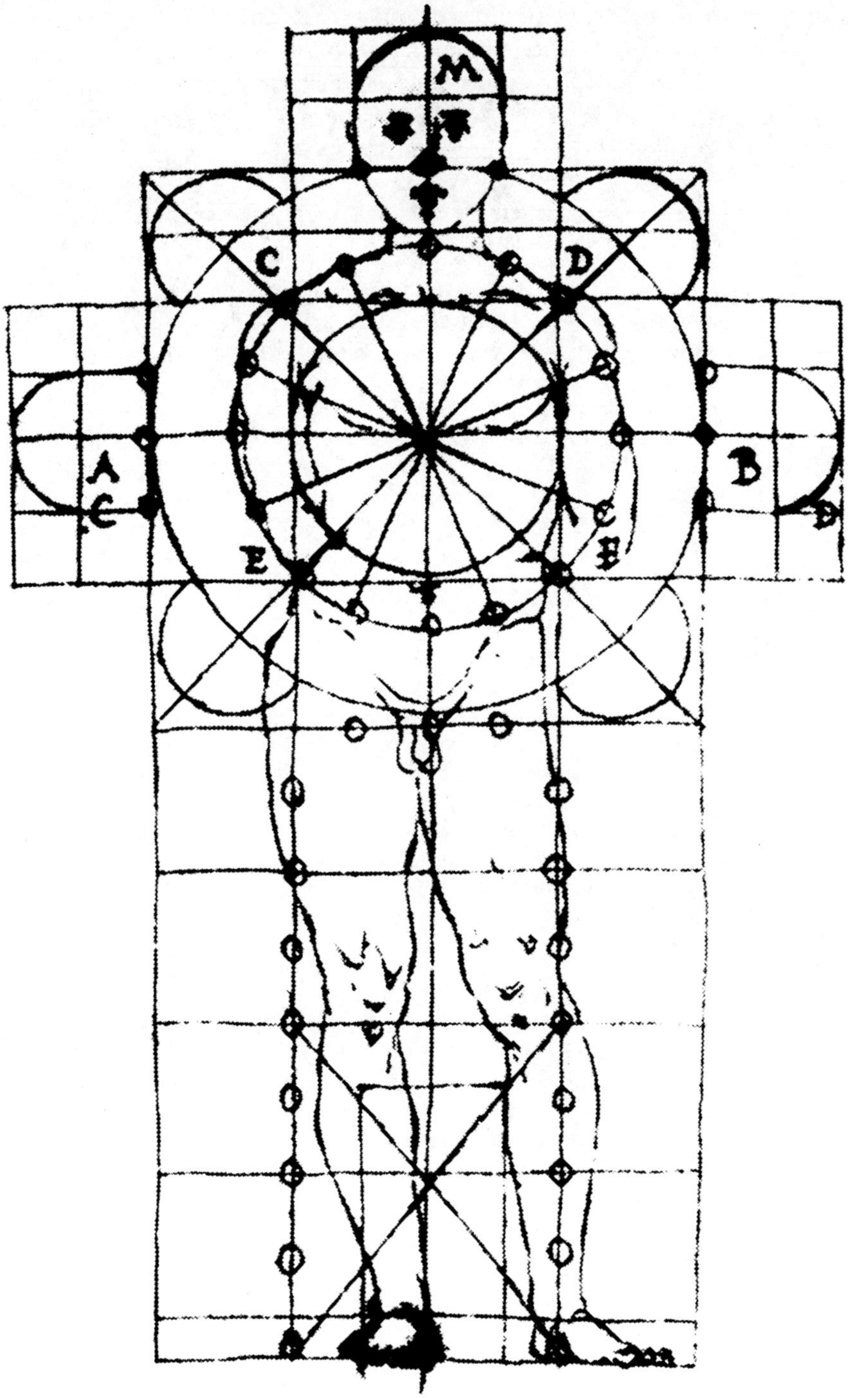

Fig. 18. Francesco di Giorgio Martini, Human figure "inscribed" over a modular grid-plan and within the floor-plan for a church (ca. 1490, *Tratatti di architettura ingegneria e art militare*).

is that one given in Francesco Colonna's *Hypnerotomachia Poliphili* (1499). Only the third book on architecture to be published in the Renaissance, and following after editions of L. B. Alberti and Vitruvius, Colonna's exotic text was, however, the first one to be illustrated. (Figs. 19, 20) As set in the narrative context of Poliphilus' "dreamed love-struggle," the hero describes how:

> I came, then, to this ancient portal of splendid workmanship, marvelously constructed with exquisite regularity and art, and magnificently decorated with sculpture and varied lineaments. I was inflamed with the pleasure of studying and understanding the fertile intellect and sharp intelligence of the wise architect; and thus I made this careful scrutiny of its dimensions, its lineaments, and its practical aspects. I measured diligently the squares beneath the columns, which were two on either side, and understood from this measurement the whole symmetry of the portal, which I will run through with a brief explanation.
>
> A four-sided figure, ABCD, divided by three equidistant straight lines and three transverse lines and three transverse lines, will become sixteen squares. Next, a figure half that size is added to it, and when this added figure is sub-divided in the same way, twenty-four squares are to be found. Such a [reticulated] figure made from threads [i.e., a grid] will also prove very useful for foreshortening in painting. Now, two diagonals are drawn across the figure ABCD, and then a straight and a transverse line, intersecting one another, so making four squares. Then in the space above the equal-sided one the four median points are marked, and the lines joining them constitute the rhombus. [And the result is] perfect symmetry of the parts....
>
> Let us now take the three preceding figures that we have drawn and reduce them to a single one, including the part added to the sixteen-squared figure, and we get this one. Now we remove the rhombus and the diagonals, the three perpendiculars and the three horizontal lines, except for the middle one, which ends where it meets the perpendiculars. In this way we obtain two perfect rectangles, one above and one beneath, each containing four squares. If we draw the diagonal of the bottom rectangle and re-draw it perpendicularly toward the straight line AB, we can easily find from its additional part the width of the arch and of the jambs. Then the line AB will be the proper place for the extended or straight beam. The midpoint of the truncated line EF will be the point from which the arched beam ends in a semicircle. This should have as much added to its downward-pointing horns as the semi-diameter of its thickness. Otherwise it will be defective, and so I cannot call it perfect; for this is how the best and most skilled of the ancients made the beautiful, exquisite and diligent designs for their arches, so as to give the arch elegance and the requisite strength, and to avoid its being blocked by the projections of the abaci.
>
> Now, under the double columns on either side, the square base, a podium, began with a plinth one foot above the level of the paved area, which was in fact the continuous boundary. From this level the reversed cymas, toruses and channels, with their astragals, rose in harmonious stages toward the podium, and thus constituted, with proper and requisite moldings, the socles or bases for the columns in antis. Next there protruded from the top of the podium the cornice with its reversed cyma and other concurrent lineaments. ... Now, adding together the entire figure of twenty-four squares, one obtains the sesquialter [4:5] proportion that gives the figure OPQT, which clearly contains in itself a rectangle and a half. Dividing that half equally by straight lines into six portions, there will result five interstitial lines and six partitions. The peak of the frontispiece occurs by rule at the midpoint of the fifth line up, from which point the lines sloped down by the proper angle until they cut the lateral extension of the corona. There, the

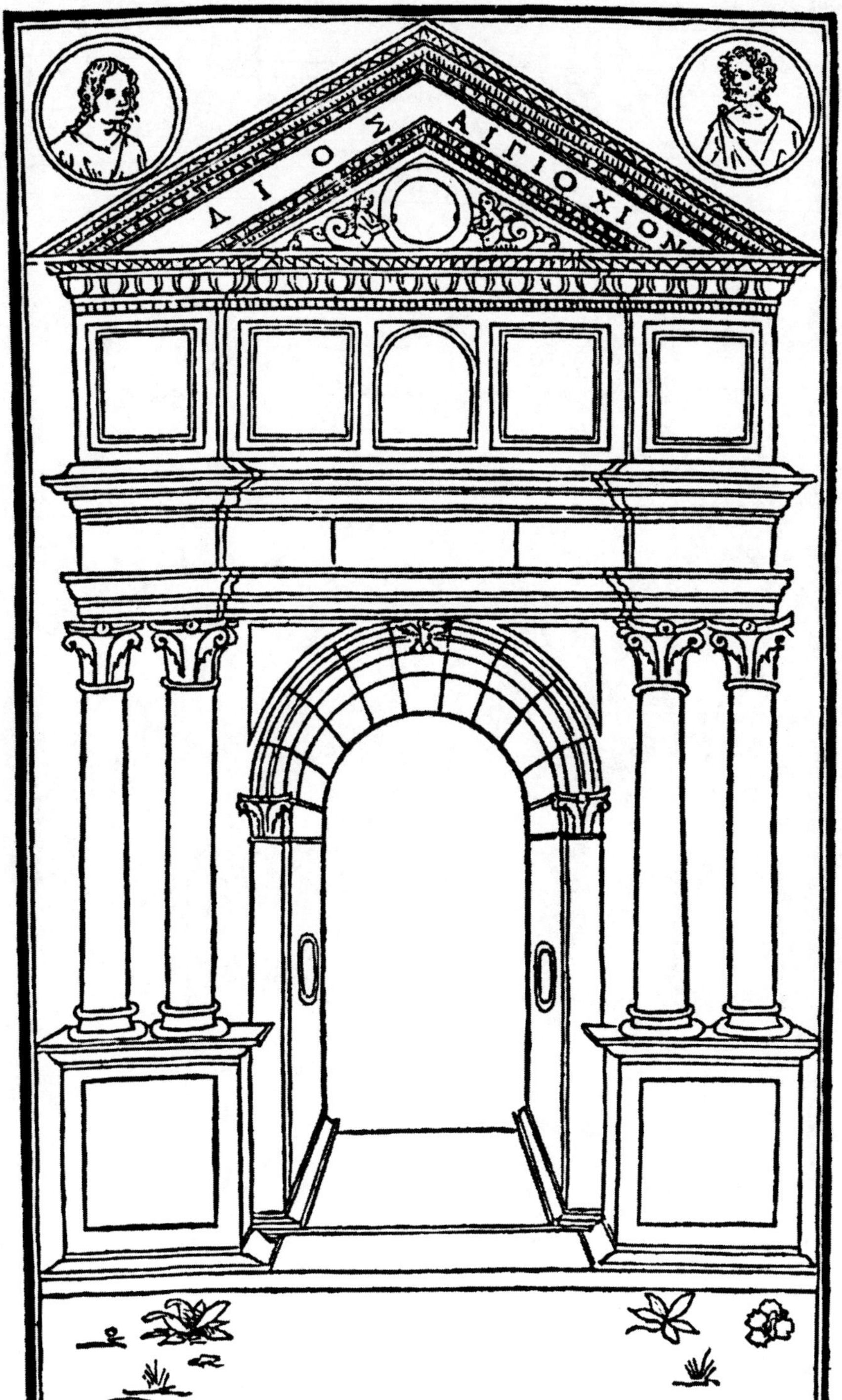

Fig. 19. Francesco Colonna, "The Perfect Portal" (*Hypnerotomachia Poliphili*, Venice, 1499).

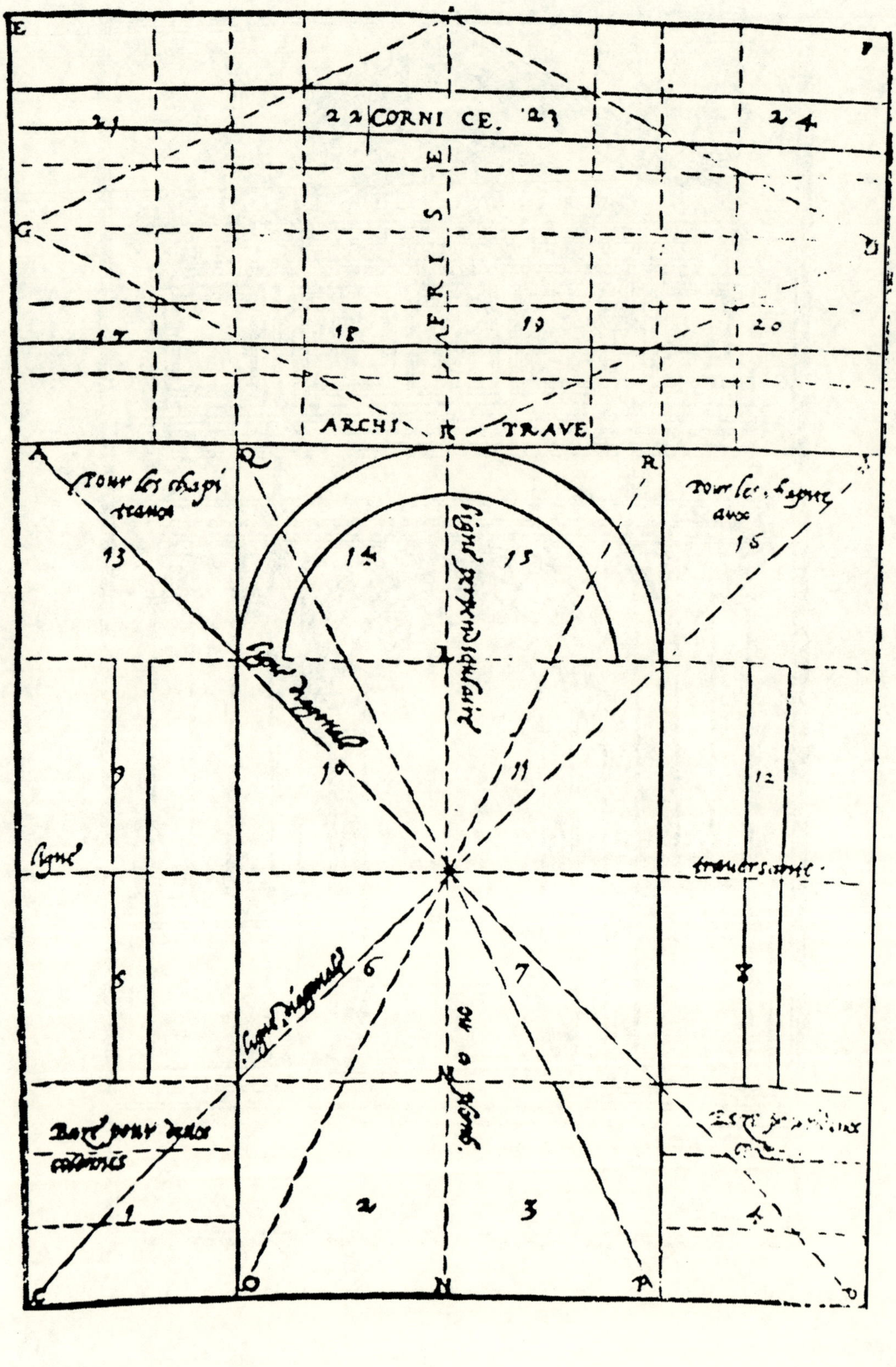
E
22 CORNI CE. 23
24
FRISE
G
18
19
20
ARCHI
TRAVE
A
R
Pour les chapi teaux
Pour les chapiteaux
13
14
15
16
ligne perpendiculaire
L
ligne diagonale
10
11
12
ligne
traversante
ligne diagonale
6
7
Base pour deux colonnes
1
2
3
O
N
P

> extremities united exactly with the cymas of the projecting corona. Finally, the frontispiece corresponded exquisitely with the lineaments of the elegant cornice: its first order was taken from the plane of the projecting square, and its last from part of the dentillated cornice, within which the angular figure was contained....
>
> The first rule that the architect must observe after the [initial] conception of the building is the square [*quadratura*], which is sub-divided to the smallest degree [within a grid] to give the building its harmony and consistency and to make the parts correlate with the whole. This is why the present portal was so beautiful in its admirable composition and invention, and why it was so particularly elegant and so faultlessly arranged, so that not even the most obscure [decorative] element in it could provoke criticism.[8]

However, perhaps the best known graphic portrayal of the architects' inscribed designs was that one published by Sebastiano Serlio in his *Architettura* (1537, with many subsequent editions and translations). His large folio volume was a practical handbook, a sort of pictorial recipe book of instant solutions for builders; as such, it was popular throughout the world, including the New World. The first book of Serlio's *Architettura* is a practical handbook of applied geometry, and the last illustration shows one how to use inscribed geometrical patterns to proportionately generate the format of a classical portal. (Fig. 21) As Serlio explains,

> If a craftsman wants to make a gate or door for a temple or church which is to be proportioned according to that place, then he must take [into account either] the width inside the church or else the breadth of the outside wall. If the church is small, and if it has pilasters or pillars inside it, he may then take the width between them [as a module] and he sets that same breadth inside an [inscribed] equal-sided square, that is, one that is as high as it is wide. Inside this square, the diagonal lines and the other two cross-cutting lines will not only show you the width of the door but also the places for, and points of the ornaments belonging to the same door, just as you see here in this figure. And if it should fall out [or happen] that you have three doors to make in a church, to that end cut three holes [or openings], and you can still retain this proportion for the smallest of them. Although the cross-cutting [process], or [sub]dividing [by intersection], is [potentially] without number, three [sets] at least should not prove too tedious [to execute]. There ends my geometry [lessons].[9]

Put in a different (hopefully clearer) fashion, first one draws an imaginary square and into this there is put the "real door," and that is initially done by pulling "inscribed" diagonal lines AB and CD; these establish the height and width of the cornice. From the same base-line, AC, another pair of diagonal lines, AE and CE, rise, so making an isosceles triangle. The intersections between the diagonals and the sides of the triangle, at F and G, mark the height and width of the door itself. If one desired more decorative elements (just as a Hispanic designer would), then one would simply add more parallel diagonals, so obtaining further intersections. As Serlio understood, inscribed subdivision by intersection is a process that is potentially without limits. And Rudolf Wittkower had observed that Serlio's "drawing seems to suggest a geometrical procedure not very different from the 'ad quadratum' method

Opposite: **Fig. 20. Francesco Colonna, "The Perfect Portal" with "inscribed figures" (*Hypnerotomachia Poliphili*, Paris, 1546).**

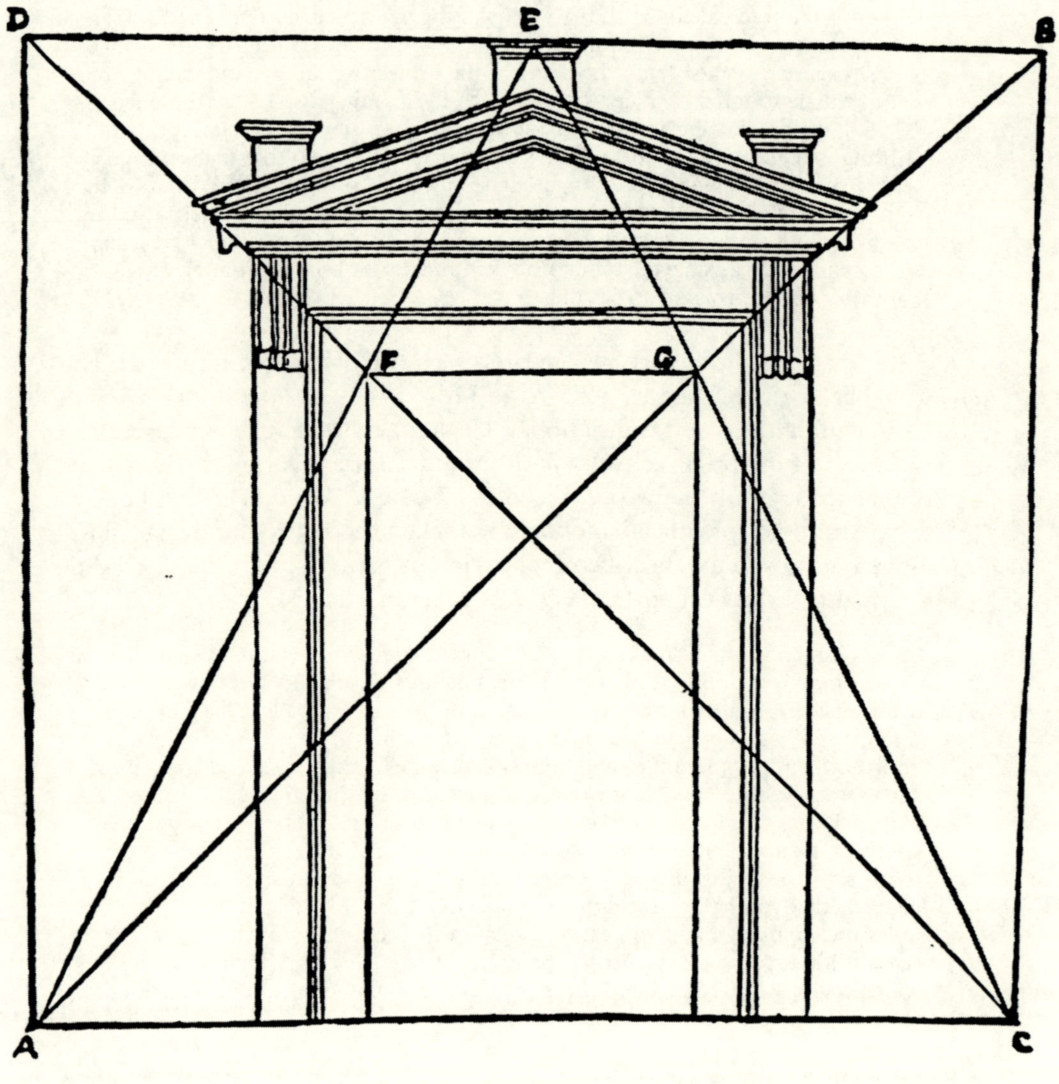

Fig. 21. Sebastiano Serlio, Inscribed geometrical patterns generating the format of a portal (Serlio, *Architettura*, 1537).

practised during the latter Middle Ages.... Medieval geometry here is no more than a veneer that enables [post-medieval] practitioners to achieve commensurable ratios without much ado."[10]

Clearly, post-medieval architectural designers used inscribed figures in the preliminary planning stage. Unfortunately, these inscribed sketches have since mostly

Opposite, top: **Fig. 22. Silvestre Pérez, Modular "inscribed figures" ordering the groundplan and elevation of the Church in Mugardos (1804). Madrid, Biblioteca Nacional.** ***Bottom:*** **Fig. 23. F. Chueca Goitia, "El trazado regulador" (regulating lines) inscribed on the façade of the Cathedral of Valladolid, designed ca. 1585, by Juan de Herrera. (Drawing given in Chueca, *Invariantes castizos*, 1947).**

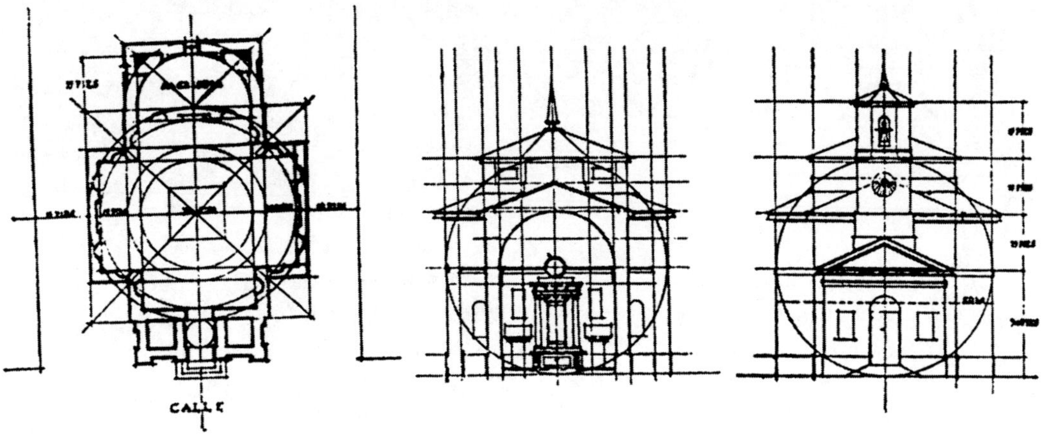
CALLE

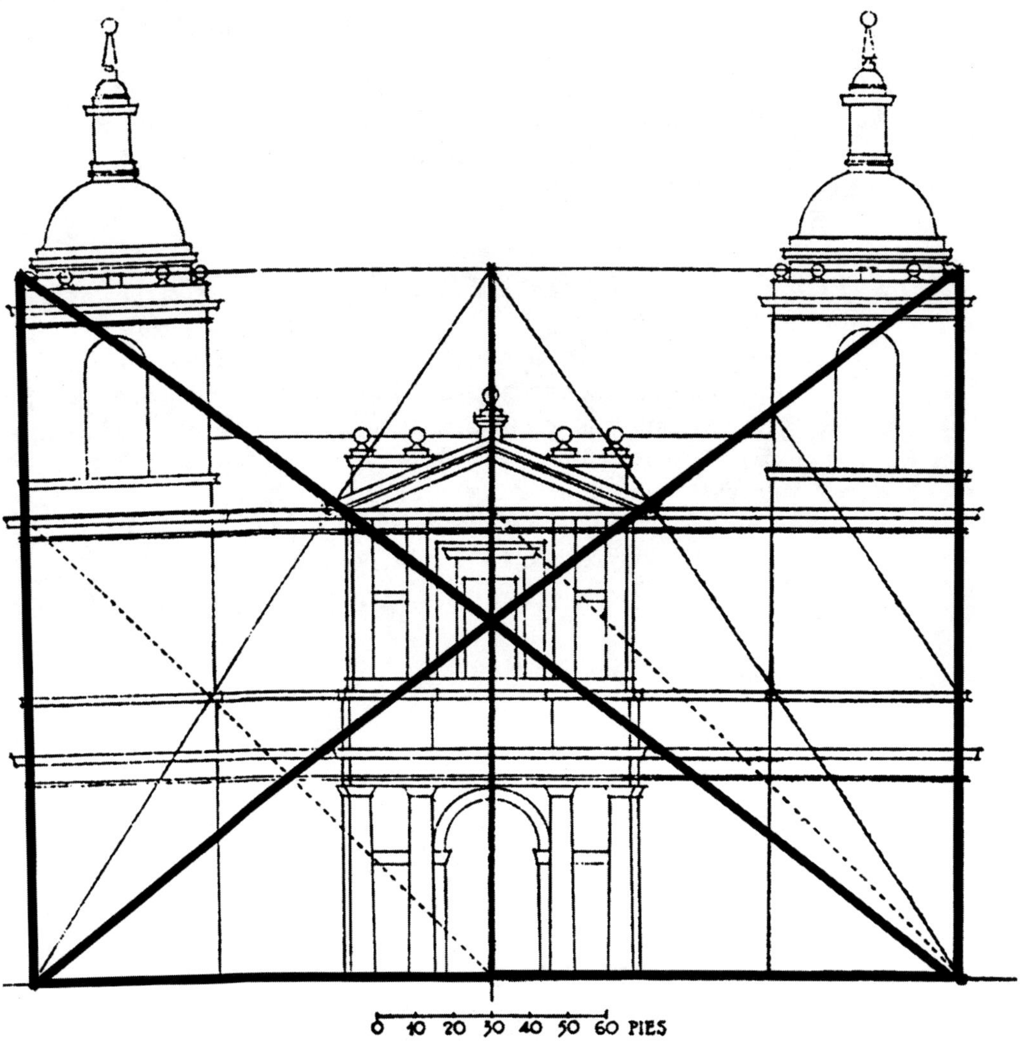
0 10 20 30 40 50 60 PIES

disappeared. But, happily, several Spanish examples have survived.[11] (Fig. 22) In this case, the architect of record is Silvestre Pérez (1767–1825), a native of Aragón, a place where *mudéjar* architecture is ubiquitous (see fig. 29). He began his planning in 1804 for the church of Mugardos (La Coruña) with a rigorously geometric conception based upon a modular system, with each squared module measuring 15 *pies*. For the ground-plan, he first combined a traditional grid-pattern with superimposed diagonals in order to establish the center for the interior; this is the spot allocated to the altar. In the elevation of the church, the critical center-point becomes the cross crowning the altar; on the exterior wall, this point corresponds to the middle of the baseline in the pediment. Chueca observed how Pérez "subjected his volumes to clearly geometric systems, so [again] showing the importance of the grid-square [*cuadrado*] as a constant factor explaining Spanish architectural proportionality."[12]

As has been noted, several modern scholars have sought traces of inscribed figures in post-medieval architectural design. For our purposes, particularly important among these is one illustrating geometrical patterns adorning the Cathedral in Valladolid, and Chueca Goitia calls our attention here to "a formula introduced [ca. 1585] by Juan de Hererra." (Fig. 23) In this case, the "formula" involved a "inscribed figure" which is wholly triangular in character. Turning to Chueca's explanation of this drawing, one reads how, in this case, "the façade is inscribed within a rectangle formed through the sesquialteran [4:5] sum of two vertical, 2:3 rectangles. As a result, it has a base of 4 and a height of 3. By means of diagonals drawn from the tips of these the specific points of the whole composition are obtained."[13]

However, it must be pointed out that the crucial center point of the entire composition was initially determined by the intersection of two *Pythagorean triangles*, and just as those were shown drawn in bold lines in Chueca's tracing. Their height then determined the bottom of the domes belonging to the paired bell-towers. Chueca also found an inscribed equilateral triangle, but this figure is wholly redundant: the paired Pythagorean triangles had already established the width and height of the central *retablo fachada*, also the inscribed horizontal line corresponding to the base of the belfry domes. More will need to be said about Herrera (and will, in Chapter 11), for he was the most important architect in Renaissance Spain (see figs. 40–51).

7

The Physical Execution of the Hispanic Façade: "Design-by-String"

As I would now generally argue, it is the post-medieval Hispanic designers' consistent reliance upon the "inscribed" *ad triangulam* procedure that distinguishes their typically "weightless" architecture from that produced in the rest of Europe, where the favored procedure literally rests squarely upon the *ad quadratum* mode. In any event, my simple mechanical-analytical, recreative procedure yields results which anyone can easily duplicate by making a proportionately marked, triangulate-module, grid-overlay using standard graphic software; that initial step is so simple that even I was eventually able to master it. The next step of the investigative procedure simply involves inserting the standard grid-pattern, treated as a "layer," into another *layer*, in this case, the grid-layer is "pasted" upon an undistorted, straight-on, digitized photograph-layer (or, even better, an accurate and measured, detailed line-drawing) of a given building or its ground-plan.

Once repeated, with various examples forthcoming (figs. 24 ff.), the computerized procedure reveals that it was the vertical intersections of the parallel diagonals—originally found with a simple plumbline—that predetermined the final positioning of each and every major or minor decorative element in the façades we examined, these being either Hispano-Islamic or Hispanic/Latin-American. We have just seen how these findings initially hold true in the cases of both the tenth-century Moslem portal-façades in Córdoba (figs. 9–11) and also in the exactly contemporaneous *mihrab* enframement in the Great Mosque (figs. 12–15) that corresponds to them in a functional and, equally, in a symbolic sense. These Islamic structures represent the geometric *fons et origo*. The next task is to see just how the same design procedure was used much later by the Spanish Christians.

As is now equally obvious to me (even though I admittedly have no written records from the world of Islam to cite), initially the entire ensemble at Córdoba had been derived wholly from symmetrical patterns—"inscribed figures"—traced out in ink on paper or vellum. The initial, wholly linear design must have been laid out flat, upon a workbench, and the able designer's tools need have been nothing more than pen, a straight-edge, and a divider-compass. This last tool seems to have been more important, for it was mainly employed to determine outer faces and proportionate intersections. The way that the simplified linear schematic had probably been translated — that is, from the drafting table onto a full-sized façade — can now be reconstructed. In spite of considerable distances in times and places, this was most likely the same simple procedure that was employed both in Córdoba (figs. 9–15) and at Tepotzotlán (fig. 1). Put in the simplest terms, for the physical translation of such façade designs—from the draftsman's reduced drawing (with no numerical dimensions being given, just ratios), thence to the towering vertical wall — the major "tool" was just a lot of string or twine. These procedures, being "artless," represent practical engineering skills at their most rudimentary, but most efficient, levels of attainment.

Being "artless," no sedentary architectural theorist during the Renaissance period would ever have deigned to describe in detail the ageless procedure so common among numberless generations of largely unlettered artisans. Nonetheless, the unquestioned utility of "DETERMINING THE PERPENDICULAR BY PYTHATGOREAN THEOREM" for U.S. Navy Construction Battalions ("Seabees") — meaning largely unlettered (especially in geometry and mathematics), also instrument-less, artisans working in the boondocks— has been recognized. And the ancient technique is thus concisely described in a 1987 government publication, a "field manual":

> The easiest and most accurate way to locate points on a line or to turn a given angle, such as 90°, from one line to another is by use of a surveying instrument called a TRANSIT. However, if you do not have a transit [and one often does not], you can locate the corner points by tape [or cord] measurements by applying the Pythagorean theorem. First, stretch a cord from your footings, and locate points C and D [defining the outer wall] by tape measurements from those footings. Now you will observe that straight lines connecting points C, D, and E, G [the far side of a rectangular ground-plan] would form a right triangle with [in this example] one side 40 feet long, and the adjacent side is 35 feet long. By the Pythagorean theorem, the length of the hypotenuse of this triangle (the line ED) would equal the square root of $35^2 + 40^2$, which is approximately 53.1 feet. Because figure EGDC is a rectangle, the diagonals both ways (ED and CG) are equal... When the tapes come taut, the joined ends will lie on the correct location for point E. If you would rather avoid [or cannot even fathom] the square-root calculation required in the Pythagorean theorem method, you can instead apply the basic fact that any triangle with sides in the proportions 3:4:5 is a right triangle. In locating point E, you know that a triangle with sides [as in this example] 30 and 40 feet long, and a hypotenuse 50 feet long, is a right (90°) triangle.[1]

Whether then or now, and as described above, on the actual building site, a given, "real" width would then be assigned to the actual length of the facing wall; this would have been only schematically presented in the original line-drawing (and two of these schemata are provided in our Seabee field-manual).

In eighteenth-century Mexico, the number values would have been given in either *pies* (feet) or in *varas* (or Spanish yards; one *vara* equals 84 cm.). From a now-known numerical measure —"width"— the actual height of the whole architectural ensemble — which at first had only been abstractly represented in a numberless, but proportionately drawn, line-drawing—then becomes nearly automatically determined. In this case, the only purpose of the initial linear schematic would have been to predetermine *proportionate relationships*, or simple ratios, between the diverse motifs—portals, cornices, columnettes, windows, sculptural ensembles, etc.—belonging to the *decoración suspendida*.

Once the lateral dimension of the site had been pegged out, light wooden scaffolding was probably then raised, with this corresponding to now-known dimensions of (proportionate) breadth and height. Once the workmen reached the top of their temporary elevated platform, two diagonal cords would then be stretched out (as seems was the case in fig. 23), with these strings running from top-left to bottom-right and the reverse, and the point of their juncture then indicated the exact center of the whole façade, typically centered on the church portal and aligned with a rose-window (fig. 1). At this time, horizontal cord is also pulled across the top of the scaffolding and pegged down, and then a plumbline is dropped down the center, through the diagonal intersections.

To determine true horizontality, a carpenter's bubble need not be not employed. As one has recently seen on some (modern) rural Mexican building sites, a piece of ordinary pipe, with a cord loosely tied at either end, serves perfectly for these simple purposes.[2] When a workman lifts the string in the center, at a point previously marked on the cord, the device falls perfectly level. From either the diagonal or the horizontal strings, all stretched out across the wall, others can be placed perfectly parallel to the first ones. The proportionate intervals between these correspond to points previously marked in the artisans' working drawing, and this is typically schematic. To determine the placement of all the vertical intersections marked on the drawing, plumb-lines (in Mexico, still often just rocks tied to cords) are next affixed to the scaffolding, and then the drawn cords are transferred to the wall as it reaches these heights. The points of diagonal intersection may be similarly predetermined by spaced marks that are put upon different lengths of twine after they have been first stretched out on the tracing floor to be so calibrated.

Whether horizontal or vertical, or even diagonal, the practical results are "string-courses"—and in the most literal sense of the term. Needless to say, this wholly pragmatic, and naturally "proportionate," layout inherently owes much more to manual facility and efficiency than to any extrinsic grasp of abstract "theory," even a once much-discussed "Pythagorean Theory."[3] In fact, for such purposes *theory* proves quite redundant and, therefore, wholly useless. As any experienced contractor will tell you, bi-laterally symmetrical plans are much easier to plan and to execute, hence are more economical, than are the disproportionate and irregular kinds. In strictly "stylistic" terms, one additionally sees how the presumably ancient string-procedure perfectly duplicates, indeed actually reinforces, the wholly linear characteristics of the schematicized drawing initially generating the planarized design (figs. 9–15).

In sum, the physical means of execution are amazingly simple—and quite infallible. By such "primitive" means: simple, but really sophisticated, workshop procedures, one can easily recreate the mundane, strictly physical, "paper-like," origins of post–Islamic, Hispanic "pseudo-architecture," a planar-geometric system seemingly wholly oblivious to gravity. The logical end result of the simple draftsman/mason's procedure was a system in which all the different elements naturally emerged quite devoid of symbolic weight-bearing functions. In short, vertically distributed "weights" do not exist on the stable horizontal plane of a drafting table, the ultimate source of all geometry-derived, abstractly proportionate, working drawings.

Since the scheme executed at, for example, Tepotzotlán seems to have come down from ancient Hispano-Moslem building techniques, brief mention may be made of the most likely context of its usage at the Great Mosque of Córdoba. Rather than "architecture" — plastic entities co-existing in a temporally determined and phenomologically "real" world — all that really mattered to the early Islamic structural designer (as is known to have been the case with all the other visual arts dedicated to Allah) were the timeless verities of *plane geometry*. As may be further suggested, the basic module (*modulus*) determining the overall proportionate system (and as executed à la Vitruvius) was likely due to a "ready-made component," namely, the Roman columns and capitals that the Isamic builders recycled (as *spolia*) within the *haram* of their Great Mosque.[4]

The geometric schemes I have extrapolated from the portals at Córdoba (figs. 9–15) were, however, certainly not "invented" in that place and time. Such techniques were probably first worked out empirically by land-surveyors in ancient Egypt around 3,000 BC (and so noted in Plutarch's *Treatise on Isis and Osiris*, as quoted in chapter 19). The most likely written means of transmission to Córdoba of the basic principles would have been the *Elements of Geometry*, as edited by Euclid. As is now known, if not an earlier one, there was certainly made available in al-Andalus a ninth-century Arabic translation of this work, by Ishaq ibn-Hunein.[5] This same source, Euclid (ca. 290–350 CE), has also been shown to have provided the basic geometric knowledge essential to medieval Christian architects.[6] Among these, one of the few to have left any written (and illustrated, ca. 1230) record was Villard de Honnecourt, a "working geometer" (*opératif*) from northern France.[7]

8

The Evidence for Pythagorean Triangulation in the Design of Tepotzotlán

With that in mind, and in the same way that the two Córdoban portals were approached, again the linear schematic is deployed, and this time it is to be applied to the portal-façade of the Jesuit church at Tepotzotlán. (Fig. 24) This imposed diagram, with two linked grids of the same dimensions (and with the source-photo—fig. 1—now having been adjusted for parallax distortions), demonstrates how the eighteenth-century Mexican building had also been based upon pre-determined systems of diagonally designed, a priori schemes of superimposed, parallel-intersectional patterns, all meeting at either 33° or 57° angles. Once again, the microcosmic leit-motiv is the ubiquitous 3:4:5 triangle, and it seems to work as well for the main façade as for the soaring bell-tower to its left (as further analyzed in fig. 28). Additionally, these are essentially the same patterns that had initially appeared in Córdoba eight (or nine) centuries earlier (figs. 9–15). The first grid-overlay (fig. 24) with triangular patterns was oriented upon the center of the façade at Tepotzotlán; then a second grid was linked to the upper right-hand corner of the first grid and extended upwards to show the repeated operations of the same geometric module upon the decoration embellishing the bell-tower (also shown in a measured line-drawing in fig. 28). In this case, the true center, or point of departure, for the whole compositional scheme seems, and logically so, to be the rose-window (*rosetón*).

Next a grid is imposed upon the central ensemble ("*imafronte*") of the church façade. (Fig. 25) In this detail, including only the first two stories but the entire width, we see how the intersections of innumerable 3:4:5 triangles centered on the rose-window determined the placements of both the height of the door below and the height of the broken cornice above the *rosetón*. The locus of the straighter cornice, between the *rosetón* and the door, was found by dividing the rectangular distance between its upper left and lower right corners by simple bi-section. Triangulation similarly determined the vertical alignments of the encrusted columns framing a stacked quartet of sculptural decorations flanking the door. Horizontal lines, aligned to intersections of further diagonal lines, determined the top of the

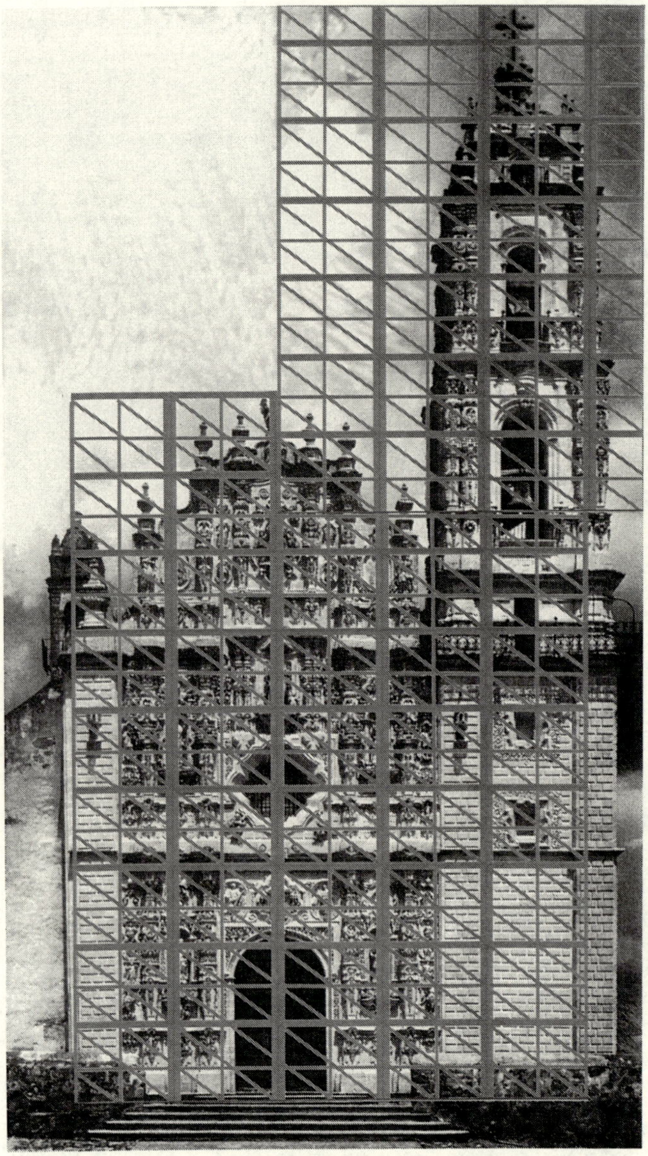

Fig. 24. Inscribed "Regulating lines" (*el trazado regulador*) ordering the entire ensemble of the façade of San Francisco Javier, Tepotzotlán. The center-point for the entire composition is the rose-window (*rosetón*). (Computer-generated grid-patterns layered over photograph: fig. 1.)

statue-niches below and the base of those above. The diagonals generated by the center-point also appear to have determined the limits of the far edges of the whole façade (1½ widths from the center) and, carried upwards, even the tips of the finials on the crowning gable.

In order to connect proportionately the bell-tower to the rest of the *imafronte* (central façade), squeezed between the two buttress-like pillars, further diagonal lines were drawn, each having an angle of 57° (again, this is the largest acute angle in a 3:4:5 triangle, with the smaller being 33°, and with both being attached to the 90° right-angle), and all these lines are anchored at regularly ascending, center-points stacked along the central, vertical axis. Sequentially, the diagonals so drawn link, first, the bottom of the portal to the far corner of the first cornice in the bell-tower; second, the center of the coat-of-arms above the portal to the second cornice; third, the center of the statue base above the rose-window to the third cornice, and the tip of the highest gable to the base of the cross at the top of the tower. Hence, both the overall height of the campanile and the dimensions of its various parts seem also to have been predetermined by intersecting, diagonal-patterns made from Pythagorean triangles arranged in parallel groups, and just as these are aligned on the portal-to-rose-window, vertical axis.

Similarly, a magnification of a section found in the center of the Tepotzotlán

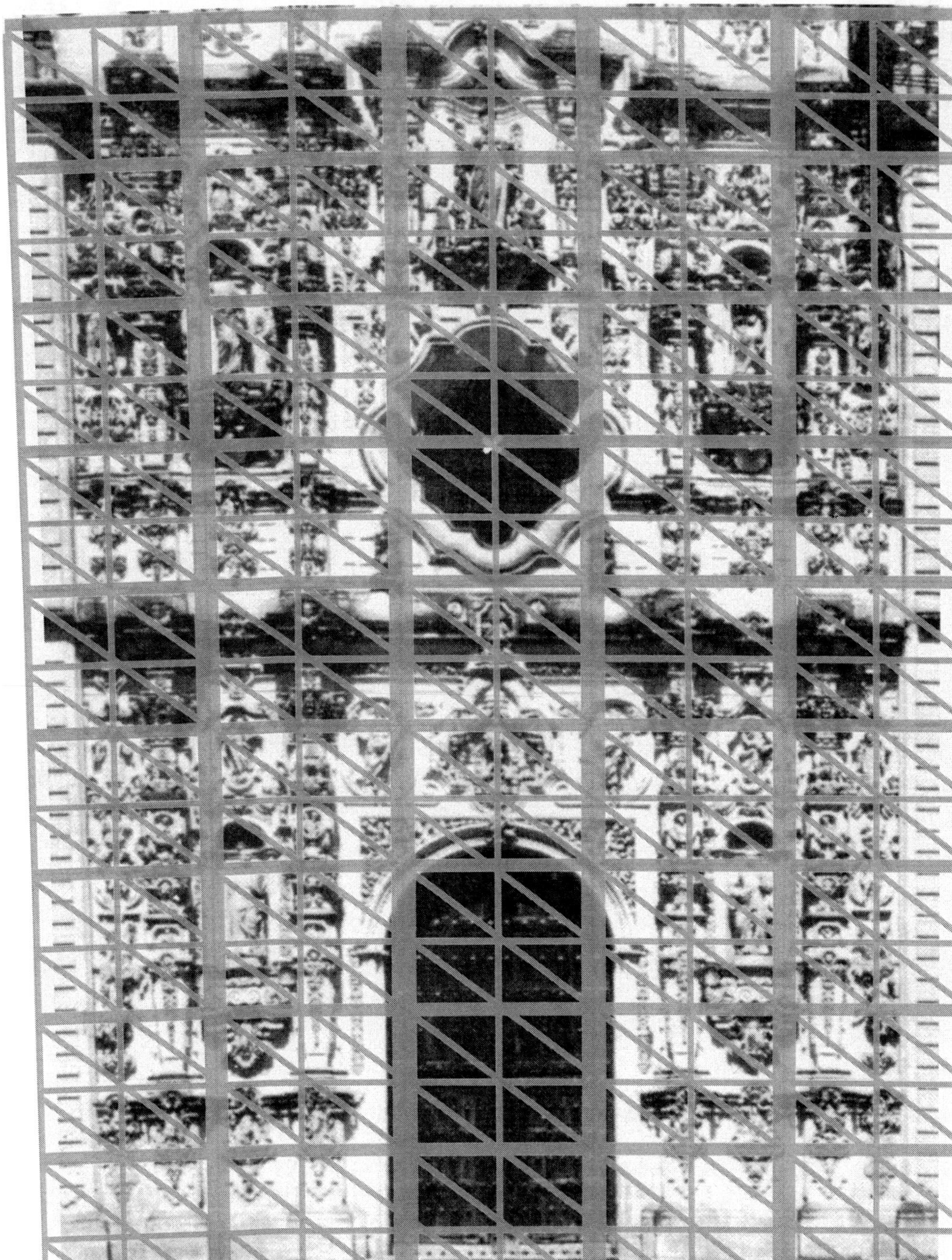

Fig. 25. "Regulating lines" operating micro-cosmically upon the ensemble of the Tepotzotlán church façade. (Computer-generated grid-patterns layered over phtograph.)

façade reveals the same inscribed operation, *in microcosmo*. (Fig. 26) The actual workings of the 3:4:5 geometric system are revealed much more clearly in this kind of detail. As we see here, as again centered upon the rosetón, the pictorial-sculptural field was flatly enframed by two rusticated, but very shallow, flanking pilaster strips that rise up to the main cornice, extending over to the bell-tower. Superimposed over a photograph of the central zone within the same *fachada retablo*, the same triangulate format reveals itself, and it now shows that, once the field had been proportionately sub-divided, the repetitive pattern was consistently applied to determine the placements of even the smallest sculptural, or quasi-architectural motifs. In actual practice, the compositional field (or canvas) is originally flat brickwork; once this is covered over by molded or skillfully modeled "*enfoscado*" (roughcast plaster), then the *imafronte* becomes a three-dimensional *fachada-retablo*.

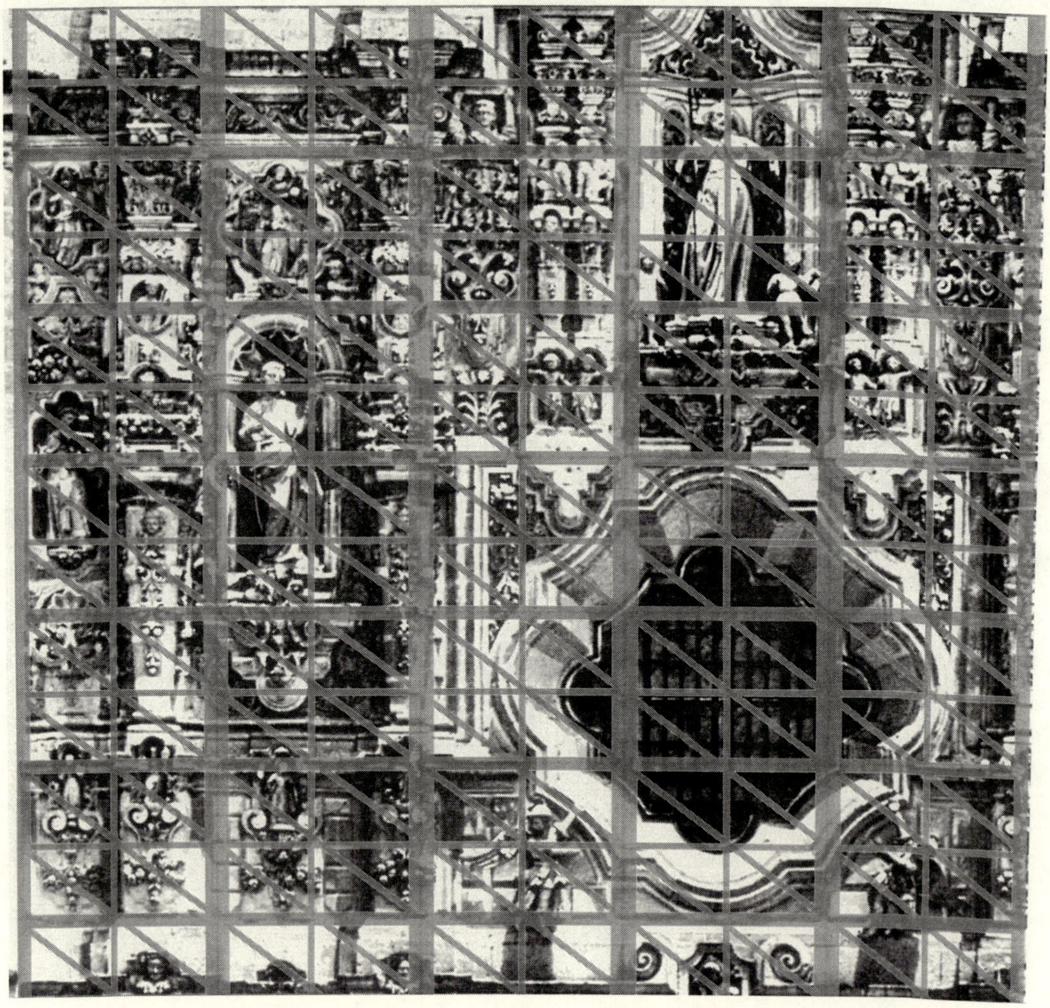

Fig. 26. Micro-detail of the "trazado regulador" operating micro-cosmically upon the center of the Tepotzotlán church façade. (Computer-generated grid-patterns layered over photograph.)

The same design principle — as "procedure" — operates throughout the church at Tepotzotlán. As at Córdoba (figs. 9–15), so too at Tepotzotlán: exterior architectural design is aligned with internalized design patterns. To further demonstrate the point, in this case, the grid may be superimposed upon a photograph of the high altar (*retablo mayor*). (Fig. 27) As we see, a succession of diagonals criss-cross over evenly spaced verticals established with a plumb-bob line, and such as these geometric patterns were initially centered upon an alignment upwardly ascending from the altar table. The resulting intersections established the proportionate placement of pilasters framing cult statures, with the determination of their respective heights and widths conforming to the triangular format. (Fig. 28) Finally, having at hand (as is usually not the case) a carefully measured drawing, in this instance of the bell-tower belonging to this church, again we find the same operation at work. The predetermined "regulating pattern" has fixed the positioning — centers, sub-centers, and their respective heights — of all the vertically aligned, decorative pilaster-strips (*estípides*). In this case, the peninsular, or "Spanish," origins of this kind of 3:4:5 triangle-module for towers is easily proven.

Our first example of a Spanish tower employing the 3:4:5 triangle as its design module is that one erected around 1315, and still serving as the bell-tower for the Church of El Salvador in Teruel (Aragón). This towering, square-base brick structure is frequently cited as a classic example of *arte mudéjar*.[1] (Fig. 29) The imposed grid-pattern again shows diagonals traversing the center-line; their intersections proportionally determine the vertical and lateral spacings for all the micro-details making up the decorative ensemble. The geographical locus for this kind of "design-by-triangulation" is, of course, *al-Andalus*. Among the most famous of the architectural attractions of modern Andalusia is "la Giralda"; now the square-based, bell-tower of the Cathedral of Seville, it was originally built, in 1198, to serve as the minaret (*al-minar*) of the Great Mosque of Seville. (Fig. 30) The same cyber-grid, as now layered over a measured, modern drawing of the ancient Islamic *al-minar*, shows the same design procedure at work as that one employed in Christian Teruel some two centuries later.

Sanitago de Compostela in Galicia is, of course, the primary pilgrimage site in Spain for devout Catholics. During the seventeenth century, the Romanesque cathedral was given a Baroque face-lift. (Fig. 31) Although completed in 1680, a grid layered over a measured drawing of the *Torre del Reloj* (Clock-Tower) reveals its hidden *mudéjar* pedigree: all the relative heights and lateral dimensions of every part of the micro-architectural design were pre-determined by the intersections of diagonals, all generated by a 3:4:5 triangle-module, and these concur with parallel vertical, inscribed lines, all of which are proportionately spaced from the central axis. These serve to define both the outer edge of the ensemble and also the enframed arch.

Given the precedent of Walls' graphic demonstration (fig. 6) that a center-point — in this case, one anchored in the middle of a courtyard (*sahn*) — had generated the entire geometry-based design of a given Islamic building complex, then it follows that the same procedure should have been employed at Tepotzotlán. In this case, the equivalent courtyard-center-point would be found within the cloister-garden (*patio*) of the Monastery. (Fig. 32) When a computer-grid is laid over a mea-

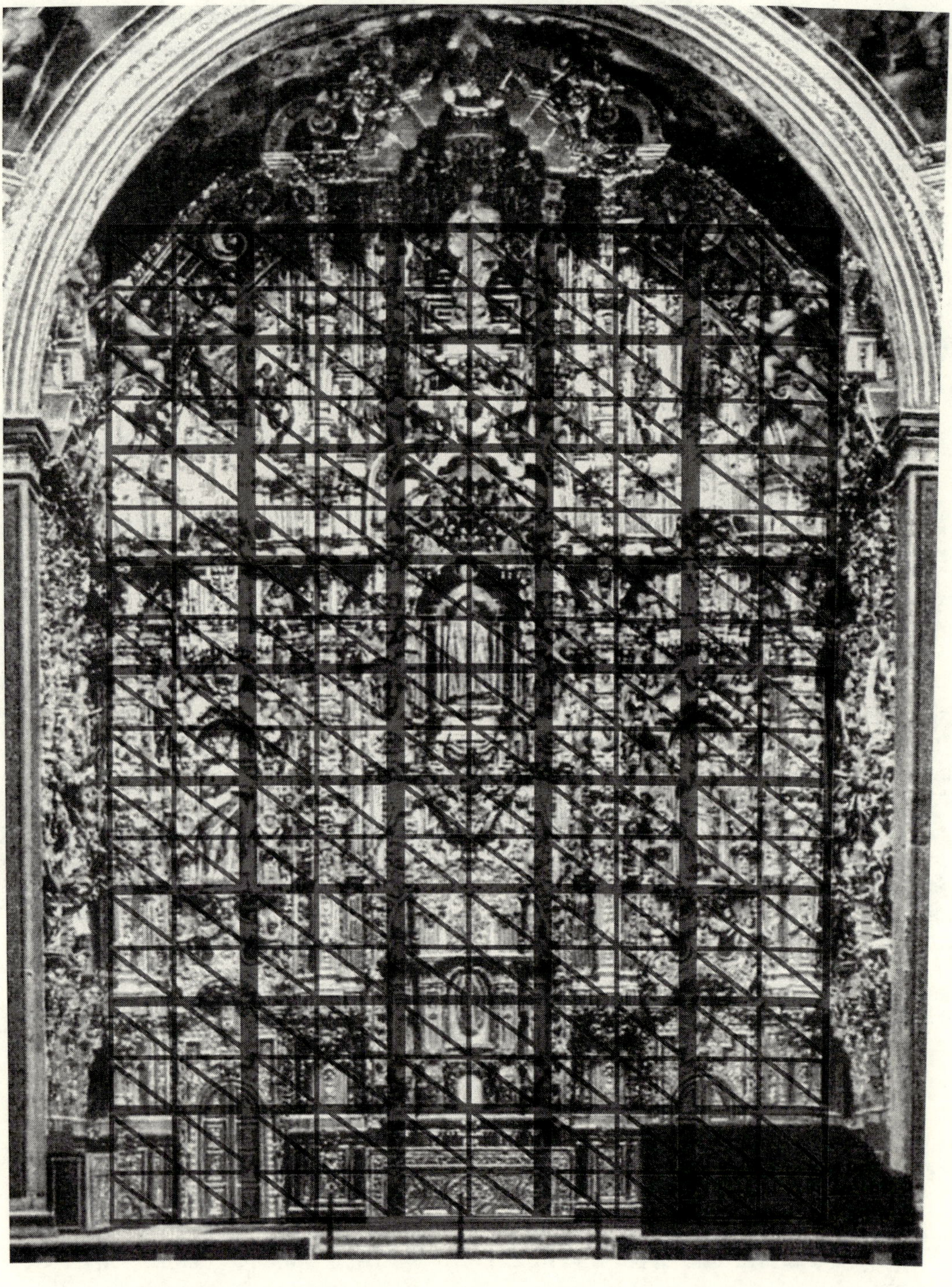

Fig. 27. "Regulating lines" operating micro-cosmically upon the *retablo mayor* within the church at Tepotzotlán. (Computer-generated grid-patterns layered over photograph.)

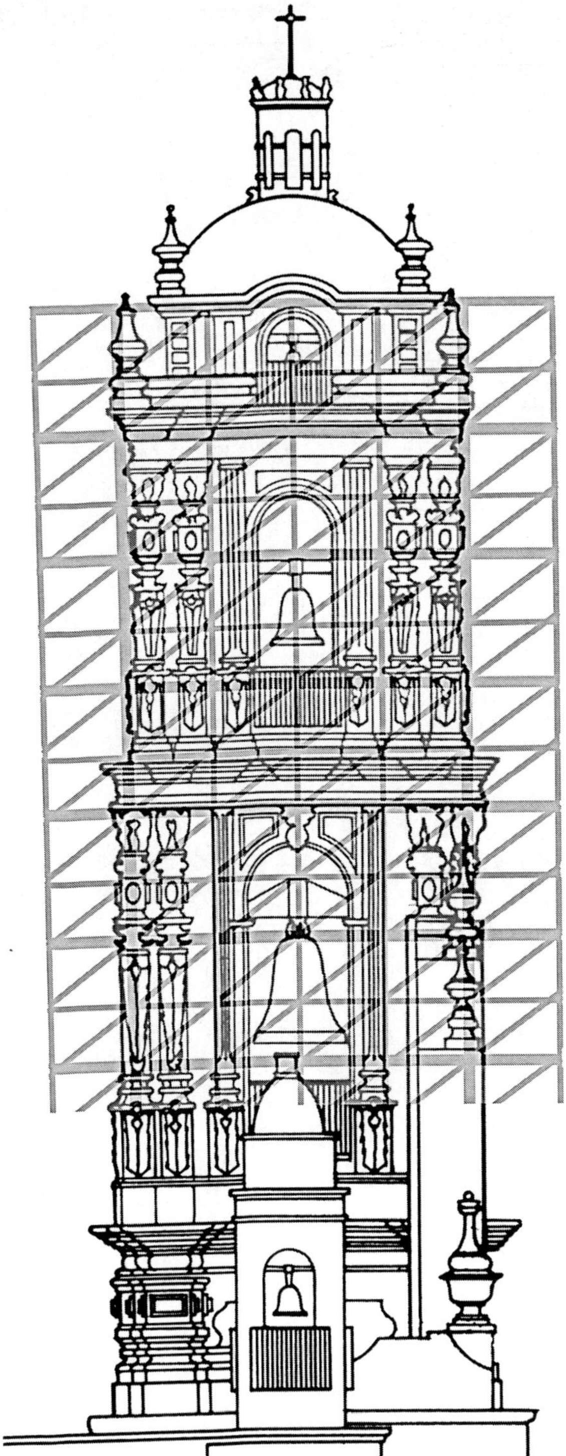

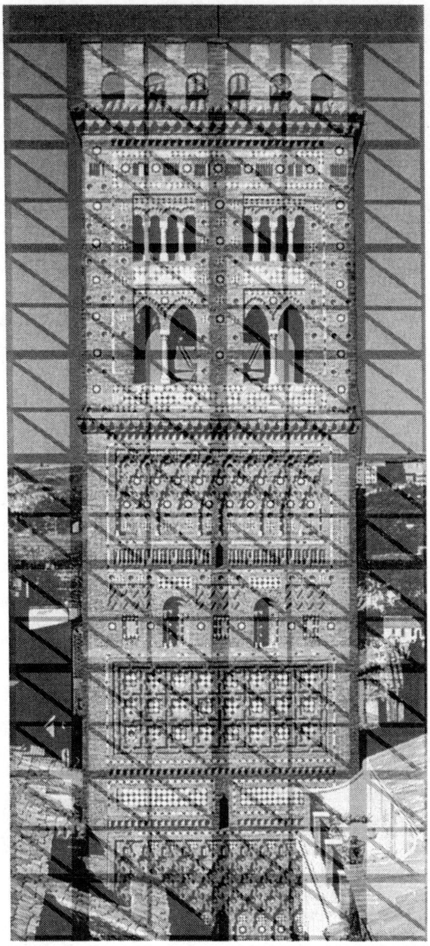

Left: Fig. 28. "Regulating lines" operating micro-cosmically upon the bell-tower of the church at Tepotzotlán. (Grid layered on a drawing given in Bottineau, *Iberian-American Baroque*, 1970.) *Above:* Fig. 29. "Regulating lines" operating micro-cosmically upon the bell-tower (ca. 1315) of the *mudéjar* church of El Salvador, Teruel (Aragón). (Computer-generated grid-patterns layered over photograph.)

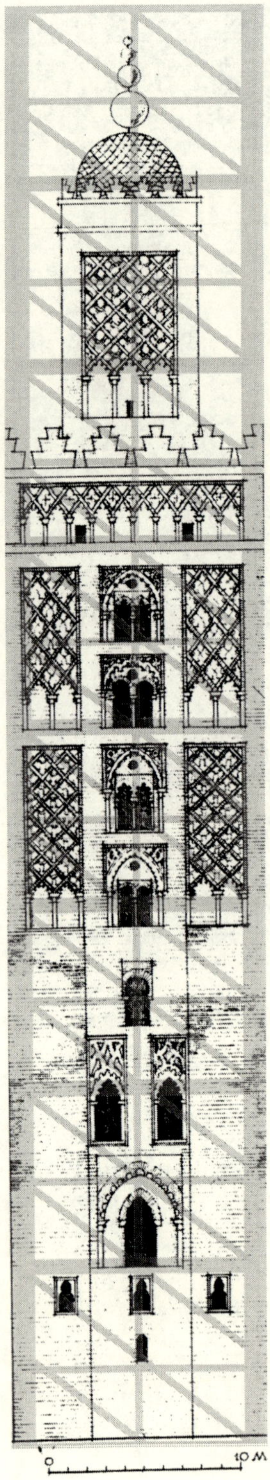

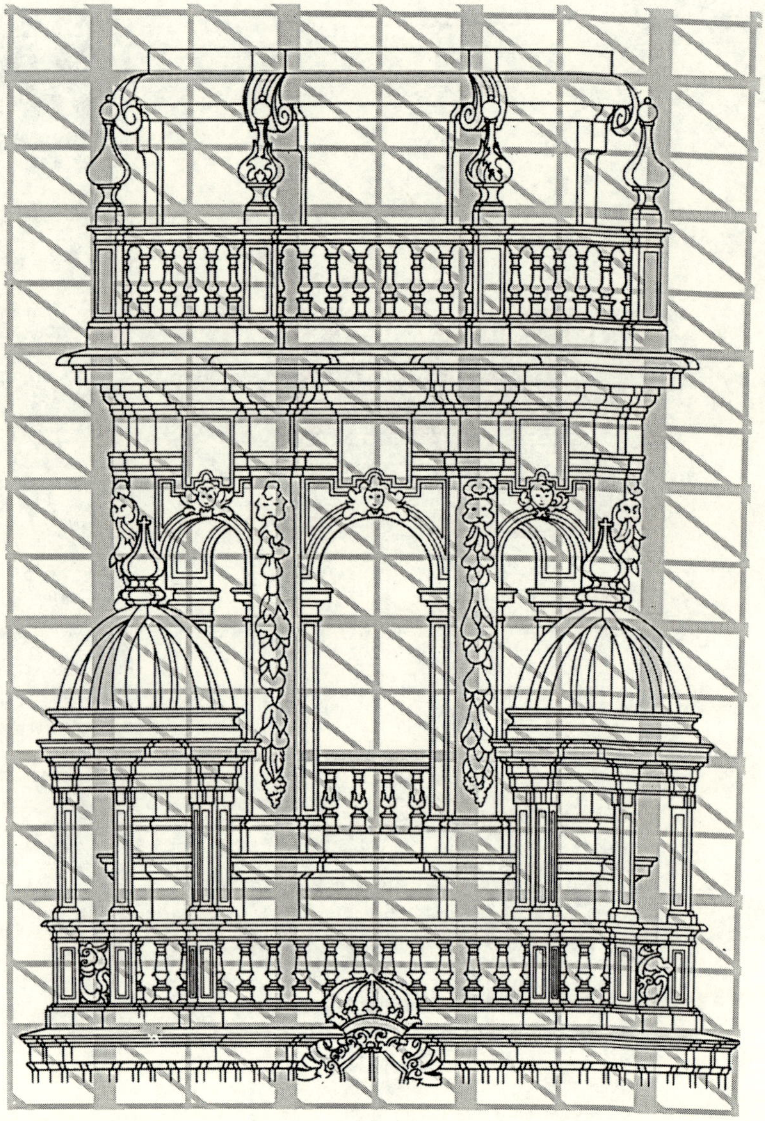

Left: Fig. 30. "Regulating lines" operating micro-cosmically upon "la Giralda," now the bell-tower of the Cathedral of Seville, but originally (in 1198) the minaret (*al-minar*) of the Great Mosque of Seville. (Grid layered on a drawing given in Borrás, ***El Islam***, 1997.) ***Above:*** Fig. 31. "Regulating lines" operating micro-cosmically upon the Torre del Reloj (1680), Cathedral of Santiago de Compostela. (Grid layered on a drawing given in Bottineau, ***Iberian-American Baroque***, 1970.)

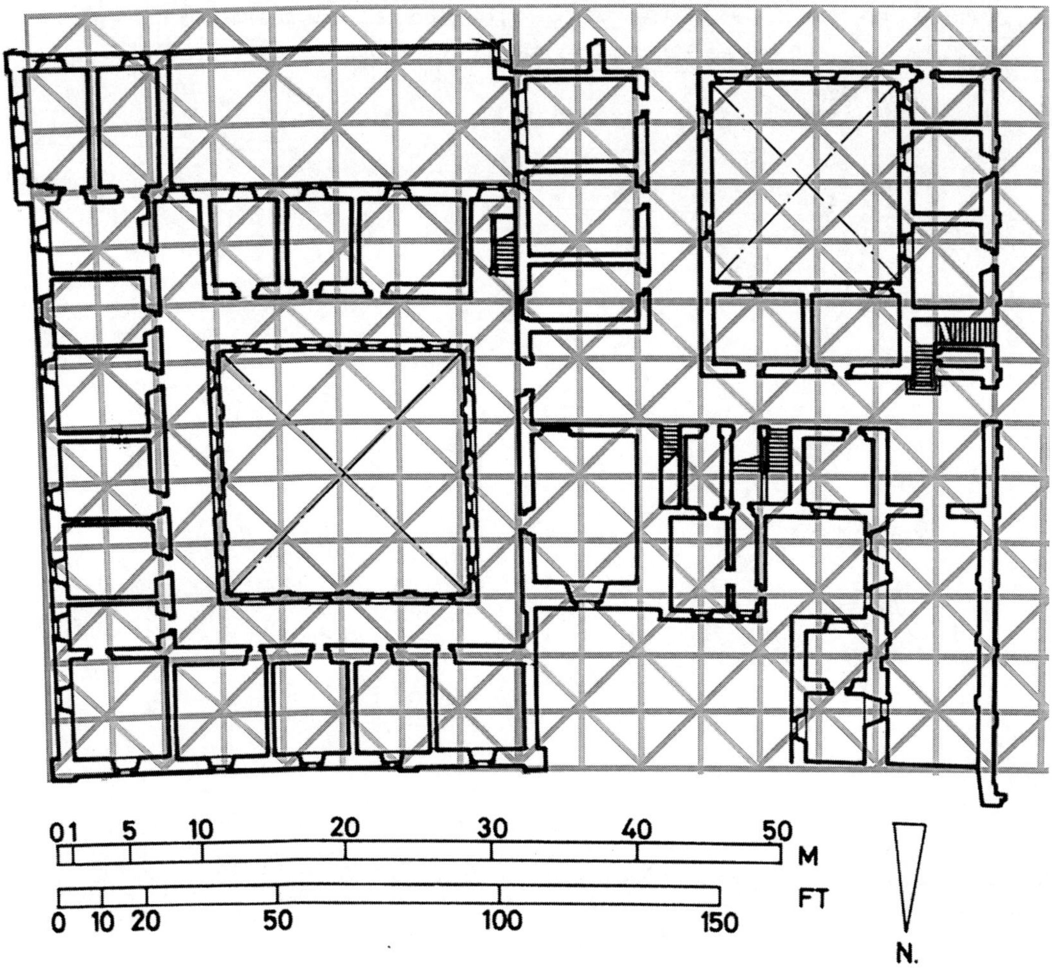

Fig. 32. "Regulating lines" (based on "lozenge-pattern" triangulation) operating microcosmically upon the ground-plan of the cloister-patio at Tepotzotlán. (Grid layered on a drawing given in Bottineau, *Iberian-American Baroque*, 1970.)

sured drawing of the cloister-patio at Tepotzotlán, the immediate result is to show that parallel lines generated by diagonal lines determined the proportionate distances between the piers in the colonnade running around the garden.

In this case however, the grid-pattern employed was based on an equilateral (3:3:3) triangle, and this departure from the other (3:4:5) module is easily explained. Christian cloister-gardens are nearly inevitably *square*; accordingly, their center-point must be derived from two diagonal intersections, so yielding four equilateral triangles. The compositional system necessarily adhered to is, therefore, the *ad quadratum* (or 3:3:3) scheme. But, overall, the proportionate results remain the same: the verticals and horizontals aligned upon the center-axis of the cloister have fixed the placement of all the other buildings in the monastery complex. This graphic demonstration shows that the axis-point of the entire Hispanic *monasterio* had been initially oriented upon the center of its open *patio* (to the left), and that the adjacent, subsequently built, monastic components radiated out proportionately from that point, then being arranged in parallel, north-south, zones, each with its own vertical axis, all as aligned upon the axis first fixed in the exact center of the squared cloister-garden.

9

Other Christian Examples of Hispanic Design-by-Pythagorean-Triangulation

Again, and in all the other Hispanic examples examined here, the basic generative motif is the "Pythagorean triangle." In the larger sense, it is, once again, simply a matter of (literally) "plane geometry." In short, it is *plane geometry*, with this meaning the specifically Islamic sort, which explains the "planarism" which Oskar Hagen had rightly cited as representing the essential formal principle distinguishing Hispanic art and architecture from its European counterparts. By the sixteenth century (at least), the originally Islamic design procedure has become thoroughly "Spanish," meaning wholly adapted to contemporary Christian usage.

As evidence for this conclusion, I submit the example of the façade of the University of Salamanca (fig. 3). Upon a nineteenth-century engraving, itself a carefully detailed and measured rendering, I again impose the triangulated, 3:4:5, grid-pattern. (Fig. 33) As this simple operation reveals, a vertical line, established in the middle of bunched colonettes placed at the center of the double portal, fixed the mid-point of the whole, tapestry-like design. The repeated intersections of a succession of drawn diagonals then predetermined the positioning of the four primary string-courses and, within these, additionally the placement of numerous vertical subunits, with these including *estípide*-pilasters framing heraldic medallions (*escudos*) and a series of pseudo-numismatic portraits of historical worthies.

A similar graphic procedure yields a new interpretation of the impressive altarpiece in the Cartuja de Miraflores near Burgos which was designed around 1496 by the Castilian master Gil de Siloé and his follower Diego de la Cruz (fig. 2). The *retablo mayor* displays at its vital center the inspiring emblem of the Crucifixion — the Tree of the Cross, a figure signifying the Host: Christ Himself— set against a geometric, or *mudéjar*, carpet-like pattern of circles and rectangles. (Fig. 34) Also at Miraflores is another stunning work designed a bit earlier by Siloé, the alabaster double-tomb of Juan II and Isabella of Portugal, commissioned by their daughter, Isabella the Catholic, and erected from 1489 to 1493 at the foot of the slightly later *retablo mayor*, with which it is aligned, in both a formal and in a spiritual sense. Framed by an octa-

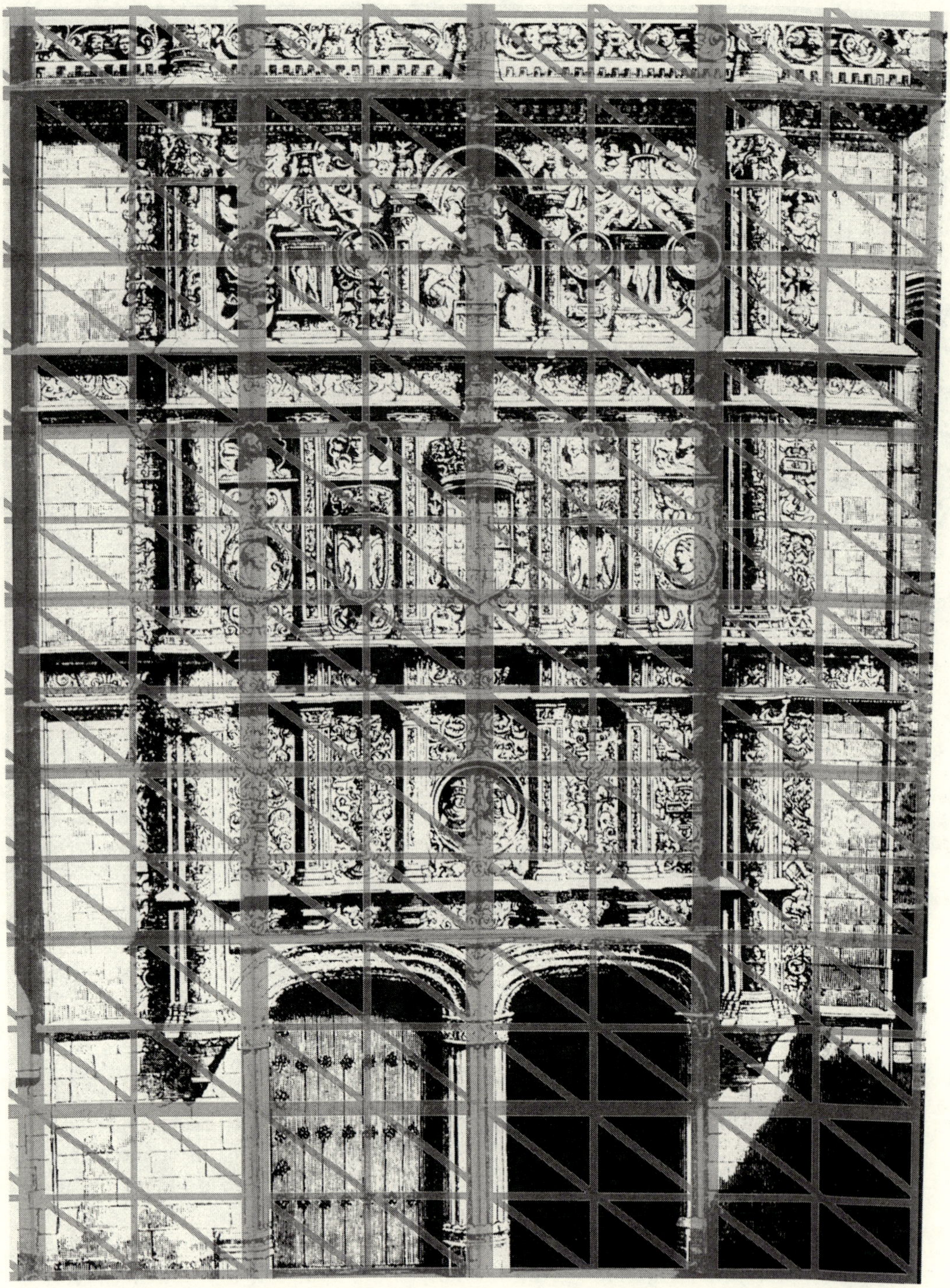

Fig. 33. "Regulating lines" (based on "Pythagorean-pattern" triangulation) operating micro-cosmically upon the façade of the University of Salamanca. (Computer-generated grid-patterns layered over fig. 3.)

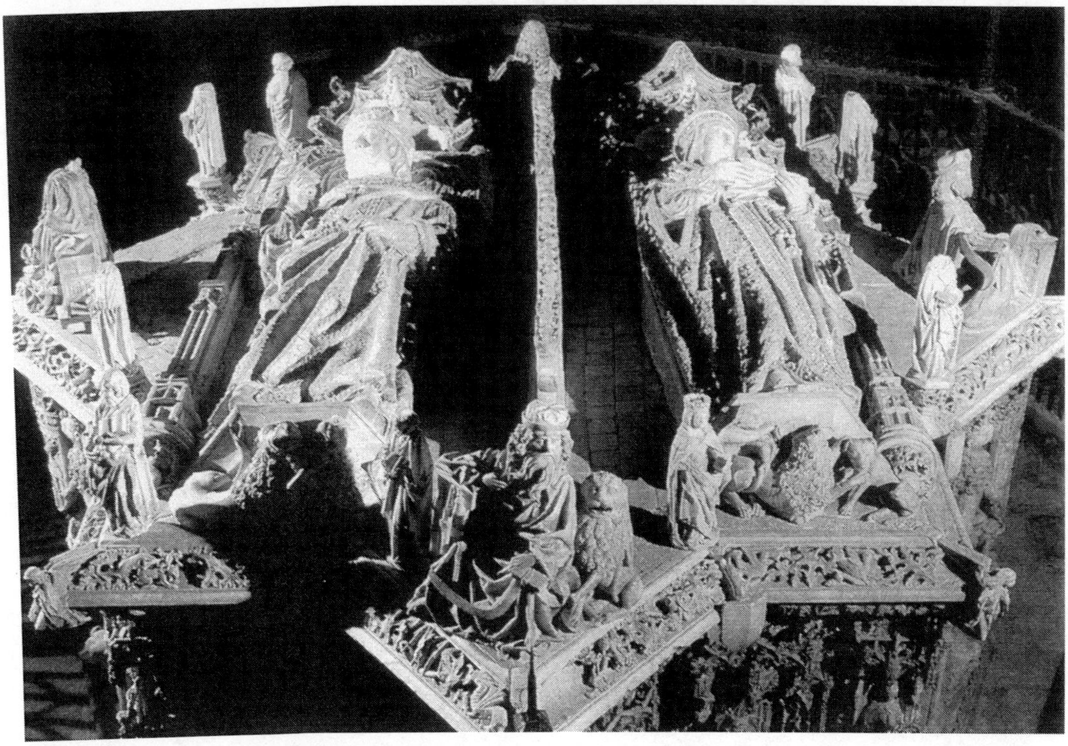

Fig. 34. Gil de Siloé, Double-tomb of Juan II and Isabella of Portugal, 1486–99. Carthusian Monastery of Miraflores, Burgos.

gon, an eight-sided, geometry-derived motif used in traditional Islamic design, Siloé's sculpted tomb marks a high point in the evolution of an Hispanic sepulchral sculptural tradition, with long-standing applications in Christian Spain. His is a truly virtuoso piece, an ensemble incorporating—besides the two intricately carved royal effigies—Biblical scenes, Evangelists, Virtues, and sixteen lions guarding the royal arms. Even more impressive than its technical accomplishment is its impelling content, revealing its liturgical function within a given ecclesiastical-architectural setting.

Since its full compositional grandeur of Siloé's tomb complex can only be appreciated when viewed from above (versus at ground level), this sepulchral arrangement also metaphorically presupposes a viewpoint from "On High," by God Himself, known by the faithful to have been figuratively situated within a Spanish *retablo*.[1] Seen from this superior perspective-point, which in Miraflores also belongs to the figure of Christ on the Cross, the recumbent figures axially aligned by Siloé towards the altar are arranged exactly like standing saints on contemporary cathedral façades, even including a telling Gothic architectural detail, the baldaquin-crown. As further viewed by the omnivoyant Deity, the compositional framework is original, a boldly proclaimed eight-pointed star, probably so because Hugo of St. Victor (among other exegetical authorities) mentions an '"eighth day," surpassing the earthly week, as signifying "Eternity" (and so also explaining the traditional octagonal shape of

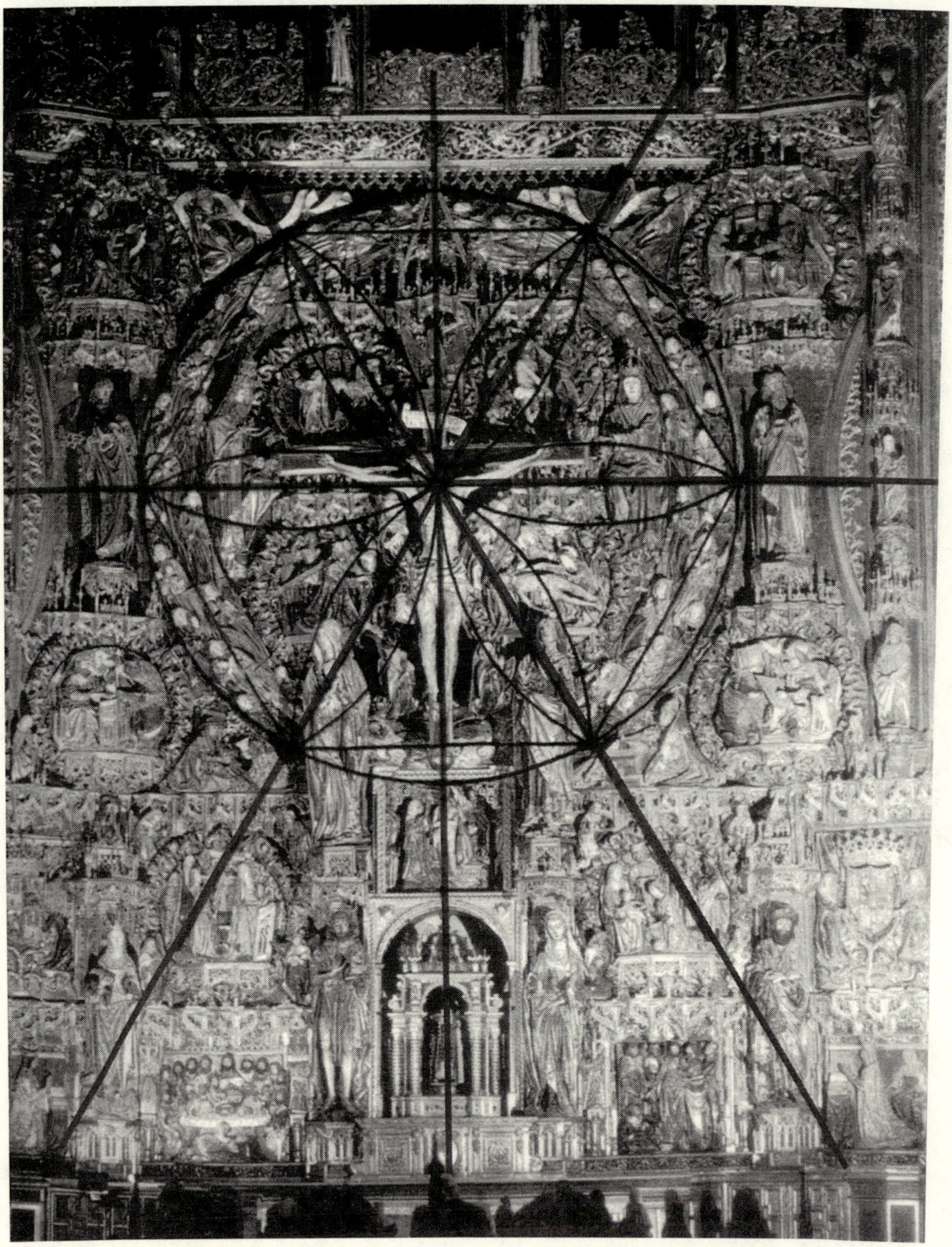

Fig. 35. Rosette pattern (from fig. 6)—generating horizontals, verticals and diagonals, with their intervals all sub-divided into Pythagorean triangles—laid over Siloé's ***retablo mayor*** (fig. 2).

baptisteries also promising eternal life to the anointed). Since Siloé's tomb is conceptually linked to the altarpiece serving as its figurative counterpart, Siloé's two-part ensemble tells us something significant about the subtle presence of a unitary Hispanic liturgical vision extending well into the eighteenth century.[2]

Following this novel exegetical explanation for Siloé's choice of a neo-*mudéjar* octagonal framework (thus providing what German art historians would call its *Rahmenthema*, a "framing theme"), a similar revelation dealing with an Islamic geometric figure adopted to Christian usage emerges from formal analysis of the Crucifixion scene framed by (or inscribed within) a rosette-pattern (fig. 2). In this case, the *rosetón* is made from stacked, circling angels, here functioning as heaven-sent "messengers" (*evangeloi*). The Christ-bearing *rosetón* is itself laid over a complex sculptural ensemble including various figural motifs inserted into boxy frames and stacked one above the other in parallel, vertical aligned micro-zones (*hileras* or *calles*). The principal, vertically ascending, axis is centered, according to liturgical logic, upon the Host guarded upon the altar in its gilt monstrance (*custodia*). (Fig. 35) To demonstrate the hidden formal logic of Siloé's composition, I have imposed upon it the Islamic geometric figure (fig. 6) which A̤. G. Walls had determined to have generated the entire composition of a given Islamic building, a *madrasa* or school of theology.

In this case, I have removed all the architectural features from Wall's drawing, and I have adopted a different operational procedure. This time, after scaling the motif to exactly conform to the size of Siloé's *rosetón* in my source photo, I then printed out the Islamic star-pattern upon an overhead transparency; this I then aligned upon my photograph (and this represents an alternative procedure to that one producing figs. 10 ff., 24 ff., and this more-or-less manual method may also be employed by researchers who wish to check my findings, then perhaps further confirming them with their own analyses of different Hispanic architectural examples).

As we have already seen, Wall's star-pattern (as in fig. 6) generates horizontals, verticals and diagonals, with their intervals all potentially sub-dividable into Pythagorean triangles. When this same hexagonal geometric figure is laid over Siloé's *retablo mayor* (fig. 35), one then discovers that the center (compass-point) of the entire composition is none other than the *Sagrado Corazón*, the Sacred Heart of Christ. The vertical axis traversing this theological compass-point is exactly aligned upon the *Custodia* bearing the *Hostia*, the transubstantiated Body of Christ given to earth-bound worshippers. The two principal, descending diagonals radiating from the *Sagrado Corazón* reach down to touch the distant figures of praying royal donors. Were two additional diagonal lines to be drawn through the exact centers of the triangles formed between what may be called the donor-diagonals and the horizontal line paralleling the arms of the Cross, and if these lines were to be extended to their full length, one would thereby locate the corners fixing the far ends of Siloé's massive sculpture-in-a-box. The triangles just described have those ubiquitous 3:4:5 proportions.

10

The Graphic Evidence for "Regulating Line-Patterns" with Pythagorean Triangulation in Hispanic Architecture

As I believe however, the detection of this simple workshop procedure, Hispanic "design-by-triangulation" and traceable at least to the tenth century in Córdoba (figs. 9–15), additionally points to specifically Andalusian origins for the designer of Tepotzotlán (figs. 1, 24–28, 31), already supposed to have been Lorenzo Rodríguez, himself a native of Guadix (Granada). In Andalusia (formerly *al-Andalus*), such weightlessly suspended, geometricizing and planarized, portal-façades were an ubiquitous visual fact of one's living architectural heritage. Across the Atlantic, the simple but venerable — and probably initially Hispano-Islamic — workshop procedure could have been as easily taught to a transplanted Spaniard as to a nearly illiterate Indian laborer. Because "Islamic" design principles of this sort certainly were never common in medieval Europe (versus *al-Andalus*), we finally understand (and "*im Detail*," as Aby Warburg would have wished) the real basis— a standard design procedure — accounting for everyone's perceptions of the decidedly unique (or "alien") quality of Spanish and Trans-Hispanic architecture.

By instrumental means of the ubiquitous straightedges, plumb-lines and levels, triangles and divider-compasses of the tradition-bound draftsman and mason, simple ratios were generated. By these means, all the positionings of the boxy compartmentalized areas and zones were pre-determined. After that point, all the eighteenth-century Christian designer in Mexico had to do was add the last, essentially incidental, details: attached architectural elements and figural sculptures. Those other finalizing details included the imposition of specific numbers, with those figures corresponding to the actual measurements (probably figured in *varas*) belonging to the site. As we have seen, these calculations were all worked out as framed within the ubiquitous 3:4:5 unit-system.

Within this predetermined geometric ordering, the final increments also included the different points for the placement of each of the various motifs current in the Neo-Plateresque canon common at the time in Latin America. Those too were a "given," and included the individual architectural motifs, with these being independently conceived *a lo romano* (as in fig. 3). As was additionally ordered by the Jesuit commissioners of the building in question, certain specified religious iconographic elements were then placed in niches upon the façade, all being situated according to the predetermined patterns. Once their relative heights and widths had been established on the work site, the individual statues of Jesuit saints were then carved in the *talleres* of skilled stonecarvers, and these workshops were either temporarily resident within the grounds of the church or perhaps located in a nearby town.[1]

Nevertheless, this seemingly anachronistic procedure was scarcely confined to Tepotzotlán. In fact, as my graphic experimentations have repeatedly demonstrated, the practice must have been nearly universal throughout Latin America. It has, in fact, been long suspected (but not yet scientifically determined) that much ecclesiastical architecture in Spanish America had been generated by predetermined linear patterns or "inscribed figures," with these now being called "*trazados reguladores.*" As it turns out, such systems of "regulating lines" still represent standard architectural practice.[2] A typical example of applications-after-the-fact of *trazados reguladores* is the regulating line-pattern drawn in 1962 by Emilio Harth-Terré and applied onto the façade of the Cathedral of Cuzco, designed around 1605 and attributed to the Spanish architect Francisco Becerra.[3] (Fig. 36) According to this diagram, the width of the façade was predetermined by a compass arc centered in the middle of the main portal. This circular figure also determined the height of the central gable. According to Harth-Terré, the proportions of the two flanking towers were predetermined by equilateral triangular sub-divisions arising from the centric vertical line, and so on, so determining the placement of all the smaller sub-divisions.

However, another imposition of the grid shows that that Becerra's design is *neomudéjar*, and that, rather than Sebastiano Serlio, his real inspiration came from the anonymous Islamic designers. (Fig. 37) As we see, a diagonal drawn from the lower left corner of the *fachada retablo* passes through the top of the hemispherical top of the main portal, and its far end determined the height of the right-hand bell-tower, and the next diagonal to the right starts from the bottom corner of the door and fixes the height of the second story in the same façade; and the next diagonal departs from the middle of the door and determines the width of the *fachada retablo* (and so forth). (Fig. 38) When the same operation is repeated upon a measured ground plan of the Cathedral, we again find that triangulation, specifically with 3:4:5 ratios, fixed the proportionate relationships between all of the major architectural features. (Fig. 39) And the same holds true when the same grid is laid over a modern scale-drawing of the elevation of the interior of the same Cathedral.

In 1969 Humberto Rodríguez Camilloni stated that the inspiration for Harth-Terré's drawing had come from Chueca's graphic investigations, as earlier published in 1947.[4] Particularly important among these is one illustrating (fig. 23) what Chueca called "a formula introduced [ca. 1585] by Juan de Hererra's project for the Cathedral in Valladolid" in Spain.[5] As we already observed, the "regulating line-pattern" is wholly triangular in character; moreover, Herrera's entire composition was indeed

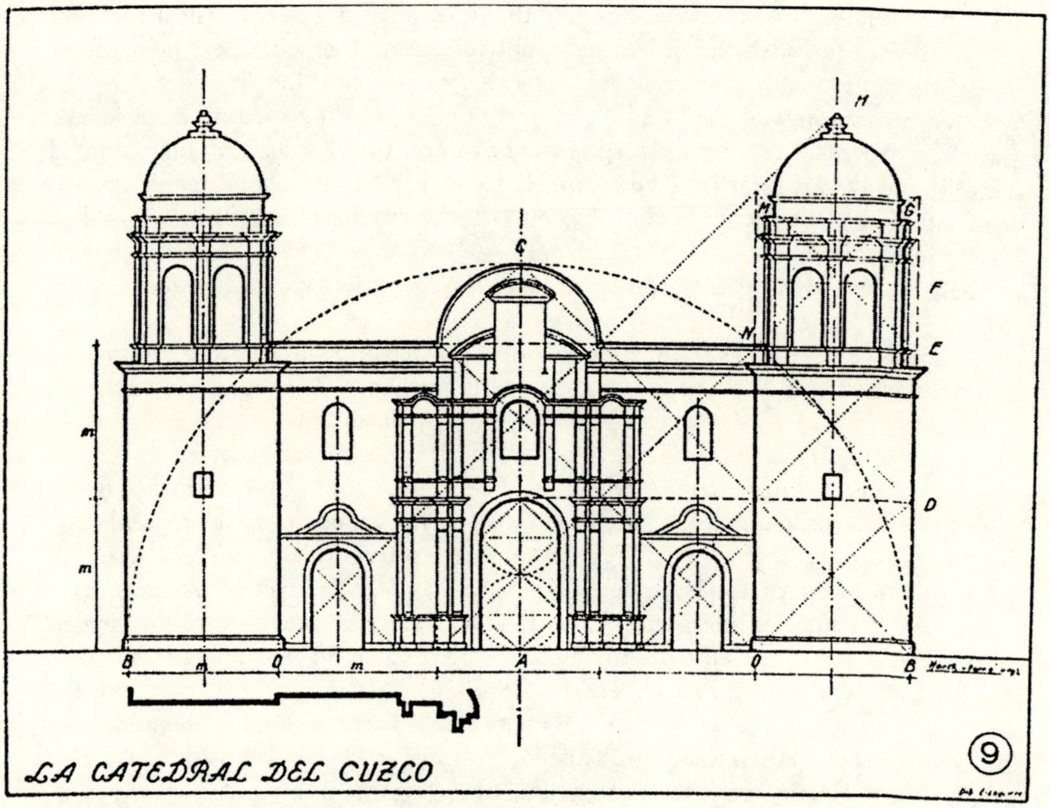

Fig. 36. E. Harth-Terré (1962), "Regulating lines" on the façade of the Cathedral of Cuzco, designed ca. 1605 and attributed to Francisco Becerra.

determined by a "formula": *the intersection of two Pythagorean triangles.* (Fig. 40) This point may be clarified further by another superimposition of the grid (this time removing the superfluous equilateral triangle posited by Chueca), so demonstrating that it was the initial 3:4:5 triangular motif which again determines all the *micro*-relationships of heights and widths.

As we now see, a diagonal drawn from the lower-left corner of the *fachada retablo* passes through the top of the hemispherical top of the main portal, there intersects with a vertical, so fixing the both the height of the second story and the position of a massive pilaster, and the next diagonal to the right starts from the bottom corner of the door and fixes the height of the second story in the same façade and also determines the width of the *fachada retablo*, and the next diagonal establishes both the bottom and the end of the lower cornice (and so forth). (Fig. 41) Final confirmation for Herrera's consistent employment of the Pythagorean motif comes from the evidence of its appearance throughout his ground-plan, and particularly in the nave, where the triangular motif established proportionately spaced footings for the massive pillars, also fixing the dimensions for the free-standing, rectangular choir (*trascoro*).

For the global solution of the anomalies of Hispanic façade design, I now pro-

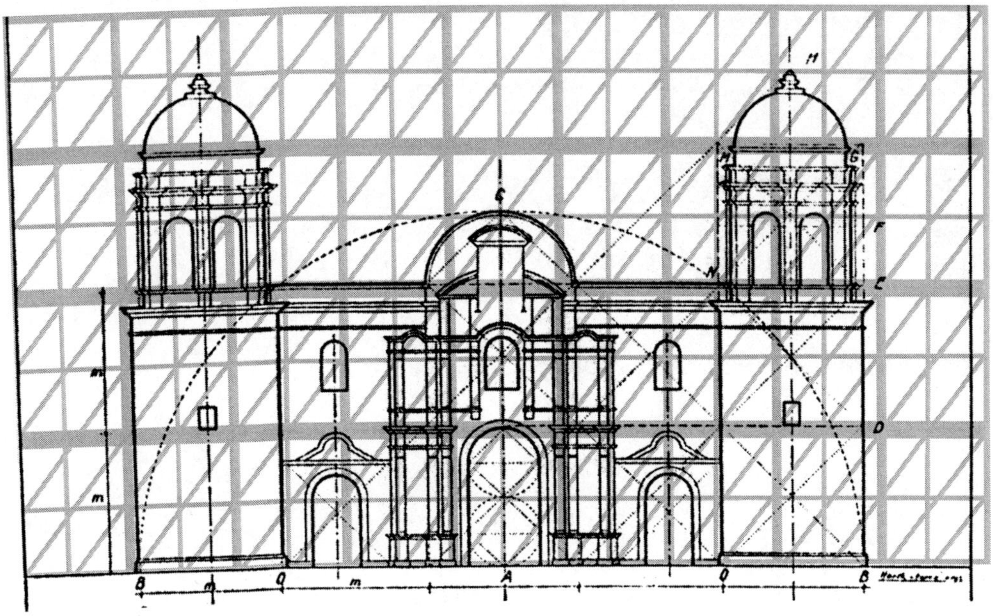

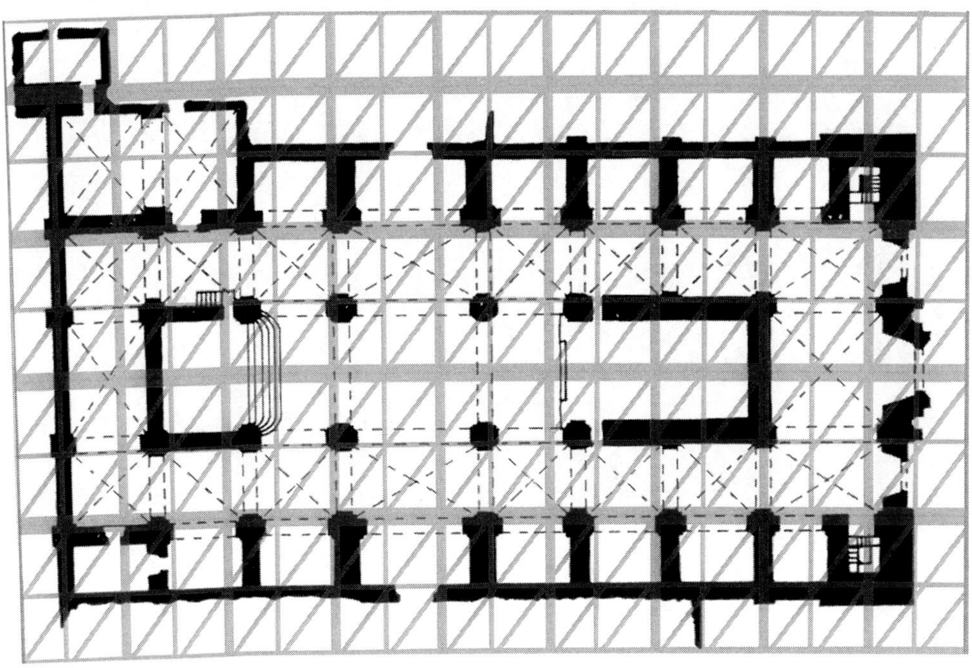

Top: Fig. 37. "Regulating Lines" generated by Pythagorean triangles and predetermining the design of the façade of the Cathedral of Cuzco. (Computer-generated grid-patterns layered over Harth-Terré's drawing: fig. 36.) *Bottom:* Fig. 38. "Regulating Lines" generated by Pythagorean triangles and predetermining the design of the ground-plan of the Cathedral of Cuzco. (Computer-generated grid-patterns layered over plan given in Kubler and Soria, *Art and Architecture in Spain*, 1959.)

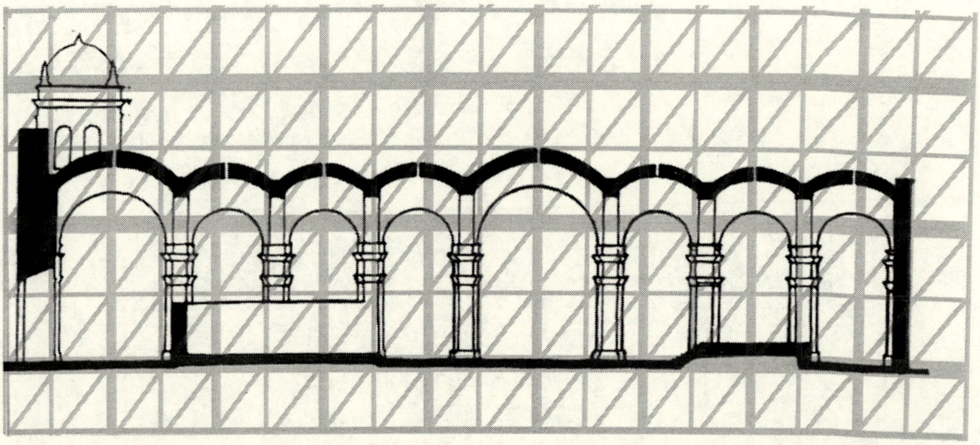

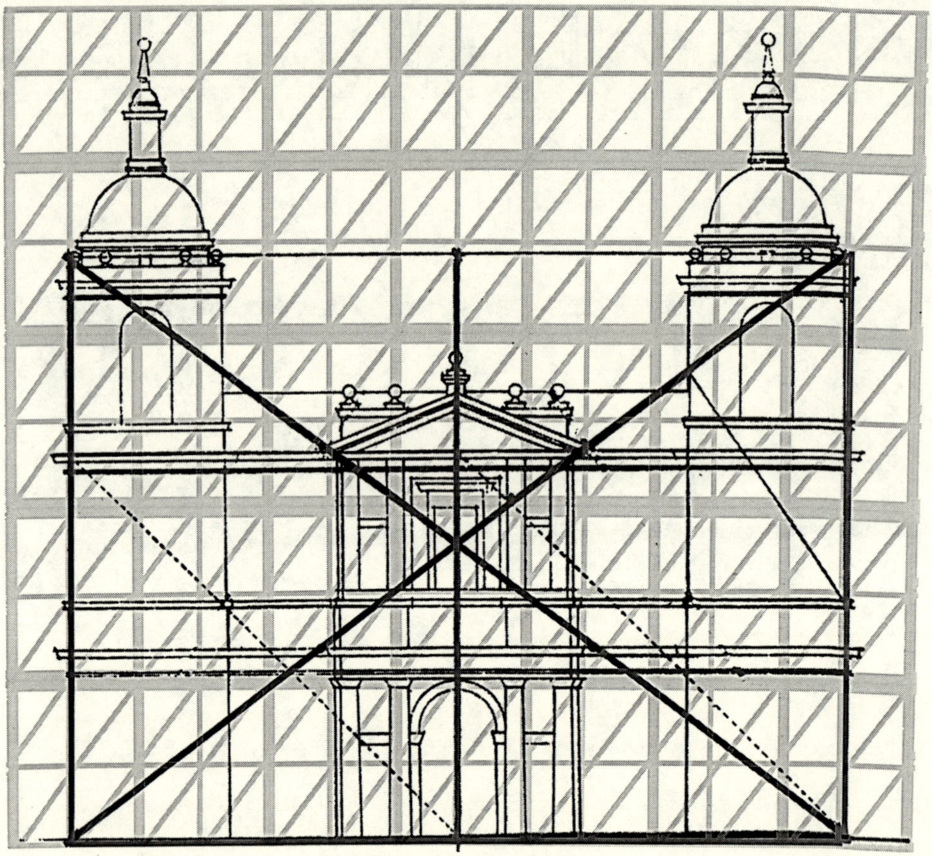

Top: Fig. 39. "Regulating Lines" generated by Pythagorean triangles and predetermining the design of the interior elevation of the Cathedral of Cuzco. (Computer-generated grid-patterns layered over a longitudinal section given in Kubler and Soria, *Art and Architecture in Spain*, 1959.) *Bottom:* Fig. 40. F. Chueca Goitia, "Regulating lines" on the façade of the Cathedral of Valladoild, designed ca. 1585, by Juan de Herrera. (Computer-generated grid-patterns layered over Chueca's drawing: fig. 23.)

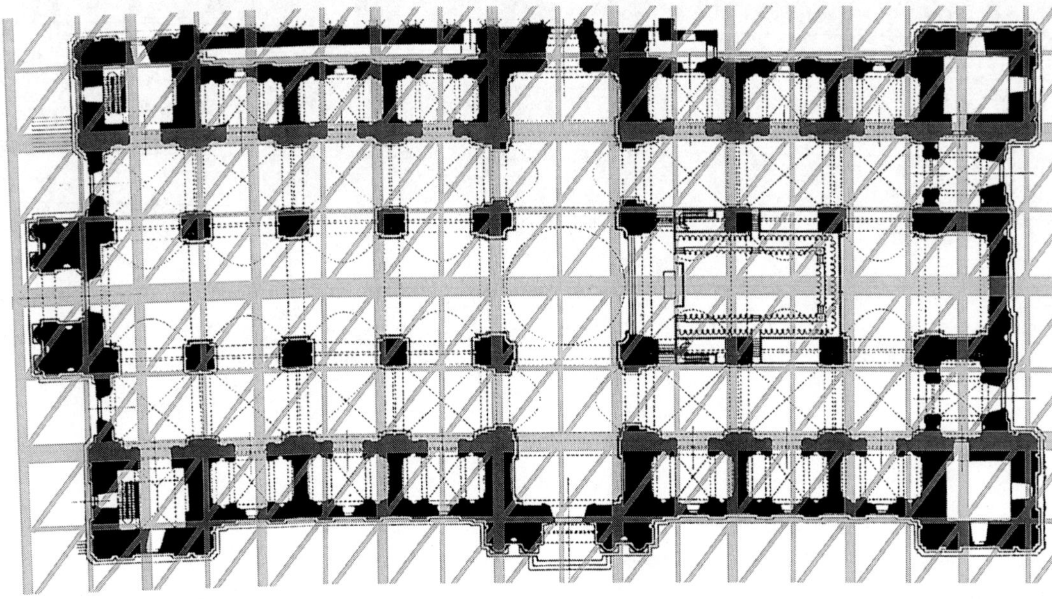

Fig. 41. Juan de Herrera's "trazado regulador" as used in the ground-plan of the Cathedral of Valladolid, designed ca. 1585. (Computer-generated grid-patterns layered over a drawing given in Bottineau, *Iberian-American Baroque*, 1970.)

pose a much more regularized, and more specifically motif-oriented system than those proposed previously by Chueca Goitia, Harth-Terré, and Rodríguez Camilloni. My specifically "Pythagorean triangulate" system fits most examples better than do those more generalized, "circular and cuadrangular," schemes previously adduced. In short, the kinds of *trazados reguladores* proposed by earlier scholars in no way account for the actual placement of both the major divisions and, additionally, even the micro-motifs included within the usually complicated *fachadas retablo* common to Trans-Hispanic architecture. To the contrary, for many (but not all) examples, 3:4:5 triangular scheme proposed here seems better to solve that problem (figs. 37–41).

The Pythagorean triangulate scheme is also faithful to the real substance of Chueca's proposal (fig. 23) for Herrera's putative *trazado regulador* (but which he did not emphasize sufficiently). As Chueca had discovered, indeed the schematic figure describing the composition of the façade of the Cathedral of Valladolid "has a base of 4 and a height of 3." With an inevitable diagonal of 5, that figure has, of course, the invariable format of a "Pythagorean triangle." I will later present, in Chapter 16, the published evidence (and which is dated 1633) showing that, in such cases, the essential tool was a *template* shaped with the same 3:4:5 ratios, specifically the one called the "*cartabón de cinco*."

11

Juan de Herrera and the Evidence for Pythagorean Triangulation in the Escorial

Unquestionably, the most important, also the most influential commission ever given to Juan de Herrera (1530–97) was the Escorial. Moreover, as we also know from reviewing Herrera's convoluted mathematical manuscript, *De la figura cúbica* (ca. 1595),[1] evidently written while he was actively involved with the design of the Escorial, this influential Spanish architect was unquestionably interested in symbolic proportions. According to my strictly graphic research, for Herrera proportionate systems were best represented by Pythagorean number systems; accordingly, he again employed the venerable 3:4:5 format at the Escorial. He, however, mainly employed these configurations in a mystical sense; in fact, he conformed to the cabalistic schemes of Raimundo Lull (and I therefore happily leave the broader interpretation of such esoterica to other scholars).[2] As I shall show, a Pythagorean-3:4:5 motif was used in the Escorial to determine both the ground-plan and the elevation of the entire edifice, just as appears to be the case with Herrera's designs (ca. 1585) for the Cathedral of Valladolid (figs. 23, 40, 41).

As mainly built between 1563 and 1584 (with the work only wholly completed in 1600) according to the wishes (and close supervision) of Philip II, the Escorial is an austere and awe-inspiring monumental structure simultaneously functioning as a palace, a basilica, a monastery, and a royal pantheon.[3] (Fig. 42) With a great basilica-church at its heart, the ground-plan for the entire complex is inscribed within an immense rectangular grid. The idea was germinating in the mind of Philip II long before he ascended the throne in 1556, and from the outset he planned it as a sepulcher-temple to the memory of his father, Charles V, the Emperor-Architect of the Universal Christendom.

As early as 1594, the Escorial was dubbed "the eighth wonder of the world," so rivaling the Mausoleum at Halikarnassos, Great Pyramids, the Colossus of Rhodes, Hanging Gardens of Babylon, Temple of Artemis at Ephesos, and all the legendary rest, also including, in the Christian tradition, the Temple of Solomon and Holy Sepulcher in Jerusalem. The Escorial is a building of enormous importance to the his-

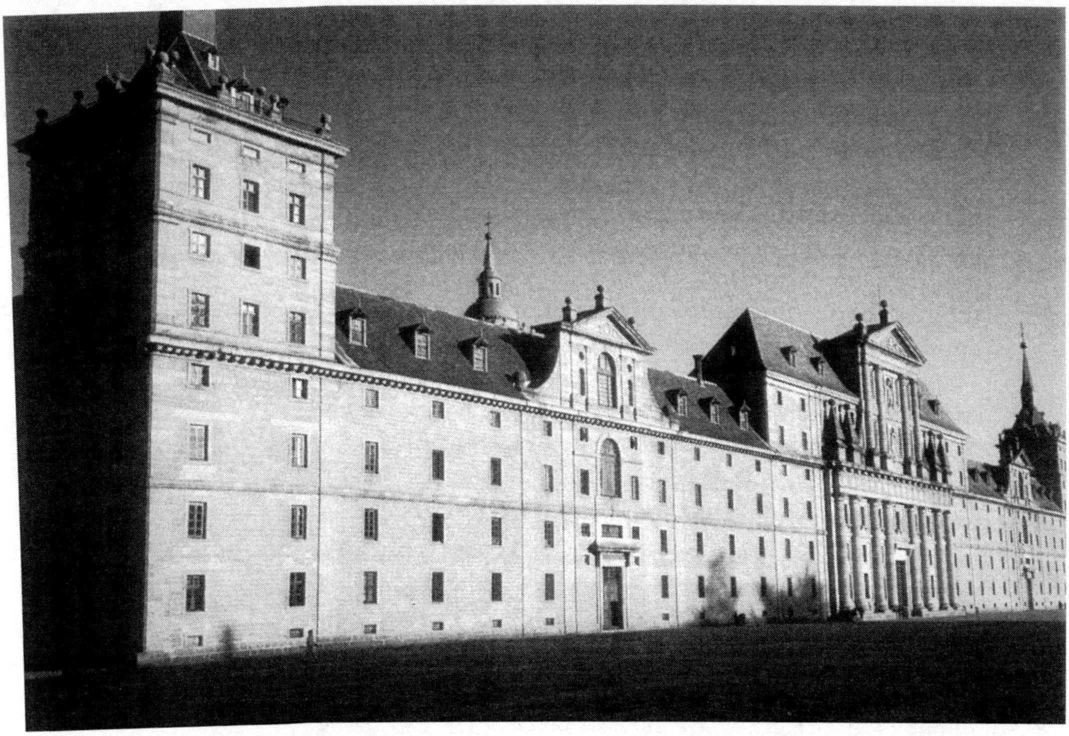

Fig. 42. Juan de Herrera, El Escorial, 1567–1600.

tory of architecture, particularly the kind embodying various kinds of political rhetoric, finding echoes in Inigo Jones' utopian project for the Palace at Whitehall, in the work of late eighteenth-century French "visionary" architects, the unadorned post-war "International Style," and, unfortunately, the fascist architectural propaganda of Hitler, Mussolini and Franco and, currently, the blank and amoral façades of powerful multinational corporations.

The ground plan was initially worked out by Juan Bautista de Toledo, who had worked with Michelangelo in the construction of St. Peter's in Rome; however, he died in 1567, before any construction actually began. The elevations, establishing the current look of the great edifice, were the work of Herrera and include massive towers with slate roofs in the traditional Flemish manner so beloved by King Philip. Construction of the great monument was largely covered by revenues extracted from the American colonies. As explained by a contemporary chronicler, José de Sigüenza, the goal was "grandeur and majesty exceeding all earlier efforts," and that was specifically "achieved by avoiding every superfluity in materials and forms."[4] Implicitly like the macrocosmic Hispanic Empire itself, according to Sigüenza, this architectural microcosm expresses "a well-proportioned body of parts that assist and respond to each other," so constituting "a whole of good measures and corresponding parts." The result was "the minimalist style" (*el estilo desornamentado*), an unadorned and mute, abstract mode. The architects of the Escorial favored the Tuscan order and the primordial rhetoric associated with it, but they saw no need to resort to any further figurative embellishment. On a scale never realized before, the Escorial is an exercise in pure geometry and Christian metaphysics.

Besides a Palace-Pantheon, a residence for monarchs dead and alive, the Escorial was (and still is) also an Augustinian monastery. In his *Historia de la Orden de San Jerónimo* (1605), Sigüenza documents not only the contribution of St. Augustine (354–430) to the influential architectural aesthetic established by Philip and Herrera but also that of Vitruvius: "the authority [was] not alone of Vitruvius ... but also that of the divine Augustine." In his *De ordine* (386), the saint established God as the great Architect of the Universe, and Augustine says that His ideas are revealed to our reason in proportion and number. Key Augustinian terms, embracing both visible form and transcendent meaning, are *equal, harmonious, similar, congruent, concordant*, and *correspondent*. Echoing both Vitruvius and Augustine, Sigüenza says that the great beauty of this building "is seen in how all its parts imitate one another ... the whole is in all the parts." Always the key terms are *reason* and *proportion*: "In all the arts correspondence and agreement give pleasure [:] correspondence loves unity and equality, whether in the likeness of equal parts, or by the gradation and ordering of unequal parts." The rigorous visual style of the Escorial thus exemplifies its ideological framework, a specifically Christian, orthodox and truly contemporary, classicism with specific moral applications of "purity."

Herrera's writings provide further evidence for the mathematical-metaphysical mentality of the Escorial. His *Discurso sobre la Figura cúbica ... según el Arte de Raimundo Lulio* (ca. 1595) is a curious work tying Euclid's essentially objective geometrical propositions to the esoteric "combinatory art" (*ars combinatoria*) of the Spanish mystic Raimundo Lull (1232–1316). Herrera's manuscript (only published in 1976) is illustrated with squared and reticulated diagrams allowing for the proportionate placement of different objects upon a plane extending deep into a literally "cubical" space. Conforming to Lullian practice, sub-squares are often identified with a letter. Like the Cabala, also popular in Spain, the formula of Lull's combinatory art makes one begin with the cipher, from the cipher comes the geometric figure, from the figure the letter, from the letter a color, from the color, a diagram, all of which point to an underlying symbolic language of proportional relationships and invisible correspondences. This is also essential to the purist "cubical art" of Juan de Herrera, court-architect of the Escorial, who also found a unique visual way of expounding what he called "divine objects" (*cosas divinas*). In his case, it serves to make pagan architectural theory Christian and transcendental. As Herrera explained one of his lettered diagrams, "the letter B signifies Generosity, C is Grandeur, D Duration, E is Power, F Wisdom, G is Will, H Unity, I is Truth, K Glory," and so forth.

Henceforth, every Spanish architect recognized that the practice of *la ciencia*, meaning the application of mathematics and particularly geometry, was the acknowledged means to achieve dignified professional status and the lucrative commissions that come with that. As George Kubler, the distinguished modern historian of the Escorial, remarked, "Whenever a King builds, Solomon comes to mind," likewise Jerusalem, the site of his great Temple. Accordingly, chroniclers of the Escorial refer to the monument as either '"Nueva Jerusalén" or "Jerusalén Celestial," that is, the

"City of God" described by St. Augustine; as was the case with the Catholic Monarchs before him, Ferdinand and Isabella, Philip II likewise bore the title "Rey de Jerusalén." The Escorial was also explicitly compared to the Temple of Solomon by Juan Bautista Villalpando, a self-declared disciple of Herrera, in a lavishly illustrated work of archaeological erudition funded by King Philip, *In Exechielem Explanationes* (1598–1604).[5]

This Jesuit scholar showed that the proportions of Noah's Ark and the Tabernacle prefigured those of Solomon's Temple, upon which there would later be erected the *Templum Domini* (actually, the circular, domed Mosque of Omar opportunely dubbed the "Temple of the Lord" by the Crusaders in 1099), itself providing the legendary form of the *Anastasis*, or Holy Sepulcher. (Fig. 43) As illustrated by the Jesuit, the Temple of Solomon is huge, was raised upon a towering podium, had a rectangular grid-plan and atrium-patios, and was even finished in the same kind of "unornamented style" that now characterizes Philip's architectural monument, the "eighth wonder" of the world. (Fig. 44) The larger point demonstrated by the Spanish scholar is that classical architecture itself, including the canonic five orders and all the principles of Vitruvius (even his Tuscan order), had all been invented in Jerusalem. Accordingly, the reductive classicism of the Escorial likewise recreates that primordial, Solomonic source and its God-given proportionality. Since the work of Villalpando was held in great respect for centuries, this was a graphic vehicle by which the austere message of the Escorial was transmitted to later architects, even those who might have no respect for the Spanish Empire and its evangelical mission.

Made under direct supervision of Juan de Herrera, in 1589 there was published a series of eleven detailed engravings showing the Escorial and its component parts. These are properly measured drawings (which are otherwise often difficult to come by in pursuing such research) and they will constitute our prima facie evidence for Herrera's consistent employment of Pythagorean triangulation throughout the great complex. (Fig. 45) The originally inscribed (or otherwise invisible), 3:4:5 pro-

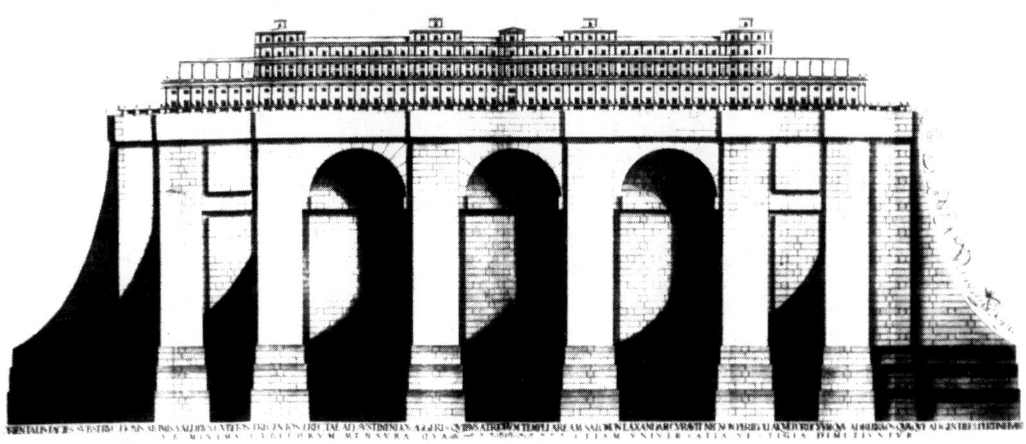

Fig. 43. Juan Bautista Villalpando, West side of the Temple of Solomon, as depicted in his *In Exechielem Explanationes* (1598–1604).

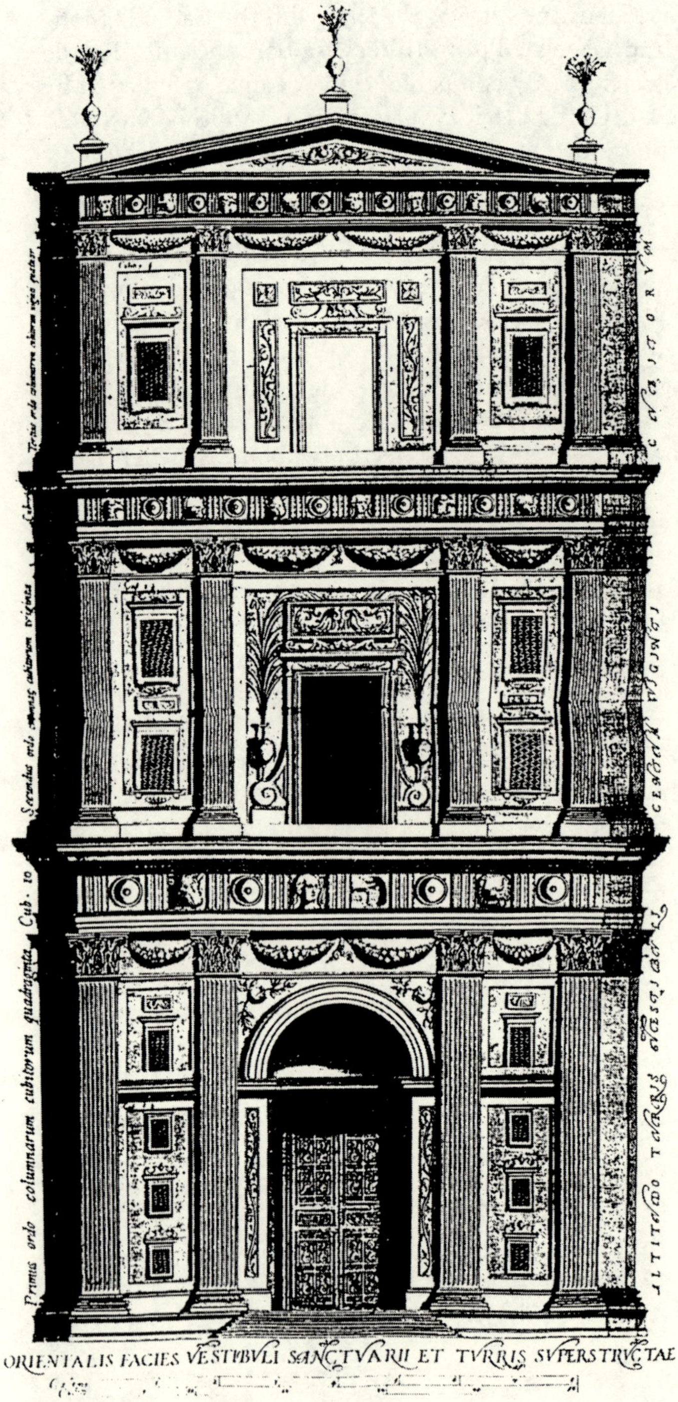

Fig. 44. Juan Bautista Villalpando, Principal façade of the Temple of Solomon, as depicted in his *In Exechielem Explanationes* (1598–1604).

portionate modules laying out the proportionate parts belonging to the complete ground-plan of El Escorial can now be recreated by again superimposing the computer-generated grid-pattern. Serving as a spiritual alignment, the central axis of the complex runs directly down the middle of the Royal Basilica. The horizonal axis is centered upon the heart of the *trascoro*. The centers of the four attached cloister-gardens are aligned upon vertical lines drawn from the central axis, with their proportional distances having been determined by triangulation. (Fig. 46) The same geometric procedure was used to determine the proportionate elevations of the interior; this is initially shown in a grid overlaid upon an elevation, showing the grand portico entrance in the west and its lateral, cloister extensions. (Fig. 47) The same operation is again performed, this time over a section running from east-to-west, through the royal residence, basilica, forecourt, and portico, and the same results are obtained: uniform proportional congruence between all the parts, and

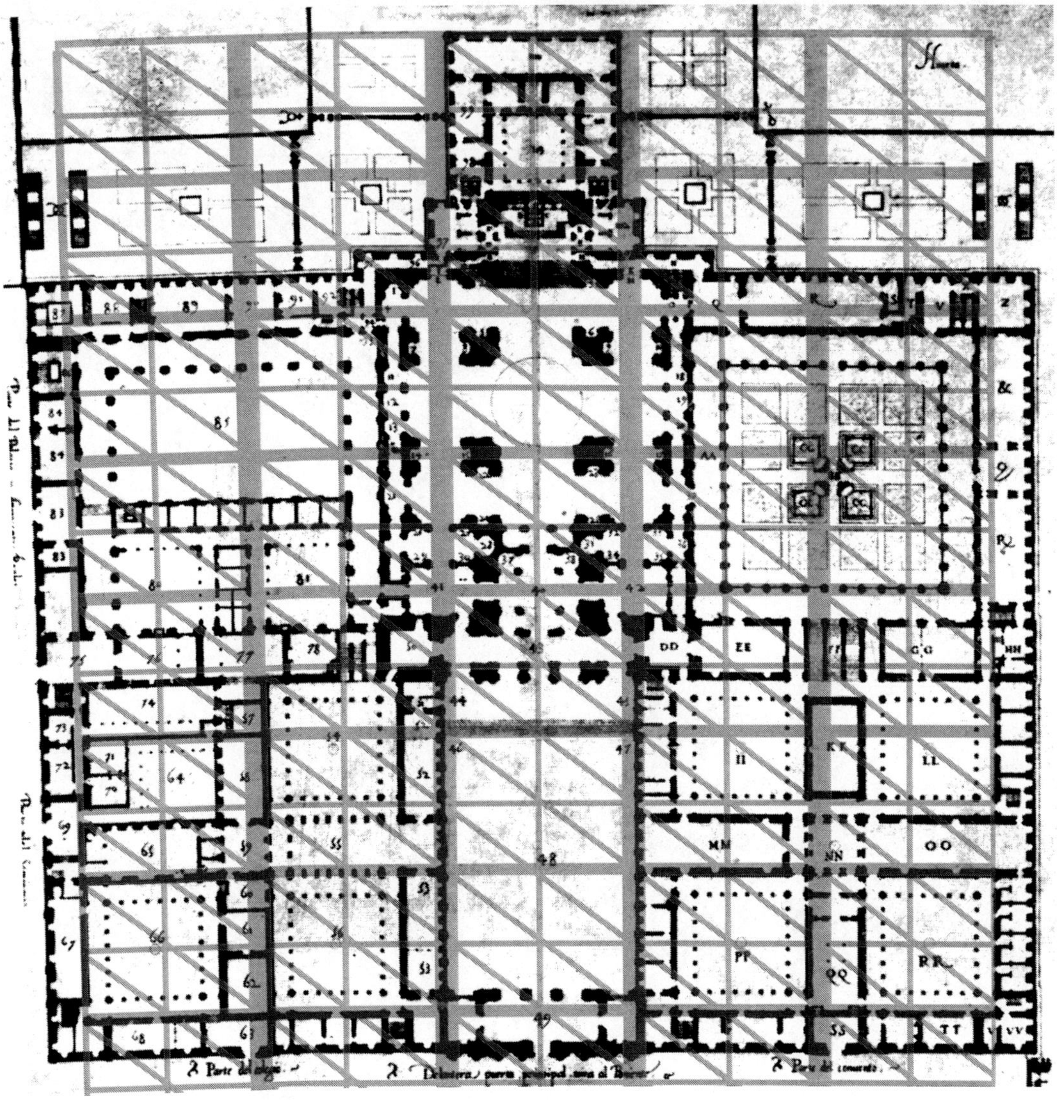

Fig. 45. Juan de Herrera, Ground-plan of El Excorial. (Herrera, *Sumario*, 1589, and now recreating its originally inscribed, 3:4:5 proportionate modules by using superimposed, computer-generated grid-patterns.)

these are so produced through the consistent application of 3:4:5 proportionate modules.

A second reference to Jerusalem is refigured *inside* the Escorial, and it recreates the New Testament successor to the Temple of Solomon, the Holy Sepulcher. The true spiritual heart of the Escorial is the Royal Basilica, and the very heart of that is its *retablo mayor*, and the literal heart of that is the monstrance (*custodia*) upon the high altar, itself containing the Host, Christ Himself in transubstantiated form. (Fig. 48) Another one of the prints published in 1589 under Herrera's supervision illus-

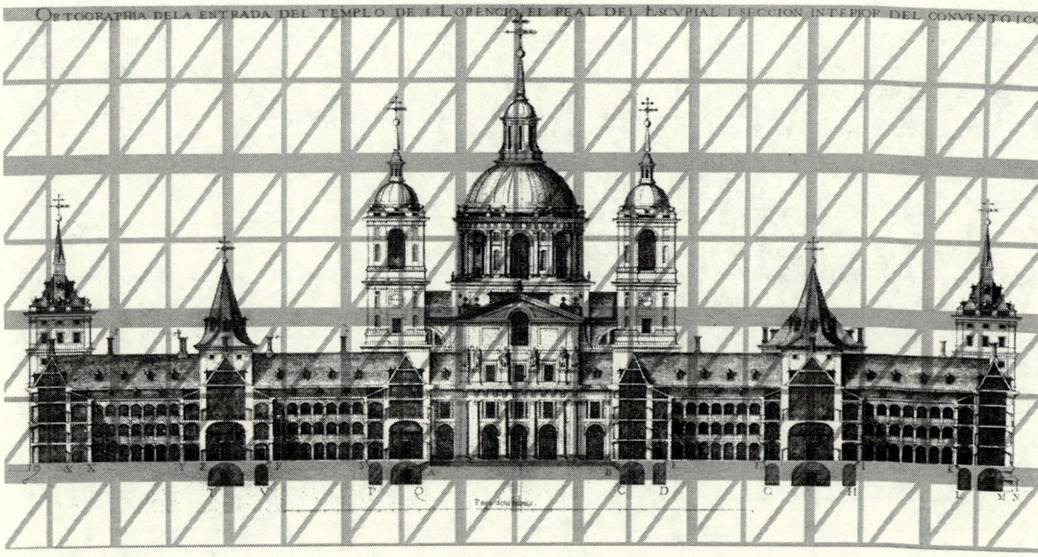

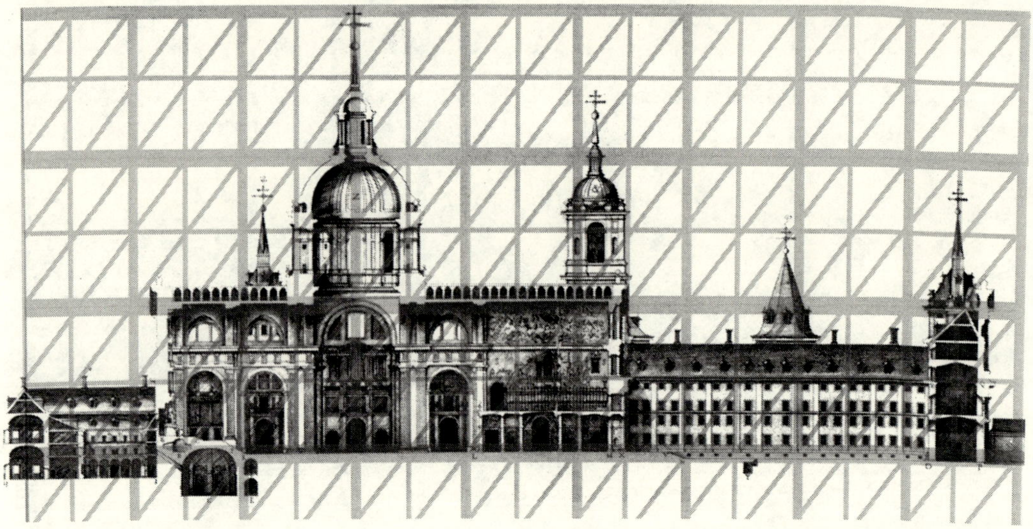

Top: **Fig. 46. Juan de Herrera, Section through El Escorial, showing the grand portal entrance in the west and its lateral, cloister extensions. (Herrera, *Sumario*, 1589, and now recreating its originally inscribed, 3:4:5 proportionate modules by using superimposed, computer-generated grid-patterns.)** ***Bottom:*** **Fig. 47. Juan de Herrera, Section (east-west) through El Escorial, showing the royal residence, basilica, forecourt, and portico, El Escorial. (Herrera, *Sumario*, 1589, and now recreating its originally inscribed, 3:4:5 proportionate modules by using superimposed, computer-generated grid-patterns.)**

trates the entire, indissoluble and scenographic, ensemble of sculpture and architecture in the Capilla Mayor; it is fittingly entitled "*Perspectiva.*" As conceived by Herrera around 1576, and later sculpted by Pompeo Leoni, two huge tomb monuments, depicting in bronze the larger-than-life-sized families of both Charles V and Philip II, are placed on either side of the great Basilica. The vision of both families is intensely

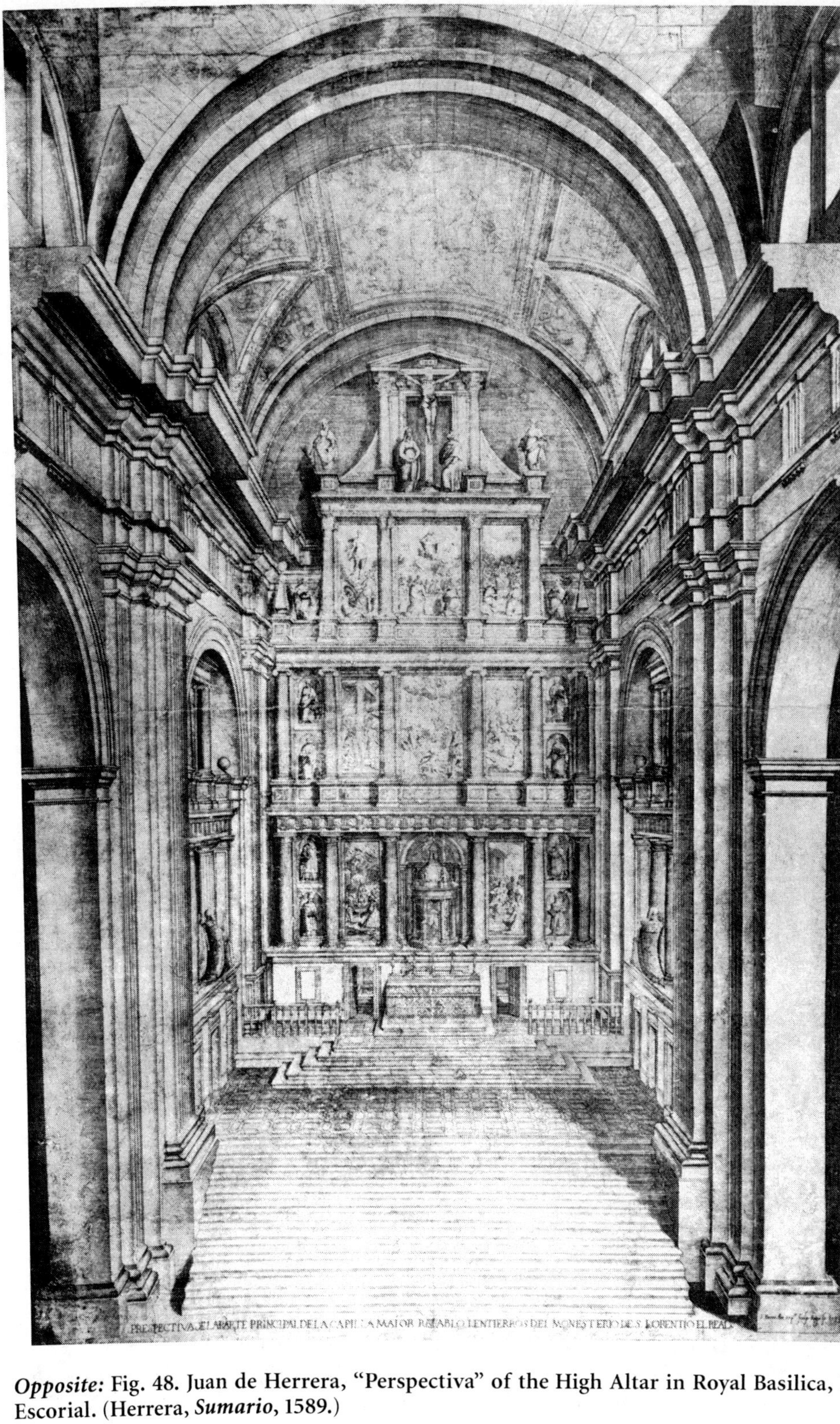

Opposite: Fig. 48. Juan de Herrera, "Perspectiva" of the High Altar in Royal Basilica, El Escorial. (Herrera, *Sumario*, 1589.)

locked onto the altar. The result, Kubler notes, is to make the sanctuary into "a triptych, with lateral wings to contain orant figures, including the donor," and he also records a contemporary calling it a "royal theater and split stage-setting" (*soberano teatro y partido coliseo*).

As shown here, and as generated by the royal vision, the vanishing-point of all Renaissance perspective is wholly spiritual, the giant *Custodia* containing a mini-*custodia*, itself the actual receptacle for the Host.[6] (Fig. 49) As is additionally revealed by a grid-overlay, the *retablo mayor* has the same triangle-derived proportion as does the entire complex of the Escorial. The horizontals produced by intersection with the inscribed diagonals determine not only the base-lines of each story but even the alignment for the arms of the cross. Moreover, here the modular employment has an obvious symbolic implication. The central axis unites the physical body of Christ above to His mystical reincarnation in the Host placed below in the *Custodia*.

The *Custodia*, a symbolic tabernacle, is both the tomb of Christ and the site of His Resurrection, the *Anastasis*.[7] Commemorative portrait-medals minted in 1563 and 1578, and depicting, respectively, Philip and Herrera, had depicted the *Templum Domini*, or Holy Sepulcher, in Jerusalem[8]; as it was so depicted, it provided the structural model for Herrera's *Custodia* (as shown at the bottom of fig. 48). Another likely visual model was Bramante's "Tempietto," an unprecedented, central plan, two-story, Doric-order, structure put into the center of the cloister of San Pietro in Montorio. (Fig. 50) This monument may be illustrated in the way that it would have been made known to Herrera, from the woodcut appearing in the Spanish translation of the third and fourth books of Serlio's *Architettura*, as prepared by Francisco Villalpando and dedicated to Philip II in 1552.[9]

Just as its traditional format proclaims, the ostensible function of the Tempietto was to serve as a "martyrium" to St. Peter since, according to Serlio, it was at this particular spot in Rome that Christ's principal Apostle was crucified. One can now posit a direct relation between this much-discussed Roman monument and the Escorial, especially given its demonstrated iconographic reliance upon the Holy Sepulchre. In short, an inscription on the Tempietto (mostly ignored by modern art historians) explicitly states that Bramante's monument was founded by the King and Queen of Spain in 1502.[10] Given the clear resemblance of the *Tempietto* to innumerable European representations of the *Templum Domini*, and especially given the honorific title of Ferdinand and Isabella as the "*Reyes de Jerusalén*," then it follows that, most likely, Bramante had been specifically encharged by them to represent the canonic sepulchral monument first erected in Jerusalem. Philip II, their grandson, unquestionably hewed to the same venerated architectural prototype in the Escorial.

Sigüenza explained that the cult of the Holy Sacrament was indeed the principal object for the "*piedad y devoción*" of Philip II and his Escorial, and thus the *Anastasis*-like *Custodia* on the high altar was indeed "the objective and center of the whole edifice." So, in Spain at least, Renaissance perspective — and just as that "*Perspectiva*" (fig. 48) is shown to be intensively focused upon a Host-bearing monstrance — is now revealed to be as much spiritual as it is geometrical. (Fig. 51) Moreover, as is additionally revealed by yet another grid-overlay, this one imposed over Herrera's folio-

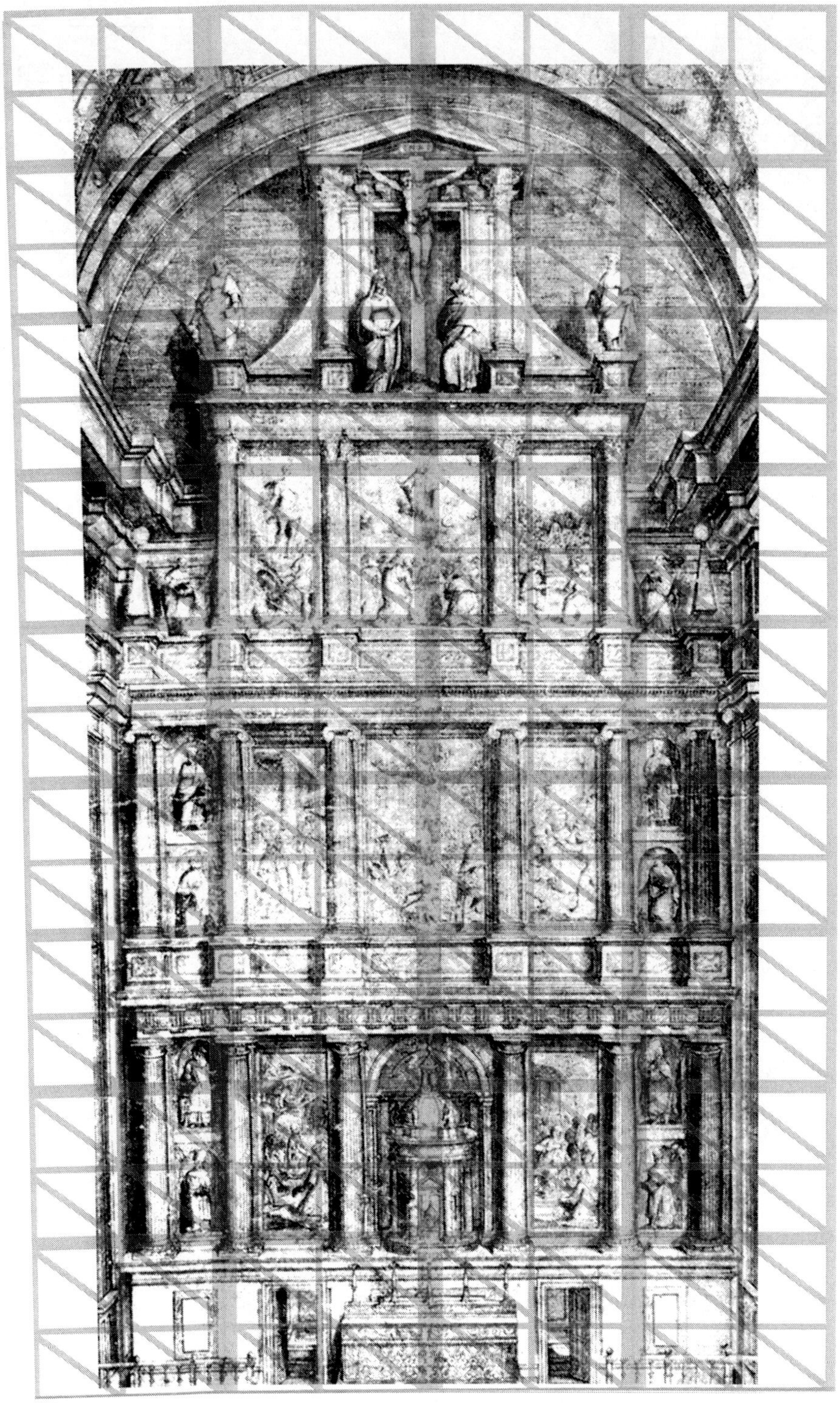

Fig. 49. "Trazado regulador" of the High Altar in the Escorial. (Herrera, ***Sumario***, 1589, and now recreating its originally inscribed, 3:4:5 proportionate modules by using superimposed, computer-generated grid-patterns.)

page size print of his *Custodia*, we find that, once again, the same *neo-mudéjar* design scheme was again faithfully hewed to. In this case, as aligned upon a central axis rising upwards through the vertically elongated *Custodia*, the center of the composition is, once again, a *rosetón*-like aperture. This graphic elevation leads to a broader conclusion. For Herrera, certainly the Holy Sepulcher (fig. 50) was itself a model of "divine harmony." As we also now recognize, it was so specifically because it, too, was based upon the venerable Pythagorean-geometrical module.

So, too, was the Temple of Solomon based upon the venerable Pythagorean geometrical module — at least, so it appeared to Juan Bautista Villalpando. (Fig. 52) When the triangulated grid is laid over his elaborate engraving of the western façade of the Temple of Solomon (fig. 44), the inscribed figures demonstrate how the proportions for the entire composition were achieved. The entire façade was first inscribed within a 1:2 rectangle (the width versus the height of the top finial); the intersection of two diagonals within this rectangular figure marked the top of the rectangular opening in the second story. As the grid reveals, all the other dimensions were established with the 3:4:5 triangle motifs.

Quickly stated, first a diagonal was drawn from the lower right-hand corner; this first establishes the height of the door and, finally, the height of the cornice in the second story. The next diagonal on the base-line fixes the width of the door, also the base-line for the for the Doric frieze. The next one marks the inner face of the vertical band of windows, with their placement having been pre-determined by horizontal lines generated by the grid. Another diagonal, drawn from the right end of the base-line for the Corinthian capitals, located the base-line for the base of the third story. In the uppermost story, a diagonal drawn from the right corner fixes the height of that zone. And so on, throughout all three stores (and the height of the central finial was fixed by yet another 33° diagonal ascending from the corner where my grid ends).

One further example, another royal palace, will serve to make the point that the Pythagorean scheme was ubiquitous in post-medieval Spanish architecture. Two Italian architects, Filippo Juvarra and Giovanni Battista Sacchetti, were commissioned to design a new Palacio Real in Madrid. This was constructed between 1738 and 1764. It was built directly over the ruins of the former royal palace, the Alcázar, which, until it burned down in 1724, had served as the residence of Philip IV and his regal followers. Notwithstanding the fact that the new architects were Italians, and that their Palacio Real certainly "looks" Italianate, nonetheless another computer overlay reveals a surprising fact. (Fig. 53) In short, the neo-classical era building had reverted to the medieval, distinctly *mudéjar* scheme of "design-by-triangulation," specifically the kind with 3:4:5 ratios. In the case of Juvarra and Sacchetti, this format was imposed upon them, presumably not by their royal patrons but instead by history. As it turns out, the Alcázar (and *al-qasr* means "fortress-palace" in Arabic) had been originally laid out around 875, and as so ordered by the then Emir of Córdoba, Mohammed I. Hence the design agreed upon in 1738 was inexorably shaped by a *sahn/patio*-centered Arabic ground-plan (like fig. 6) worked out over eight centuries earlier in a then–Moslem Madrid, which was then called *Magerit.*[11]

There are, of course, other clear instances of employment of the 3:4:5 triangu-

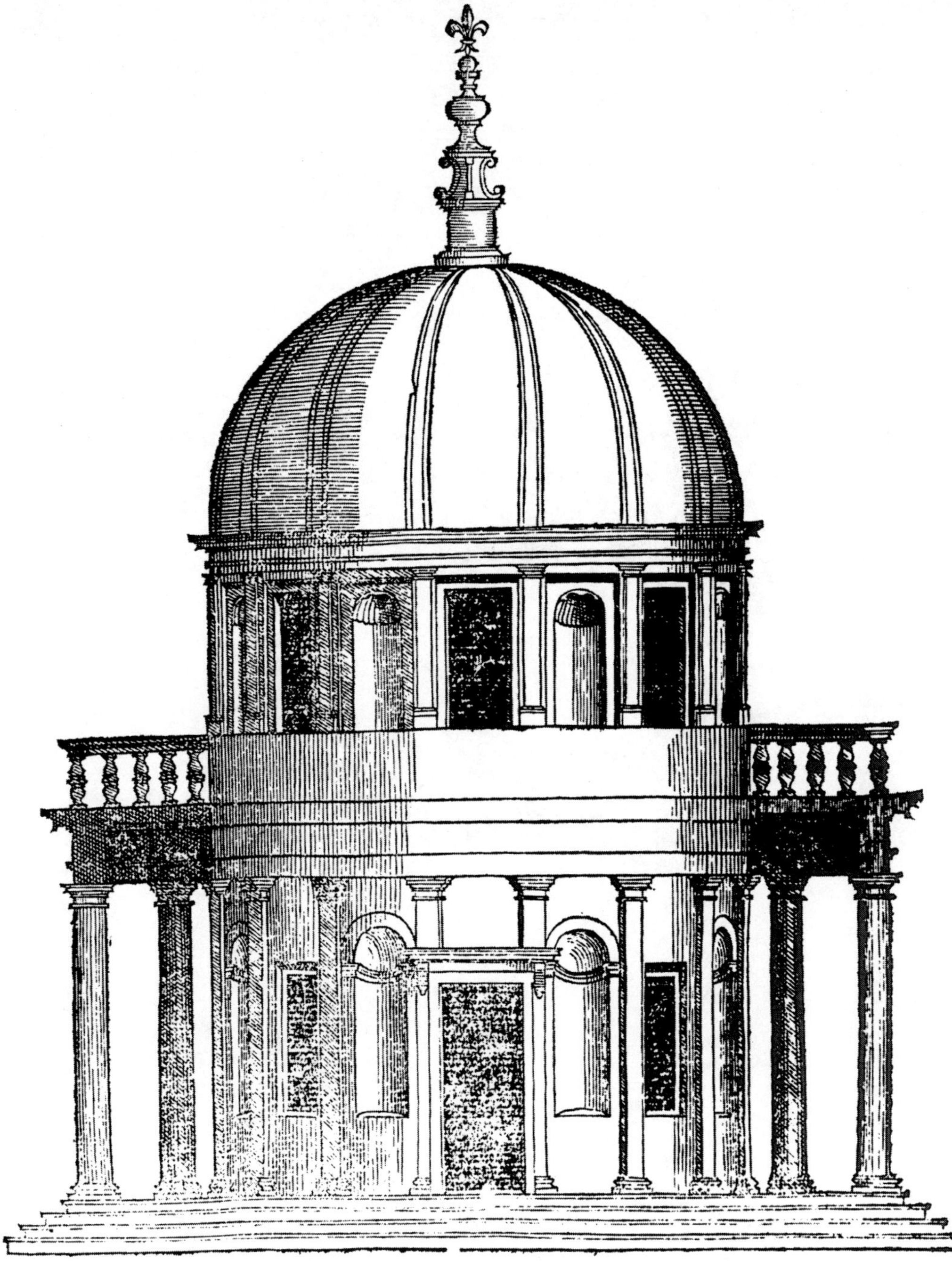

Fig. 50. Donato Bramante, Tempietto. S. Pietro in Montorio, Rome, 1502. (As illustrated in *Tercero y Quarto Libro de Arquitectura de Sebastián Serlio*, 1552.)

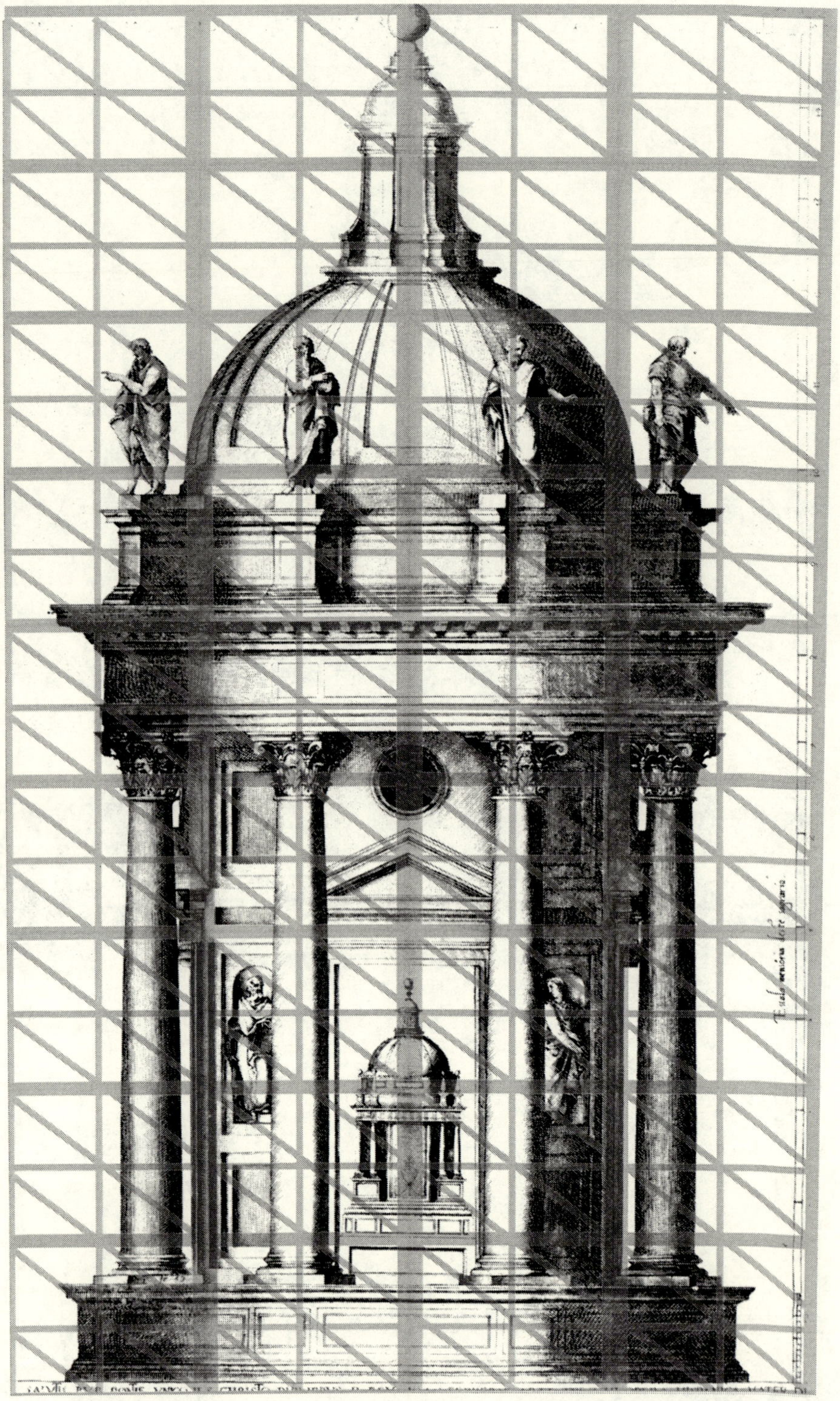

Fig. 51. "Trazado regulador" of the ***Custodia*** placed on the High Altar in the Escorial. (Herrera, ***Sumario***, 1589, and now recreating its originally inscribed, 3:4:5 proportionate modules by using superimposed, computer-generated grid-patterns.)

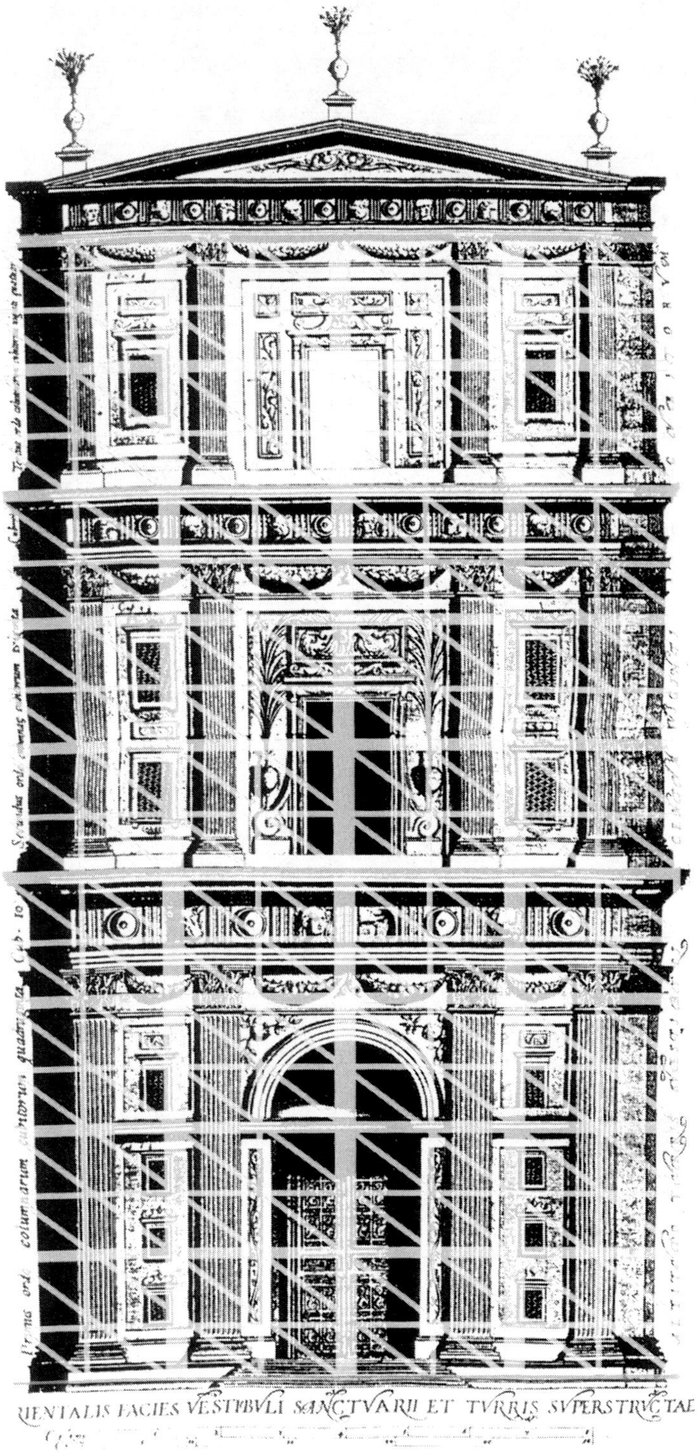

Fig. 52. Principal façade of the Temple of Solomon (Villalpando, ***In Exechielem Explanationes***, and now recreating its originally inscribed, 3:4:5 proportionate modules by using superimposed, computer-generated grid-patterns.)

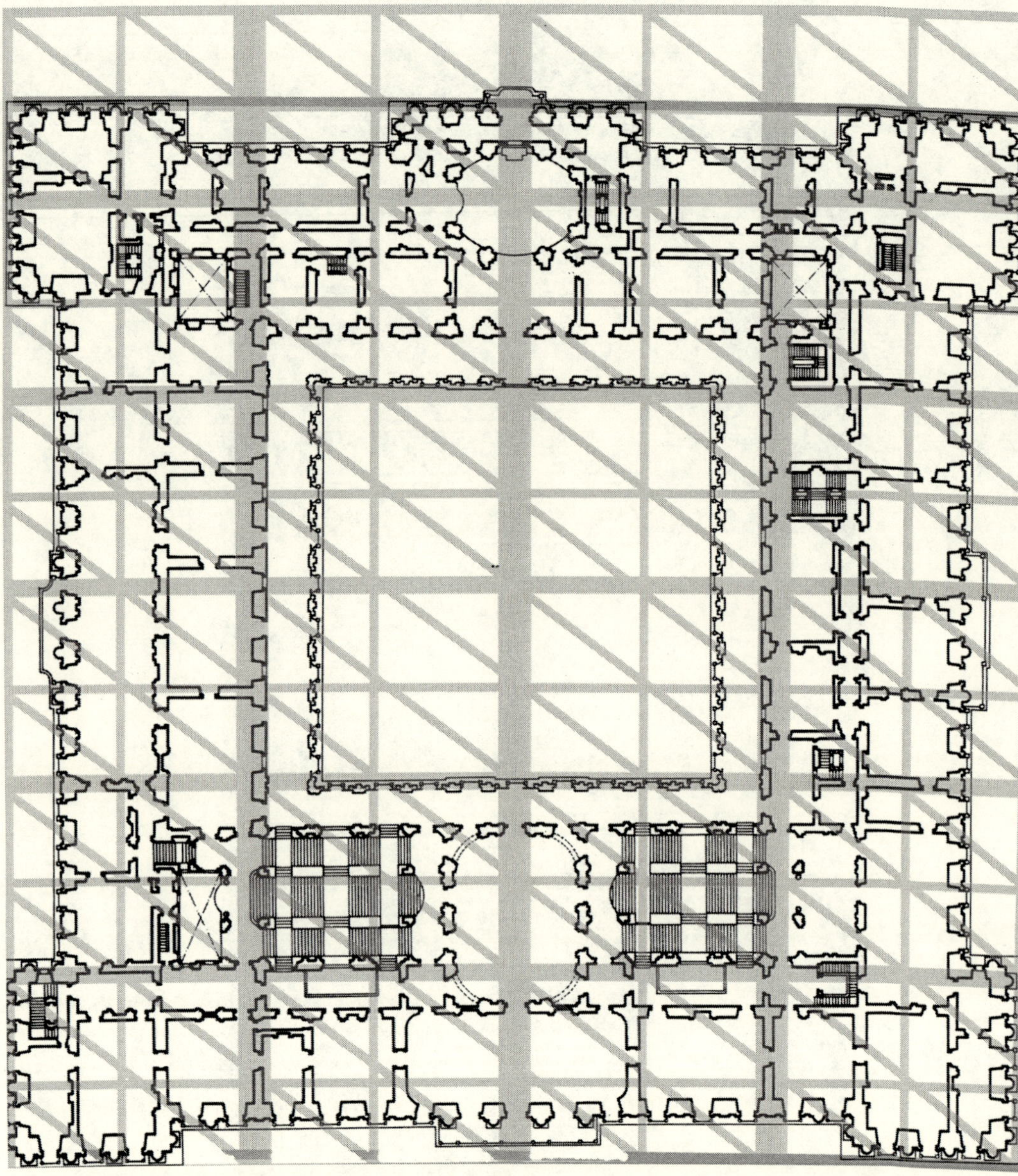

Fig. 53. F. Juvarra and G. B. Sacchetti, "Trazado regulador" in the ground-plan of the Palacio Real, Madrid, 1738–64. (Computer-generated grid-patterns layered over a drawing given in Bottineau, ***Iberian-American Baroque***, 1970.)

lar scheme in peninsular Spanish architecture, ones which I have already checked by laying the grid, as printed on an overhead transparency, over various examples, and the best graphic evidence is provided by recently made, measured drawings (as in figs. 28, 30, 31, 36, 38, 40, 41, 53).[12]

12

Ocotlán: Another Mexican Example of "Design-by-Triangulation-and-String"

Juan de Herrera, a royal architect seemingly obsessed with Christian number-symbolism, is one thing, and a rustic builder charged with designing a humble parish church on the civilized fringes of Mexico is quite another. A major problem with the previously posited patterns of superimposed "regulating line-patterns" proposed by the likes of Chueca Goitia, Harth-Terré, and Rodríguez Camilloni is that those schemes seem much more appropriate to chair-borne theory. As such, they are obviously not quite so adaptable to the traditional, "hands-on," procedures of known artisan workshop practice.

Evidently "ancient," the 3:4:5 system is, in any event, demonstrably traditional within Spain; unmistakably, it has been so used there, and since (at least) the tenth-century Córdoban designs (figs. 9–15). Moreover, graphic proof for the complementary employment in Spain of inscribed geometrical figures ordering both ground-plans and architectural elevations can be dated as late as 1804 (see fig. 22). According to our findings (figs. 24–28, 32), in eighteenth-century Mexico, the traditional design scheme had additionally shown itself to be still adaptable to even the most primitive working conditions, that is, the kind then typically obtained in Spanish America.

As further evidence for the ubiquity of a *trazado regulador* system specifically employing Pythagorean triangle, we can submit several additional examples. Upon each of these we had again superimposed the *a priori*, plane geometry-derived scheme, and the interested reader may repeat the experiment by reference to those published photographs upon which our digitized grid schematum had been laid.[1] Since we have already provided sufficient graphic evidence to make our case for the local phenomenon, only one additional example of Latin-American architecture designed with inscribed, specifically 3:4:5 triangulate, regulating line-patterns needs to be discussed and illustrated here.

The Sanctuario Church at Ocotlán (Tlaxcala) is the site where devout pilgrims gather at a spot where the Virgin was said to have made an apparition around 1540;

towards the end of the eighteenth century, a new shrine was erected to honor her visitation. (Fig. 54) From the atrium gate, a splendid vista greets the eye. The snow-white, turbulent surface of the massive façade is compressed between bright-red, pillar-like tower bases. The mathematical precision of the flanking, *neo-mudéjar*, hexagonal tile-patterns contrasts with the measured complexity of the *fachada-retablo*. The mysteriously suspended, paired belfries repeat the sculptural vigor of the façade; sitting upon the base of two inscribed triangles, their apex is anchored in the *rosetón*, and the base-line runs through the bottom cornice of the clock placed directly above the portal. The vertical axis is 3 units high; the horizontal length is 4; and the diagonal is 5 units.[2]

As was the case at Tepotzotlán (figs. 24–26), the formal (and spiritual) center of the composition is the *rosetón* containing a stone replica of Our Lady of Ocotlán. (Fig. 55) This time eliminating the bell-towers, and concentrating wholly on the centered altarpiece-façade, our cyber-grid is again laid over our source-photo. A doubled triangle generated from this centric point determined the height of the main portal below the rosetón, and another, flipped on its horizontal axis (as in fig. 7), fixed the height of the pilaster framework. Another such triangle determined the center-point of the gable at the top of the building, containing a clock chiming the hours for devotions. The result is that, through triangulation, three spiritually significant points on the façade are vertically aligned and proportionately distanced from one another. The diagonals drawn downwards from the center point determined: (1) the mid-point of the second story of the composition; (2) the center of the capitals in the first story; (3) the base-line of the sculptures on this story; (4) the axis of the terminating column on this row. The upward path of the same diagonal line similarly determined the top of the third story of sculptural decorations, and also the height of the finials at either end of the entire composition, and so forth. As may be expected, the imposition of a smaller 3:4:5 grid-pattern reveals the geometrically predetermined source of nearly all the other intersections on this *fachada retablo*.

In the next step, when our grid is now laid over a measured drawing of the ground-plan of the entire complex at Ocotlán, the same triangle-determined proportionate relationships are produced throughout the architectural complex, including a monastery with an open patio and various out buildings. (Fig. 56) In this case, the essential centric point, with this generating the lateral proportions, becomes the central axis of a terminal octagon, the *camarín*, sheltering a massive, gilded *retablo mayor*. Now we may repeat the same procedure, this time over a measured drawing of the Sanctuario itself. (Fig. 57) The single nave opens on to three bays; the nave is quite wide, spanning almost fifteen meters. In the fourth bay, the nave walls curve inward to form a triumphal arch, with this being transformed into a huge shell, a nine-meter rectangular crossing, closed with an octagonal dome. As the measured ground-plan reveals, the cross-ribbing in the bays was determined by 3:4:5 triangulation and, as now shown by our grid-overlay, the same module also fixed the footings for the attached piers, also the squared ends of the flanking bell-towers outside the church.

Obviously, the 3:4:5 layout will not be found in *all* Hispanic buildings. Nevertheless, although previously ignored, it is amazingly common. The interested reader

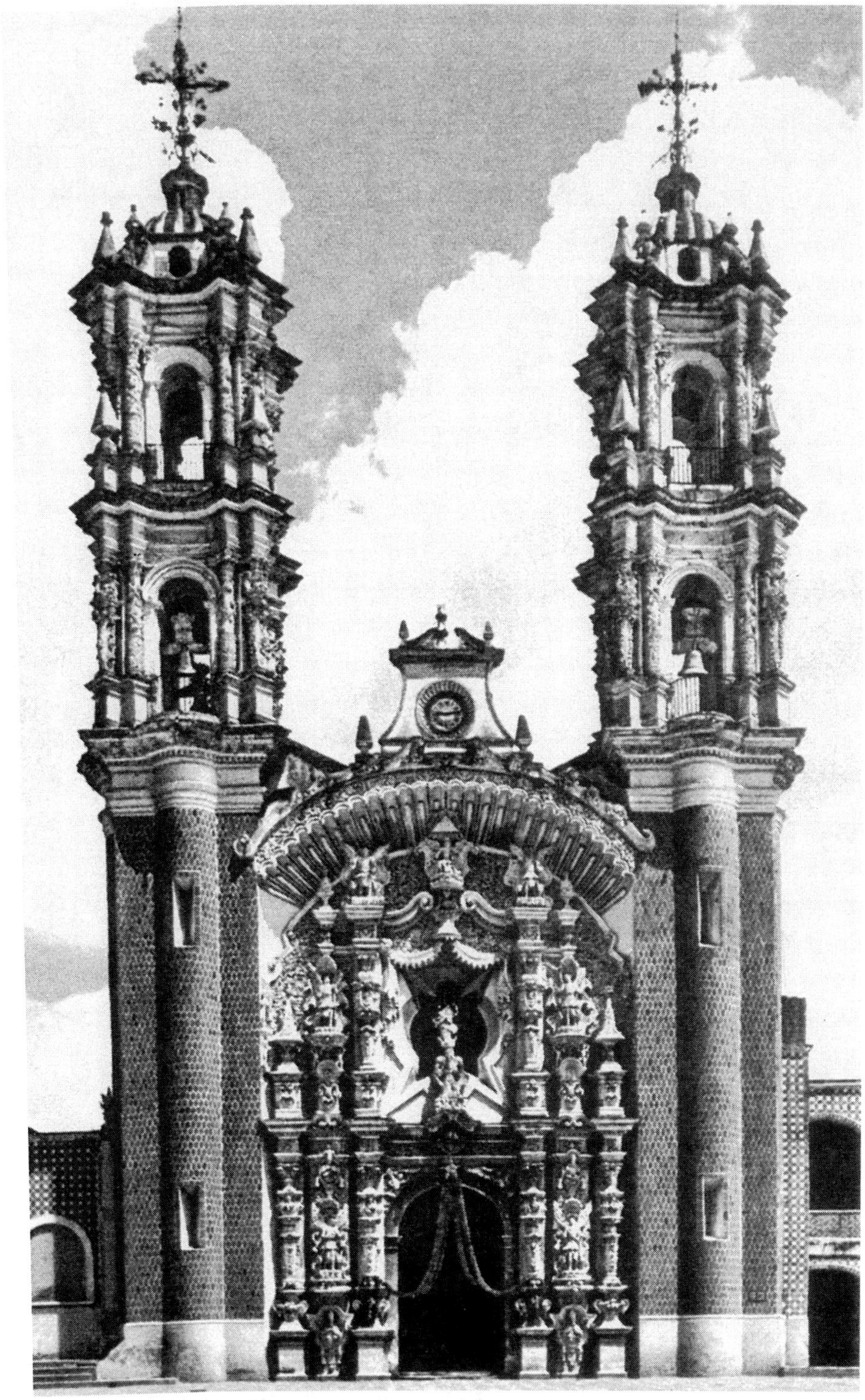

Fig. 54. Façade of the Santuario church at Ocotlán (Tlaxcala), ca. 1780.

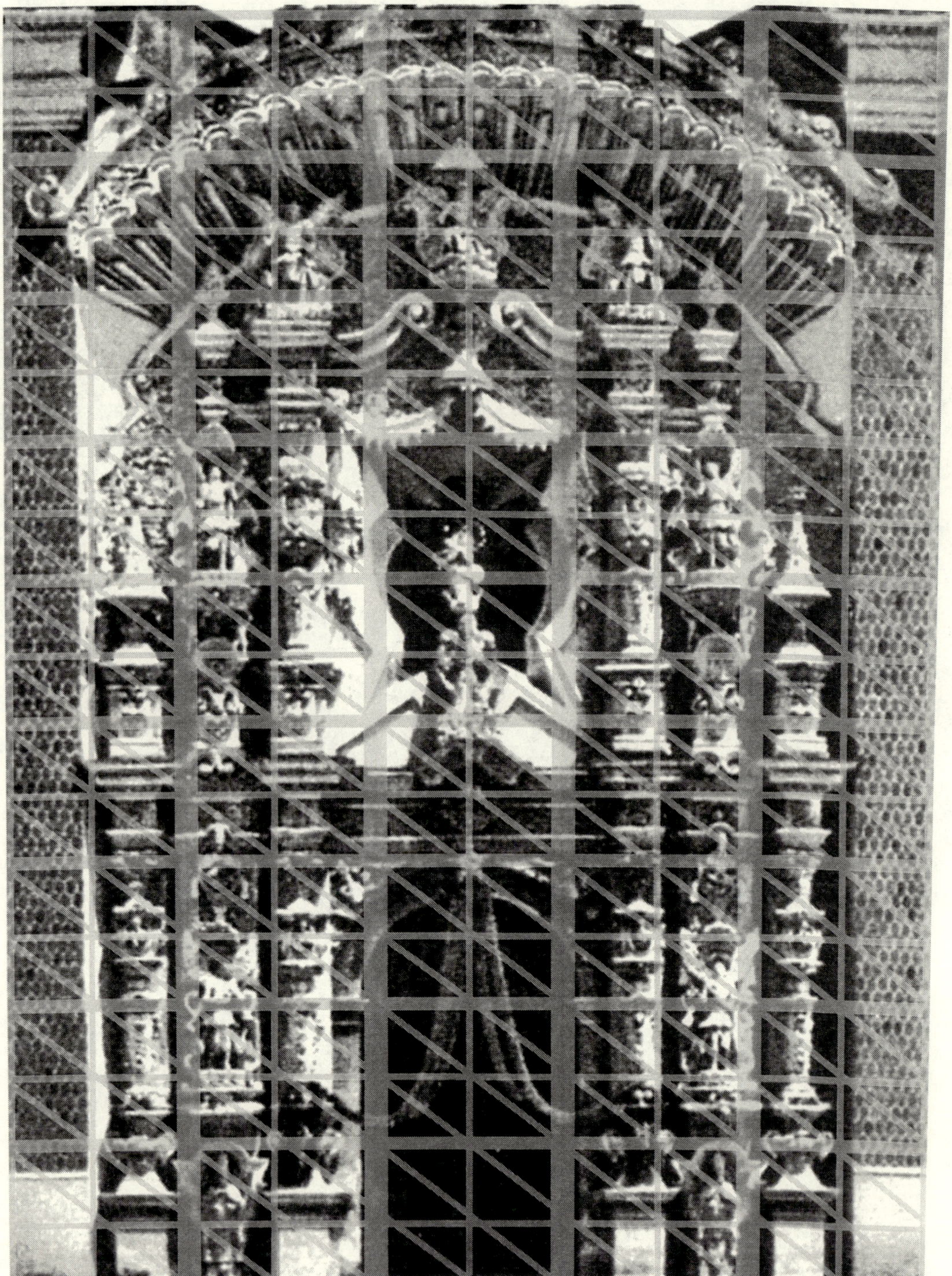

Fig. 55. "Trazado regulador" patterns used to design the façade of the Sanctuario church at Ocotlán. (Computer-generated grid-patterns layered over photograph.)

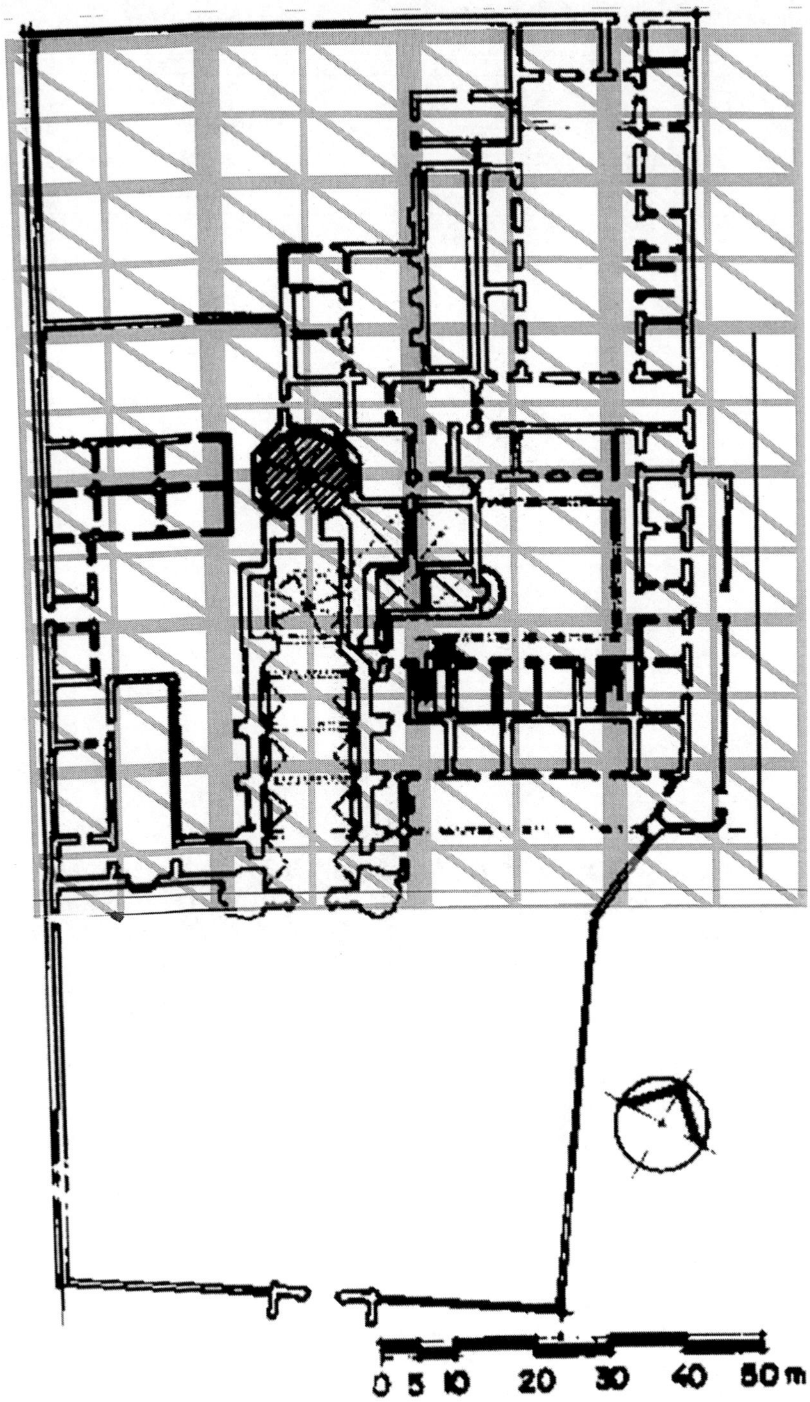

Fig. 56. "Trazado regulador" patterns used to design the ground-plan of the church-monastery complex at Ocotlán. (Computer-generated grid-patterns layered over a drawing given in Mullen, *Architecture and Its Sculpture in Viceregal Mexico*, 1997.)

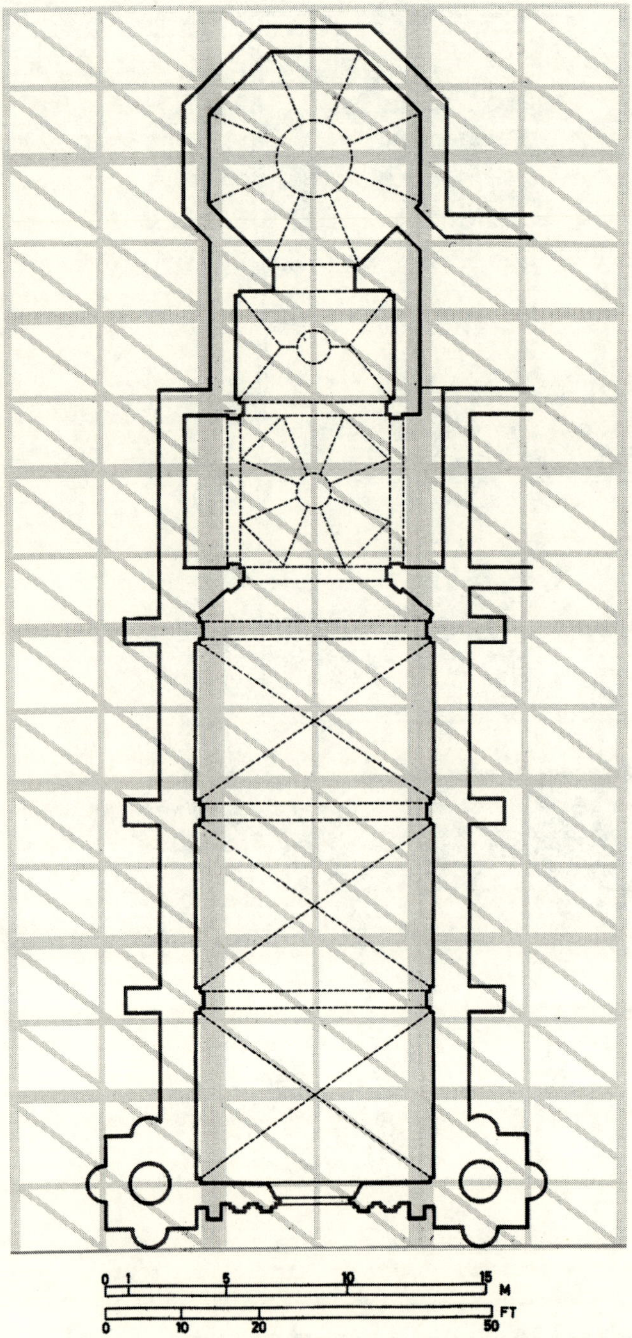

Fig. 57. "Trazado regulador" patterns used to design the ground-plan of the Sanctuario church at Ocotlán. (Computer-generated grid-patterns layered over a drawing given in Bottineau, *Iberian-American Baroque*, 1970.)

may now refer (in a note) to a number of other examples of Trans-Hispanic architecture which I find to be similarly based upon comparable 3:4:5 regulating line-patterns.[3] However, these examples do not by any means exhaust the potential corpus of 3:4:5 plans in Latin America and Spain. In any event, because these wholly geometric observations can be objectively duplicated, or even expanded upon by any *other* interested investigator similarly utilizing a digitized grid-overlay, these findings are, therefore, *scientific.* My conclusions are "scientific" inasmuch as they represent what scientists, rather than art historians (once optimistically calling themselves *Kunstwissenschaftler* = "art-scientists"), call "repeatable results." Whereas DNA and genome research has produced those *repeatable results,* "cold fusion" never did: *Quod non erat demonstrandum.*

Simply put, the strictly visual evidence points to the survival, and as late as the eighteenth-century in Latin America, of: (1) a once universally applied, early-medieval workshop procedure of "design-by-string"; (2) said designs were all initially derived from the "design-by-triangles" procedures; and (3), uniquely in the case of the Hispanic (vs. European) architectural designers, those imputed procedures clearly, and consistently, produced the same kind of 3:4:5,

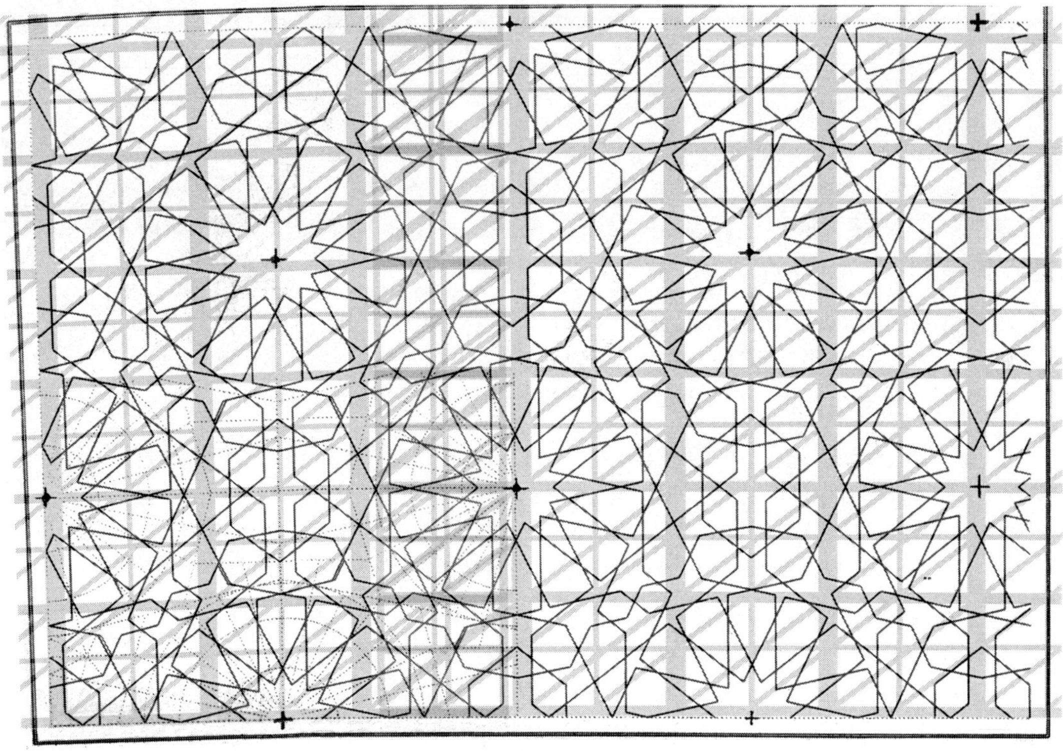

Fig. 58. J. Bourgoin, "Underlying (or Non-Inscribed) Heptagonal Design-Modules Determining the Generation of Arabic Star-Patterns," *Les Eléments de l'art arabe: le trait des entrelacs* (1879), Plate 169 (with Bourgoin's dotted-line under-drawing here enhanced with computer-generated grid-patterns of non-inscribed, 3:4:5 triangle-modules).

ad triangulum, organizational leit-motiv as that one which we (following Chueca's lead) have identified as having first appeared on Spanish soil in Moslem Córdoba (figs. 9–15).

Now the graphic evidence may be presented which shows the persistent Hispanic predilection for the 3:4:5 triangular module seems to have been a local peculiarity, a distinctly *mudéjar* departure from standard (or orthodox) Islamic practice. In short, I find that the geometrical framework for the so-called Pythagorean triangle is a *heptagon*, a seven-sided figure.

To the contrary, and as was determined long ago by the scrupulous graphic research of Jean Bourgoin, as published in his large-folio volume *Les Eléments de l'art arabe: le trait des entrelacs* (1879), besides being logically derived from templates employing inscribed triangles (as shown in fig. 7), and as specifically designed to be displayed in or upon architecture, traditional Islamic (meaning non–Hispanic) decorative schemes were mostly derived from the *equilateral triangle* (as was shown in fig. 8). Subdivided, the equilateral (3:3:3) triangle produces a 3:5:6 module, with these modules in turn producing those geometric figures called *hexagons*, *octagons* and *dodecagons*, even *pentagons*.[4] For instance, a basic Islamic module like the *hexagon* is inscribed within a square, the edges of which have been clipped, so producing four 3:5:6 triangles, and the interior of the hexagon is made up of six 3:3:3 triangles and, bisected, each of these triangles yields two 3:5:6 mini-triangles. Moreover, the *hexagon* is also a "natural" figure; since even bees produced hexagonal modules to make their beehives, and so the hexagon was often used as a design module by architects, particularly the modern ones with theoretical inclinations.[5]

To the contrary, *heptagons*, the kind based upon the 3:4:5 triangle module, are rare in strictly Islamic (vs. standard Hispanic or *neo-mudéjar*) usage.[6] To demonstrate the paucity of heptagonal figures in Islamic decoration, the reader can easily corroborate this conclusion by laying a transparency inscribed with a grid having those figures over good photographs or linear tracings of Islamic designs from places other than Andalusia. In any event, the specific operations of the Pythagorean triangular figure within a standard heptagon is easily demonstrated with our computer-generated grid-patterns of non-inscribed, 3:4:5 triangle-modules, here laid over one of Bourgoin's six heptagonal drawings. (Fig. 58) In short, the essential vertical alignments of the major heptagonal star-patterns, all proportionately arranged in parallel rows, were again pre-determined by diagonal intersections, and those diagonals establish innumerable triangles having those distinctive 3:4:5 ratios.

13

The Historical Evidence for "Design-by-String" and with Grids

Whether one is working with a tenth-century Islamic façade or with a *fachada retablo* from the eighteenth century in Mexico, the evidence by which to deduce the underlying design procedures is necessarily visual and empirical. The vagueness of the surviving written testimony making mention of the preliminary design stages of architectural projects, particularly its working drawings, may be exemplified by the explanation given in on the first page of Leon Battista Alberti's *De re aedificatoria* (1489):

> Let us therefore begin thus: the whole matter of building is composed of lineaments [*lineamenta*] and structures All the intent and purpose of lineaments [the initially drawn design scheme or schematic outlines] lies in finding the correct, infallible way of joining and fitting together those lines and angles which define and enclose the surfaces of the building. It is the function and duty of lineaments, then, to prescribe an appropriate place, exact numbers [*certum numerum*], a proper scale [*modum*], and a graceful order for whole buildings and for each of their constituent parts, so that the whole form and appearance of the building may depend on the lineaments alone. Nor do lineaments have anything to do with material, but they are of such a nature that we may recognize the same lineaments in several different buildings that share one and the same form, that is, when the parts, as well as the siting and order, correspond with one another in their every line and angle. It is quite possible to project whole forms in the mind without any recourse to the material, by designating and determining [on paper] a fixed orientation and conjunction for the various lines and angles. Since that is the case, let lineaments be the precise and correct outline, conceived in the mind, made up of lines and angles, and perfected in the learned intellect and imagination.[1]

A bit more helpful is a builder's manual published in 1639 by Fray Laurencio de San Nicolás. In the brief (two folio-pages) chapter in his *Arte y Uso de Arcitectura* dedicated to the way to "sketch upon paper any building" ("trazar en papel cualquier edificio," Cap. XXI), the author states that "it is suitable that every plan is laid out with right angles, although some use circular figures. However, the sturdiest buildings are those based on right angles." He next briefly describes what appears to be the preparation of a squared grid as an essential step preceding the entire design process.

> In order for the novice designer to become adept at sketching both elongated and square ground plans, he will [first] generate angular [patterns] with lines drawn over blank paper, and where ever he wishes; so doing, he creates right angles in the intersection of the lines. In drawing [further] parallel lines, their opposing angles will also be at right angles [so setting up a grid scheme]. The first thing to do is to divide a [given] straight line into given lengths; these divisions are called modules [*módulos*] by Vitruvius, and by us they are commonly named the scale [*pitipié*]. This is the regulating unit, according to which every part of the building must be drawn.[2]

As previously noted, we have little or no textual materials that relate to the actual working conditions and procedures encountered at building sites. Especially, we have little practical information regarding the devising of those schematic preparatory sketches (*lineamenta*) and I find no other explicit mention made in any architectural manual of the obviously essential employment of a drawn grid. Nonetheless, we do know that grids had been employed by architectural designers since the time of the ancient Egyptians. The modern historian's dilemma was explained long ago by Eugene-Emmanuel Viollet-le-Duc in his *Entretiens sur l'architecture* (1872). In his ninth lecture on architecture (extensively quoted in the Appendix), he remarked that

> In the Middle Ages, however, [and] incontestably, and probably [also] in those of Classical Antiquity, certain [practical] methods were adopted [by designers], according to which the proportions of buildings were [pre]determined. We have but little [written] information on the subject. The loss of [artisan] traditions, the utter decay of official instruction [in architecture], have allowed the clues, which formerly guided architects in the labyrinth of those mysterious branches of knowledge and with which the professional guilds were formerly so familiar, to slip from our hands.[3]

It has been over a hundred-and-thirty years since Viollet-le-Duc wrote those lines. Modern historians of architecture have since tried to fill in the gaps in our knowledge "of those mysterious branches of knowledge."[4] Again, the best way to solve the problem is essentially, and simply, to ignore official architectural *theory*. In the first place, the process we are attempting to reconstruct must antedate any kind of written architectural theory (and we may even include Vitruvius in this disclaimer). Therefore, it is equally obvious that our primary historical examples—either an Islamic mosque or a Mexican church—were certainly not designed to serve as abstract exemplars of architectural "theory." Neither is it likely, for that matter, that their design peculiarities are due to any overbearing concern with esoteric "number symbolism."[5]

These buildings were, furthermore, designed by humble masons—and not by any theological scholars, and certainly not by erudite "architects." In the case of Córdoba (figs. 9–15), when the term "*architectus*" had not yet been invented,[6] and, at Tepotzotlán and (certainly) at Ocotlán (figs. 24–28, 54–57), no "*arquitecto*" was likely to have been available at the site. Instead, the designer would be called an "*alarife*," or master builder.[7] When, back in the homeland, Spanish designers had earlier resorted to *la geometría*, they did so for strictly pragmatic purposes.[8] But whenever they did employ geometric formats, they were past masters of the science, for theirs was a *scientia*—geometry—which they had first been taught in Islamic *al-Andalus*.

In such cases, obviously the geometric procedure was used for wholly practical ends and, therefore, its understood purposes were not at all "theoretical." As François Bucher reminds us, like Vitruvius, the pragmatic medieval designer's

> primary tools consisted of a straight-edge, a square, a compass, and a divider. Rules, including the construction of any inscribed polygon, were based on the use of these instruments. Thus, we almost always deal with [practical] design precepts that had lost any [direct] connection with theoretical geometry or mathematics.... The geometric canon [as applied] was at first based on modular addition, and then [in the late Gothic period] it was replaced by modular inscription or rotation, thus producing proportions containing irrational numbers.... The standard rectangle, as well as the golden section, were geometrically constructed, at least from the twelfth century onward. The geometric [or manual] systems ordered the grid upon which the plans, elevations, and details evolved.[9]

Another eminent student of Gothic architecture, the late Paul Frankl, explained (in 1960) that, since standards of measurement were not uniform across medieval Europe,

> The architect, therefore, placed no value at all on plotting his drawings in a determined [numerical] scale. That does not mean that he could get along without them entirely. He had at least to indicate the length in feet for the first measure that he prescribed, and it was then, so to speak, the preordained fate of the building, whether a "foot" in that locality was, for example, 29 cm or 30 cm. But, for the further course of his work, it was not important to be able to express every distance between two points of the structure in arithmetical quantities; *he had only to determine the interval geometrically*. Of course, he had to see to it that his geometrical figures could be enlarged to any desired size by a purely geometrical method, and this method is, in general terms, that of the transfer of proportions and angles. *The angles remain the same in the transfer [from the original drawing] and, likewise, the proportions*.... One could not choose any desired dimensions, but was restricted to those that could easily be transferred by the aid of simple geometrical figures.[10]

This same investigator also provides us with a clear explanation of the way that the Gothic designer—a *magister operis*, or works-foreman, and not the early modern types of "*architectus*"—transferred his plan from the drawing board onto an immensely bigger façade. As Frankl tells us, but as we had already deduced must have been the case hundreds of years later on a Mexican site, it is all done with *string* in the Middle Ages. According to Frankl, typically in medieval practice,

> The magister confines himself in his drawing to a few lines that exactly correspond to what the builder has to do on the site. The builder, that is, marks off the outer square first, for example, a side length of seventy feet, describes, at the end points, right angles, and then he marks off from there the seventy foot [or any

> other distance] sides. Then *he runs two strings along the diagonals*; first, to test the right angles and, second, to determine, at the same time, the center. Next, he divides one of the sides of the square into two equal halves, *then he connects the center of this side with the center of the square by means of a string*, and then marks off this length on the diagonal, from the center out. The point so obtained is one of the corner points [in this example] of the "garden," or for the open court [designed proportionately, just as one may design a façade]. *He finds the other three points by the same simple method, and he can now connect these four points by strings*. Thus the mason or stone-mason is enabled to begin setting the stones.[11]

As explained by Frankl, the procedure is absolutely clear cut and unmistakably identified as a matter of customary workshop routine. Unfortunately, Frankl was not able to cite any contemporary text that actually described the standard procedures belonging to "design-by-string." Similarly, we ourselves have never been able to find a written statement like this—that is, none composed by either a mason, or even coming from a proto-architect—that so clearly defines the old "design by string" methodology. Nevertheless, all the available evidence (mostly visual, and here empirically interpreted with the aid of a computer) indicates that this must have been the procedure that was nearly always utilized in traditional artisan workshops in order to generate proportionate designs. Evidently, the simple techniques of design-by-string were so well known that no builder ever felt the pressing need to write them down for posterity. In any event, the entirely "oral" nature of the transmission of architectural "craft secrets" in Spain during the Renaissance has been repeatedly emphasized by Antonio Bonet Correa.[12] Certainly, the inexpensive string patterns have no grand "theoretical" justification that would have interested a lettered architectural *amateur*.

Nonetheless, there is a one practical medieval artisan's manual that actually does explain the "design by string" procedure in some detail, and it does so in very much the same way as did Frankl. The great advantage of this text is that its author shows the great importance of—besides cords drawn vertically or horizontally or diagonally—the wielding of divider-compasses in order to create proportionate distance-points. In this case, however, the writer was not an architect, but instead a painter. As one reads in a famous craftsman's handbook, called *Il libro dell'arte* ("Book of Craftsmanship"—and not "art") and written around 1390 by Cennino d'Andrea Cennini,

> In the name of the Most Holy Trinity, I wish to start you on painting. Begin, in the first place, with working on a wall; and for that I will teach you, step by step, the method which you should follow.... When the plaster is dry, take the charcoal, and draw and compose according to the scene or figures which you have to do; and take all your measurements carefully, snapping lines [or cords] first, getting the centers of the spaces [by drawing strings diagonally from the four corners]. Then snap some [more lines], and take the levels from them. And this line [of string], which you snap through the center to get the level, must have a plumb bob at the foot [of the cord]. And then put one point of the big compasses on this [center] line, and give the compasses a half turn [i.e., swing an arc] on the under side. Then put the point of the compasses on the middle intersection [or vertical axis] of one line with the other, and swing the other semicircle on the upper side. And you will find that you make a little slanted cross on the right side, formed by

> the intersection of the lines. From the left side, apply the line to be snapped, in such a way that it lies right over both the little crosses; and you will [now] find [and without the aid of a spirit-level] that your line is horizontal on a level. Then compose the scenes or figures with charcoal, and always keep your areas in scale and regular [so suggesting the use of a preliminary grid-drawing].[13]

Another passage from *Il libro dell'arte* is also interesting inasmuch as (1) it refers to the same string procedures being used for the layout of simulated buildings appearing in frescoes; and (2) it describes a system of architectural spatial illusionism based on orthogonal lines which are be drawn in receding patterns. The second factor indicates the presence of a simple draftsman's procedure around 1390 that is (almost) like the scenographic layouts that were to become common in geometry-based "perspective" renderings of architecture drawn in straight lines after 1425, that is, just two-score years later. According to Cennini,

> If you want to do buildings [in a fresco painting], get them into your [small preliminary] drawing [*bozzetto*] in the scale you wish, and snap the [string-] lines. Then lay them in with [under-paintings, in earth colors].... Then take a long ruler, straight and fine; and have it chamfered on one edge, so that it will not touch the wall ... and [with this hand-rest] you will execute those little moldings with great pleasure and delight; and, in the same way, bases, columns, capitals, façades, fleurons, canopies, and the whole range of [architectural details belonging to] the mason's craft, for it is [also] a fine branch of our [craft] profession, and it should be executed with great delight. And bear in mind that they [i.e., the architectural details] must follow the same system of lights and darks [*chiaroscuro*] that you have in the figures. And draw in the buildings by this uniform [string-derived] system: [this means] that the moldings which you make at the top of the building should slant downward, from the [nearest] edge next to the roof; the molding in the middle of the building, halfway up the face [at about eye-level], must be quite level and even; the moldings at the base of the building underneath must slant upwards [and back into space], in the opposite sense to the upper molding, which slants downwards.[14]

Cennini's text — strictly a practical handbook, and hence not really meant to be a "theoretical" treatise — apparently demonstrates that *all* kinds of medieval artisans — whether painters or, especially, builders — were evidently all well acquainted with the "design-by-string" procedures.

We may now return to consider the manner in which the grids actually used by the Hispano-Islamic designers for the ornamental gates at Córdoba (figs. 9–15) might have actually looked. One of these must have been the "lozenge-pattern" grid (figs. 8, 58), the standard kind drawn on either paper or parchment. What follows is admittedly a hypothesis, one proposing yet another kind of grid, one having three dimensions and functioning as a portable tool. However, given the absence (as noted above) of any written Arabic sources, with none describing either any of their clearly standard, design procedures or even the simple workshop instrumentation actually employed to execute these intricate decorative schemes, one really does *need* to reconstruct these methods, also their rudimentary tools. Unquestionably, even if we have no texts describing them, they actually *did* exist. In any event, we do know that the use of a grid by artists — particularly sculptors, and certainly by architectural designers—was ubiquitous, and it has been so since at least the time of the ancient Egyptians.[15]

As we know from the archeological evidence, one sort employed by the Egyptians was the squared kind of grid drawn on papyrus (fig. 4). That is also the same kind of ink-drawn grid that we do know Islamic architects actually used (just as is shown in fig. 5). As I would argue, another kind was made from *strings* or *cords*, with these being stretched over a slim wooden armature. Rather than a "drawing," it was a *tool*, and its light weight meant that it was portable; hence, like plumb-lines and set-squares, it could easily be taken to working sites. On the actual building site, the Arabic artisan would have looked *through* these squared patterns—just as one might today look through one of those lattice-screen windows that are still to be seen throughout the Moslem world. Then the foreman would, and just as contemporary surveyors did (and still do), direct a worker to fix the intersections for the centers of his star-patterns (as in figs. 8, 58) upon to the wall or wooden panel providing the actual field for the decorative ensemble. Once those critical center-points were marked on the wall or panel, the rest was pretty much a matter of routine execution: the shaping, cutting, and setting of the individual tiles or wooden segments making up the geometry-derived composition, itself potentially one of infinite extension.

Much later, Albrecht Dürer called the perceptual effects resulting from employment of such a simple device, the kind based on reticulation-by-string, a "*Durchsehung*" (literally a "seeing through").[16] We do actually have in hand a verbal account of the portable stringed grid, and rather a famous one at that. Rather than Islamic, the author of our description is European, none other than Leon Battista Alberti. "*Perspectiva*" was first described by Alberti in 1435 (and his manuscript was later published in 1540 as *De Pictura*). He explains that it was *all* done by means of an "intersection," and "a single straight line forms the diagonal of connected quadrangles." The actual means used to find the intersections was just "design-by-strings," and it was literally used to create a squared grid-pattern (and one assumes that it was also essential to Alberti's *lineamenta*). According to Alberti, "I believe nothing more convenient [to create perspective] can be found than the grid [*velo*], which I call the intersection among my friends, and whose usage I was the first to discover." Then he describes the *velo*-grid's simple construction as follows:

> It is done like this: a *velo* is loosely woven of fine threads. It is divided up by thicker threads, these dyed whatever color you please; you make as many parallel, square sections as you like, and then it is stretched over a frame. I set this velo up between the eye and the object to be represented, so that the visual pyramid passed through the loose weave of the velo.... It always represents the same surfaces unchanged, for once you have fixed the position of the outlines, you can immediately find the apex of the triangle you started with, which is extremely difficult to do without the intersection.... A further advantage is that the position of the outlines and the boundaries of the surfaces can easily be established accurately.... You can situate precisely all the features on the panel or wall which you have similarly divided into appropriate parallel intervals.[17]

Although Alberti states that "I was the first to discover" squared grid-patterns, this claim is clearly erroneous (see figs. 4, 5); in any event, he refers to the "*grati-*

cola" (grid) elsewhere in his text. As John Spencer remarks, "In the fifteenth century [or much earlier] the reticulated net was probably used in conjunction with squared paper [a grid] to obtain accuracy in drawing, and then later transferred to panel or fresco by means of the same [drawn or woven] squares."[18] More specifically, William Ivins has put Alberti's supposedly unprecedented invention into its proper historical context, the routine workshop procedures of humble artisans:

> He made a model of the lines of his vision by stretching strings from the eye-hole to the corners of the squares on a checkerboard.... To get the needed information all that was required was to set a position for the picture plane and then to take measurements of the places at which the lines of vision ran through it. To do this with a foot rule would have been possible, but difficult, laborious, and impractical. What was needed was a graphic method. Therefore, Alberti, taking a trick from the carpenters and stone cutters, had recourse to a template, i.e., a thin piece of board cut at one end or side until it just fits a contour.... He could place this over the strings (or lines of vision) at any position he wished.... With the template in this position, he marked on it the height at which the rows of strings passed through it.... It was quite as simple and ingenious as anything an intelligent carpenter might do.[19]

In effect, we are dealing with a simple sighting device. Specifically, we are looking for the historical evidence of much earlier sighting devices employing intersecting strings or threads. One of those instruments was used by the Greeks and Romans who called it a *diopter*, or "two sight." A sophisticated surveying instrument possibly invented in the fifth century BCE, the pole-mounted diopter functioned rather like a modern theodolite. Polybius (*Histories* X: 42–47) recommended use of the diopter, by which he meant a sighting tube or device with grid-pattern "sights." Pivoting on a circular plate, gearing enabled the cross-hair sights placed on top of the diopter to be turned in various directions and exact angles could be read off the inscribed top disk.[20] However, a more likely historical source for Alberti's *velo* comes from commonplace military equipment, namely the crossbow. This weapon was described by the Chinese as early as the fourth century BCE. Some two centuries later, the accuracy of crossbows was greatly improved by the use of the world's first grid-sights for aiming. These sights were made of fine horizontal and vertical wires crossing each other to form square grid-patterns.[21] Lightweight and with proven usefulness, the familiar crossbow sights would have easily adapted themselves to peaceful, craftsman's purposes.

Although a device most likely already a commonplace artisan's tool in medieval European workshops, Alberti was evidently the first person to *verbally* describe the string-grid, sighting device. Certainly Alberti was the first person to use a string-grid for an unprecedented, specific purpose. That was the artistic novelty which we call "perspective," meaning a strictly graphic way of determining the *illusionistic* effect of receding, orthogonal lines belonging to actually perceived buildings with (originally) horizontal string-courses and colonnades. (Fig. 59) In his practical manual explaining artistic procedures, *Underweysung der Messung* (1538), Albrecht Dürer had even illustrated the simple utilitarian device, and his print usefully includes the other essential element, a previously drawn grid-pattern. The procedure is simple:

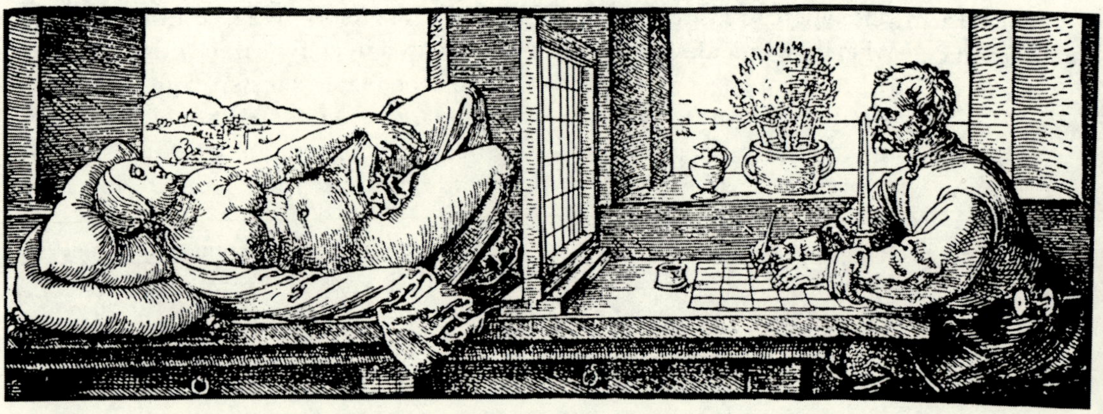

Fig. 59. Albrecht Dürer, "Artist Drawing a Woman with the Aid of a String-Grid and a Grid-Pattern," woodcut, 1525. (*Underweysung der Messung*, 2nd ed., 1538.)

the artist sights through a fixed eyepiece, then he directly coordinates what is "seen through" the string-grid to what has been already drawn over the grid-pattern.

Any one so fortunate as to have actually seen Islamic architecture in the flesh, no matter whether its function is religious or just domestic, knows just how ingenious and intelligent the Moslem carpenters, masons, and tile-setters actually were. Besides the squared grid, surely they were equally familiar with the string-grid, the kind allowing the artisan-designer to "see through" it towards a distant wall, the kind awaiting the transfer of his previously drawn design. The Islamic artisan's source-drawing (evidently looking like figs. 8, 58) would have shown the proportionately spaced center-points essential to the forthcoming, elaborate decorative ensemble, the kind capable of infinite extension. Taking repeated sightings through the reticulated net, as though looking down the barrel of a crossbow, the next step is to have a helper physically fix the center-points, as established in the designer's grid-drawing, upon the wall or panel. (Fig. 60) Panofsky describes just how the evidently traditional workshop procedure (and just as it was again graphically demonstrated by Dürer) was typically done:

> The human eye is replaced by the eye of a big [threaded] needle, driven into a wall, to which is fastened a piece of string with a sight at the other end; the operator [or artisan] can then "take aim" and mark them on the grid [or on a wall] with the [distant] situation determined, not by the position of his eye but by that of the needle. The [string-grid] apparatus eliminates the human eye altogether; it consists, again, of a needle driven into the wall and a piece of string, but the piece of string has a pin on one end and a weight on the other. Between the eye of the needle and the object is placed a wooden frame, within which every point can be determined by two movable threads crossing each other at right angles. When the pin is put on a certain point of the object [or, in reverse, in the drawing] the place where the string passes through the frame determines the location of that point within the future picture [or decorative panel]. This point is fixed by adjusting the two movable threads and is at once entered upon [or drawn from] a piece of paper hinged to the frame; and by a repetition of the process the whole object may be transferred gradually to the drawing sheet [or vice versa: from the drawing sheet to the wall or panel].[22]

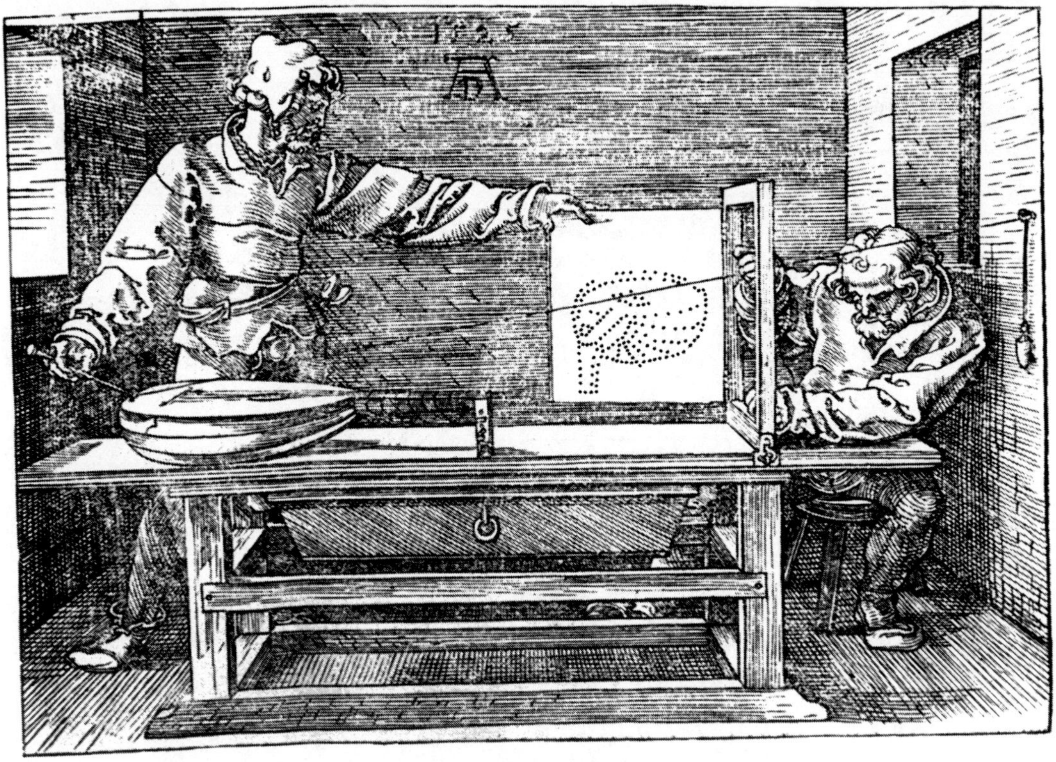

Fig. 60. Albrecht Dürer, "Draftsman Drawing a Lute with the Aid of Drawn Strings and a Grid-Frame," woodcut, 1525. (*Underweysung der Messung*, 1st ed., 1525.)

Perhaps the most elaborate depiction of a design procedure employing a stringed grid, and as specifically used in an architectural context, is that one provided in Andrea Pozzo's *Perspectiva et architectorum* (1693). (Fig. 61) In this case, the specific task is to transfer to the distant, curved ceiling of a church nave a design initially worked out on a relatively small piece of paper prepared with a grid pattern. Pozzo's procedure is succinctly explained by Martin Kemp: "His perspectival design was first drawn to scale on a sheet of paper divided into squares. A matching grid of strings was suspended across the nave at cornice level. A further string was then used to take 'sightings' through the suspended grid from the viewing position, *O*, in such a way that the grid was projected into the curved surfaces of the vault. The design was then transposed line-by-line from the squares of the drawing to the curved equivalents on the ceiling."[23]

Again, we have no written evidence from the Islamic world to back up this hypothesis of Islamic "design-with-string-grids." Indeed, we have no evidence by which to back up any *other* hypothesis. In fact, we have no other evidence explaining Islamic design procedures, nothing, that is, but the physical remains of innumerable Moslem decorative ensembles. Since these are considerable in number, all magnificently done, and all quite consistent in their traditional formats, then "Occam's Razor" demands a simple explanation for how it was all actually done. In

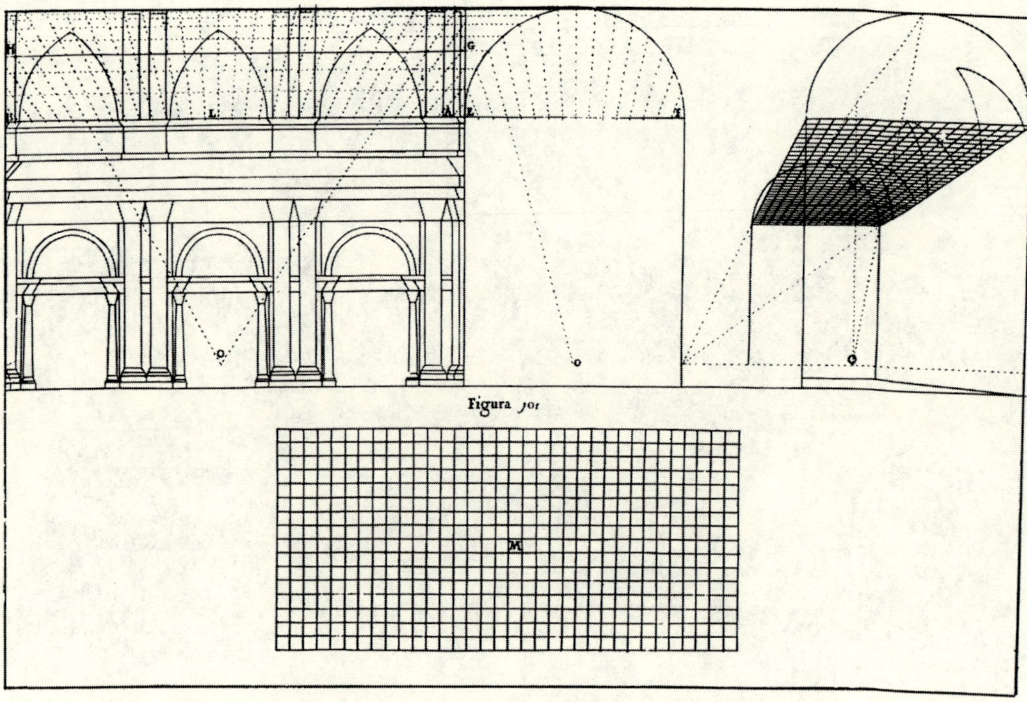

Fig. 61. Andrea Pozzo, "Use of a String Grid to Project an Illusionistic Design onto a Curved Vault." (Pozzo, *Perspectiva et architectorum*, 1693).

fact, the explanation given here is so simple and logical that I myself could set about (and I certainly am not a professional, or even a competent, craftsman) the execution of a typically *morisco* decorative panel of ceramic *lacerías* to be fixed on to the brick walls in my very own backyard (or *patio*).

For want of earlier, pre–Cennini, written documentary evidence specifically referring to the humble design-by-string scheme, also to its instrumental sibling: the *velo*-grid, we can only strongly presume that the same simple system prevailed centuries before in the design of classical temples, for those were obviously designed according to proportionate relationships. In a complementary sense, unquestionably the layouts belonging to newly-founded Hellenistic cities were made to conform to the "grid," "gridiron," or "chessboard" format.[24] Since Antiquity, the grid-plan has served as the colonial urban scheme par excellence. This was notoriously the case in Spanish-American urban planning, and that fact alone provides further evidence for an all-pervasive "grid mentality," the kind we have already shown to be also inherent to the Hispanic *fachada retablo*. In Latin America, we also know for a fact that Hispanic city-plan designs exhibiting the "*damero*" (chessboard) and "*parrilla*" (grill) formats were literally laid out with set-squares and hemp cords—"*a cordel y con la regla*."[25]

We actually have in hand a sixteenth-century description showing how the practice of instant urbanization was actually employed in Mexico. As was explained by Friar Toribio de Benavente ("Motolinía") in his *Historia de los indios de la Nueva*

España (1541), the town of Los Angeles was founded *ex nihilo* in Tlaxcala province in 1530. Although it was planned to be used by the Spaniards, Toribio says that Indians traveled from various places to assist the European colonists; moreover, they did so voluntarily: "todos vinieron de buena gana para dar ayuda a los cristianos." To carry out their building tasks, the natives brought with them laths, ropes and measuring cords (*latas y ataduras y cordeles*). The Spaniards brought with them a plan for the town, and this had been prepared previously by a local stone mason—"ya traían hecha y sacada la traza del pueblo [hecha] por un cantero que allí se halló." As witnessed by Toribio, after attending mass,

> Then, without further delay, the Indians cleaned up the site and, having stretched the cords [*echados los cordeles*], they divided up forty parcels destined for forty settlers. And because I was there, I can say that it seems to me that there were no more than those [forty Spaniards] who first settled the town. Later, but on the very same day, the Indians set about to erect houses for all those residents for whom the parcels had been marked out. They worked so speedily that all the houses were finished within a week's time.[26]

Moreover, it may be believed that this grid-inspired system of design-by-string harks back to the very beginnings of monumental architecture, in Phaoranic Egypt. Hence, we may also assume that, centuries later, the very character of "classical" architectural design itself was also predetermined by those already long-standing but now mostly forgotten, wholly empirical traditions of grid-inspired design-by-string. Ancient procedures had certainly produced distinctively proportionate and harmonious designs and, as I now argue, those mainly were due to the wholly pragmatic, artisan-shaped means of their immediate execution. Granted that, then all written architectural "theory" (and from whatever period) that ascribes symbolic valences to any buildings endowed with such proportionate schemes is clearly speculation rendered *a posteriori* to the chronological locus of the specified edifices. Preferable, therefore, are *a priori* analyses derived deductively, the kind then repeatedly applied to the simple, wholly visual, *facts* attesting to the actual means of the physical execution of given architectural constructions. The big gap in our knowledge is not "theory," but instead routine *artisan* practice and procedures.

14

The Non-Visual, or Textual Evidence for Hispanic "Design-by-Triangulation-and-String"

Unfortunately, there is a paucity of architectural writings from the New World; a notable exception is the work of Friar Andrés de San Miguel (to be quoted below: figs. 67–70). The only other example known today to scholars is an anonymous, also un-illustrated and un-dated (ca. 1800), manuscript with 47-folio pages, "Arquitectura mecánica conforme a la práctica de esta Ciudad de México."[1] Probably written by a reasonably informed amateur, this reads something like a notebook, into which the unknown author randomly set down diverse observations in a rather disconnected way. No matter; his text sporadically gives us invaluable hints as to the actual techniques employed by Mexican workers on building sites. Even better, these casual observations confirm one's initial hypotheses about how "design-by-triangulation-and-string," including the deft wielding of templates (*escantillones*), was actually done in the rustic Mexican outback.

According to our anonymous architectural chronicler,

> The starting point for a building is the floor. Its value is measured in square *varas*. [A preliminary step is] staking the site [*estacamento*]. After outlining the foundation [*cimiento*] with powdered lime, a trench is opened, leveled, and straightened with a mason's square [*escuadra*] and cords [*los hilos*]. All this is done in conformance to the schematic design prepared by the overseer [*todo esto se hace por el Mapa que tiene hecho el Maestro*]....
>
> Various terms [are] used by the master builders. The workshop [*taller*] is the site where masons [*los canteros*] do their work. The boss [*el principal*] is the one who pays the others; he takes on the role of the teacher-master [*maestro*] and he is called an overseer [*capataz*]....
>
> Vertical lines have [first] to be inscribed [*rayar*] upon a wall; their placement is determined with a plumb-line [*plombo*] and the placement of the [parallel] horizontal cords is set with a level [*nivel*]. A cathetus [*cateto*: the acute angle of a triangle] is used to determine a perpendicular line which descends [diagonally] from top to bottom, and not in any other direction [*y no cualquiera otra*]....

A template [*escantillón*; from *escantillar*, an architectural term, alternatively meaning to measure from a point, or to trace lines upon a wall, or even to mark out with a template] represents the modular measure for the given object [*la medida de la cosa*]. It is a general term; thus to apply a template [*dar escantillón*] means to ascertain the [proposed] thickness of a cut stone, a beam, and so forth. An arcade [*danza de arcos*] represents a regular succession of arches, or one may inscribe [*rayar*] a series of *danzas de arcos*, whether these are to be placed on one side or on four sides [of a building].

Those [Churrigueresque] façades [*portadas*] presently often built here [in Mexico City] are commonly referred to as sculptures [*obras de talla*] and, truly, they have become nothing more than side-altarpieces [*colaterales*] set up on the street [i.e., they are implicitly recognized to be "*fachadas-retablos*"].

The order in which the architect should proceed [in his design] is as follows. The design [*planta*] for the façade should be made by a skilled illustrator [*pintor*], but this work must be done under specific directions given by the master to assure, for example, that all the proportions are made to correspond within the first and second registers, and so forth [throughout the composition]. Better yet, one hires a master joiner [*ensamblador*] to draw the plan, or the elevation [*el alzado*], as some call it. *Ensambladores* are those craftsmen who make altarpieces. It is said that they are familiar with designing the elevations for façades, also in the use of moldings suitable to be fitted with vertical extensions [*voladas*] and other projecting elements [*proyecturas*].

Before beginning the work, the architect needs to examine and correct all the other components [in the plan]; if he knows how to design himself, he can envisage [the completed state of] the plan which the joiner brings to him. He will check to see if it is agreeable to the eye and will know if there are any defects to correct. For example, it may not be in a proportional relation to its cornice. Or perhaps the pedestals are not appropriate to the order designated for them; this is important if they support Doric columns [rather than Ionic or Corinthian orders], and other things of this sort [must be taken into consideration].

Having corrected the design after having assured himself about all its details, he then calls in the stone sculptors and confers with the overseer of the masons about whatever is needed in order to begin the work. This way, they will work together closely, carefully following the drawn designs showing [both] the elevation [*la planta de perspectiva*] and the layout of the sculptural decoration [*y la iconográfica*], and they will not exceed the limits set in the designer's master plan....

The master should [always] have with him a large, well-made wooden level [*nivel*] along with his plumb-lines [*plomadas*], all carefully made, [and] drawing paper and parchment [*papel de marca y pergamino*] upon which to draw the designs for buildings [*para la delineación de las casas*], [also] a carpenter's square [*escuadra*] and hemp cords [*cordeles*] to level and square whatever he is working upon. And, given the likeness of this profession to the work of surveyors [*el oficio de agrimensor*], he should have with him his bubble-level [*nivel de agua*] and its stand.... A good protractor [*abujón*], with triangles along its edges for calibration with the geometric scale, all this he has to have with him.... He should have his *cordeles* marked in *varas*....Another well known tool is called the *salta regla*. It is made from two movable rulers connected at a point in the manner of a compass; with this, any angle can be measured and transferred to the stonework.[2]

This anonymous notebook provides us with invaluable documentation for the actual workshop techniques and procedures current in eighteenth-century Mexican architectural practice. Such simple methods and rudimentary tools must have been used for centuries. What one had originally hypothesized as the most logical way it would have been done is here shown to have been, in fact, actual practice. Another important issue resolved by this document is the utter lack of "theory" belonging to such design procedures. It was all a matter of practical expediency. That, and nothing else — except for an expressed desire to produce the best optical effect.

15

Vitruvius: Theoretician — or Artificer?

Before the Renaissance, the *only* complete and available manual dealing with the design and execution of monumental buildings was the *De architectura, libri decum* that had been composed by Marcus Vitruvius Pollio around 25 BCE.[1] In spite of its usually being treated as a prime example of "architectural theory," and in the light of what we have just learned, we may now believe that this famous text now can profit from a semantic re-examination in the light of now-forgotten craft traditions. It is, of course, a commonplace that Vitruvius—himself an experienced architect—was very much concerned with "*geometria*." His geometric bias is unquestionable.

Nevertheless, we are also usually told today that Vitruvius primarily viewed geometry as the physical manifestation of superior organization principles, that is, mainly as what we today call "theory." Those well-known Vitruvian compositional principles are, respectively (*De architectura*, I. 2. i), "Order," "Arrangement," "Symmetry," "Propriety," "Economy," and, above all, "Proportion [*Eurythmia*]." Again, these familiar design factors are customarily presented today as "theoretical principles," and they have been consistently so discussed since the Renaissance.[2] This mostly theoretical bent is best demonstrated by Leon Battista Alberti in his treatise *De re aedificatoria* (ca. 1450; first published in 1486). In this work, the first book wholly devoted to architecture since Antiquity, Alberti meant to outdo Vitruvius. However, rather than addressing himself to humble artisans, Alberti writes in Latin, and his elitist audience is the future patrons of architecture, Renaissance princes and merchants. To define proportionality, Alberti coined a new term, *concinnitas*, and its application is explained by him (in Book IX, ch. 5) in a wholly abstract, hence essentially unrealizable, fashion:

> It is the task of *concinnitas* to compose parts that are quite separate from each other by their nature, according to some precise rule [left undefined by Alberti], so that they correspond to one another in appearance. That is why when the mind is reached by way of sight or sound, or any other means, *concinnitas* is instantly recognized. It is our nature to desire the best, and to cling to it with pleasure. Neither in the whole body nor in its parts does *concinnitas* flourish as much as it does in Nature herself; thus I might call it the spouse of the soul and reason....

> Beauty is a form of sympathy and consonance of the parts within a body, according to definite number, outline, and position, as dictated by *concinnitas*, the absolute and fundamental rule in Nature. This is the main object of the art of building, and the source of her dignity, charm, authority, and worth.[3]

If this passage would be read today to a building contractor, obviously he would have no idea how to set about in practice to design, for instance, a series of doors and windows having a proportionate relationship. He would not have a clue how to begin his task. In the light of this, as applied to Vitruvius, the "theoretical" epithet now seems somewhat of an anachronism; certainly, it obscures the ancient, largely pragmatic, or strictly "workshop," origins of his weighty design axioms, particularly once those became invested by Renaissance theorists like Alberti with essentially non-practical, usually rhetorical, arguments about the "decorum" of applied ornament.

Let us again review the text of this unique historical document. As Vitruvius had stated in a wholly practical manner, mastery of mathematics (*geometria*) "affords many advantages to architecture, mainly by teaching us proportion through the use of the measuring compasses [*primum ex euthryramis circini tradit usum*] since these are tools that make it easy [for us] to lay out site-plans for buildings [*e quo ... in areis expendiuntur descriptiones normarumque et librationum et linearum directiones*]." We recall that he further stated (I. 1. iv) that for all kinds of preliminary workshop planning there are to be inevitably applied "carpenter's square, the set-level, the compasses, and the plumblines"—*norma, librationum, circini, lineas*— and that is because these simple tools facilitate both design and execution.

In Vitruvius' own words (I. 1. i), geometric "theory [*scientia*] is the ability to demonstrate and rationalize [that is, *a posteriori*] the manual productions of dexterity [*res fabricatas*, which are themselves derived *a priori*] within the intellectual principles of proportion [*rationis proportione*]." Moreover (I. 1. iv), "the difficult problems of [architectural] design are solved by geometrical rules and methods." Therefore, "geometry also furnishes much assistance to architecture. It teaches use of the ruler and compass, and thus it facilitates the laying out of building on their sites by the use of set-squares, levels, and plumb lines." Although admittedly prone afterwards to "rationalizations" (*rationis*), proportionality really originates from — meaning, was initially (or *a priori*) "demonstrated"— in extremely simple workshop practice. Again, our conclusion hold that the procedures ensuring geometric *proportione* were, in short, tool-derived: "*per instrumentum*," hence, the rationalization (*scientia*) only came long after the fact of a long-standing, pre-*scientia* practice.

In short, proportions are derived by Vitruvius in a very practical manner. He explained (I. 2. ii) that "the selection of module-units [*modulorum*] arise from the members of the work itself, and, starting from thse individual parts of the members, one constructs the whole work to correspond accordingly to those modules—*ex partibusque separatis ad universae figurae*." Therefore (VI. 2. i), "There is nothing to which an architect should devote more thought than to the exact proportions of his building [and] with reference to a certain part [or design module] selected to serve as the standard." On a much more practical — and literally instrumental level, Vitruvius advises (I. 2. ii) that, at all stages of the operation, "layout and design demand competency in the manipulation of compasses and rulers; by these [instrumental] means, working plans—laid out upon flat surfaces—are obtained." To again repeat the same statement in Latin: "*Ichnographica* [the initial design worked out on a flat sheet of paper] *est circini regulaeque modice continens usus, e quo capiuntur formarum in soliis arearum descriptiones.*"

A review of the basic points laid down by Vitruvius again reveals that a simplified linear drawing is the initial step. This example of "*ichnographia*" is a proportionate ground-plan that naturally and inevitably arises from the manipulations of compasses and straight-edges. The next stage described by Vitruvius (I. 2. iv) is called the "*Orthographia,*" explained as the representation of "the elevation, which is an image of a vertical façade [*erecta frontis imago*]; it will be lightly sketched in to show the outlines of the work for future stages [*operis futuri figura*]." Those "future stages" are those actions later done at the actual working site, where the initial *frontis imago* is be applied, and then a few, minor adjustments will arise due to local exigencies.

As it turns out, the only other old text known to me which had described in detail the actual physical process of transferring the *ichnographia* to the *orthographia*—the necessarily planarized "picture of the front of the building"—was contained in Cennino Cennini's *Il libro dell'arte* (and as was quoted in chapter 13). In any event, Vitruvius was, of course, continuously known by builders since Antiquity (and evidently by Cennini as well).[4]

All that remains now is to demonstrate Vitruvius' knowledge of our standard 3:4:5 design module. Granted his later authority (and translation into Spanish), then his comments on this traditional design-by-triangles scheme may be presumed to have been known to most architects, even to those rustic designers working in Spanish America. As Vitruvius explained in the Introduction to his Book IX, there are "many useful theorems" to help the artisan make measurements in the field, and these are "to be achieved by lines correctly drawn," particularly those configurations otherwise "inexplicable by arithmetic." In this case, we are dealing with *geometry*, the kind involving "diagonal lines drawn from one angle to another angle by using a square," and by "taking this diagonal line as the length, you describe yet another square," as many as you want in fact, and with these you produce triangles. Above all (liber IX, Proemio, v–viii), it was

> Pythagoras [who] showed that a [right angle] set-square [*norma*] can be formed without the aid of an artisan [*sine artificis*]. Thus, the result—which carpenters only reach very laboriously, but scarcely with exactitude—can be demonstrated from the reasoning and methods of his teaching. First, we take three measured rulers [*regulae*]—with one three-feet long, the second four, and with the third one being five feet in length. Then these three rulers are then to be joined together with tips touching each other; so doing, they make a triangular figure [*trigon*]. So joined, they will also form a right-angle [*norma emendata*].
>
> Now, if a square figure is extended from the original length of each one of these rulers, the square on the side of the three-foot-long ruler will have an area of nine square feet, while that of four feet gets sixteen; and the one with a five-foot length now yields twenty-five square feet. And it turns out that this cubed area [which may be either horizontal: a "ground-plan"; or, if vertical, it becomes an "elevation"], which was made up from the two squares, respectively the three and four feet in length [and containing 9 and 16 squares respectively], is equal to that of

> the square described on the [third] side, which is five feet long [and it now shows 25 squares; see fig. 63, left side]. When Pythagoras discovered this fact, he had no doubt but that the Muses had inspired him to the discovery, and it is said that he very gratefully offered sacrifices to them.
>
> This theorem [*ratio*; actually just a practical artisan's technique] affords a useful means of measuring [in advance, and away from the building site] many things, and it is particularly serviceable [for example] in the building of [vertically-ascending, evenly stepped] staircases in buildings [*scalarum aedificationibus*]. It assures that the steps will be inserted at the proper [or proportionate] levels [*uti temperatas habeant graduum librationis*]. Let us suppose that the [vertical] height [or elevation] of the story, from the floor above to the ground below, is to be divided into three parts.
>
> Five of these parts [forming the diagonal] provide the correct string-courses for the stairway [*scalis scaporum*], so establishing the proper longitudinal inclination [*iusta longitudine inclinatio*] for the stringers [i.e., the sub-divided horizontal elements and upright columns, inserted within the frame or scaffolding]. Now dividing into three equal parts the height between the floor above and the ground below, let four parts be set off [with plumb-bobs] from the perpendicular diagonal [*a perpendiculo*], and have the fastenings for these parallel strings be tied down there [*et ibi conlocentur inferiores calces scaporum*]. The design will thus be proportionately adjusted [*temperatae*]—so also will be the sub-divisions of the several parallel steps comprising the staircase itself [*et graduum ipsarum scalarum erunt conlocationes*]. A drawing of this ensemble is attached [in Vitruvius' ms., but it is now lost to us].

This description, one written two-thousand years ago, effectively matches the one published two millennia later in the Seabees field-manual for inexperienced artisans working in the boondocks with only the most rudimentary tools at hand (and as it was quoted in chapter 7). Long before, Pythagoras' practical triangulation method was praised by the likes of (besides Vitruvius) Plutarch (*Moralia* 5. 90), Cicero (*De natura deorum* 3. 36), Porphyry (*De vita Pythagora*e 36), and many more.[5] This description of the Pythagorean 3:4:5 triangle was, of course, included in all the later translations of Vitruvius' *De architectura*. The grandest, also most lavishly illustrated of all these was the Italian edition prepared by Cesare Cesariano, and published at Como in 1521.

Somewhat oddly, Cesariano illustrated the passage found in the second chapter of Vitruvius' First Book, where the ancient author discussed the generation of ground-plans, elevations, and perspective views of any given buildings (and just as was quoted above), with an extraordinary illustration: the "Gothic" structure of the Cathedral of Milan. (Fig. 62) Rudolf Wittkower explained that such a depiction (see also fig. 16) of the Vitruvian-inspired "reign of triangulation, of the square, and of the circle, with the latter procedures two absolute essentials of classical and Renaissance aesthetics," served to "demonstrate the validity of Late Gothic geometry, the '*Germanico more*' (as Cesariano put it), within a classical system."[6] And, as Paul Frankl further explains, the "Pythagorean," or 3:4:5 triangle, was indeed the modular basis of the design of the Cathedral of Milan, likewise laid out using "strings" or cords:

> Their choice is understandable as soon as one observes [fig. 62] the triangle that now results from a base of 32 *braccia*, the width of the nave, and an elevation of

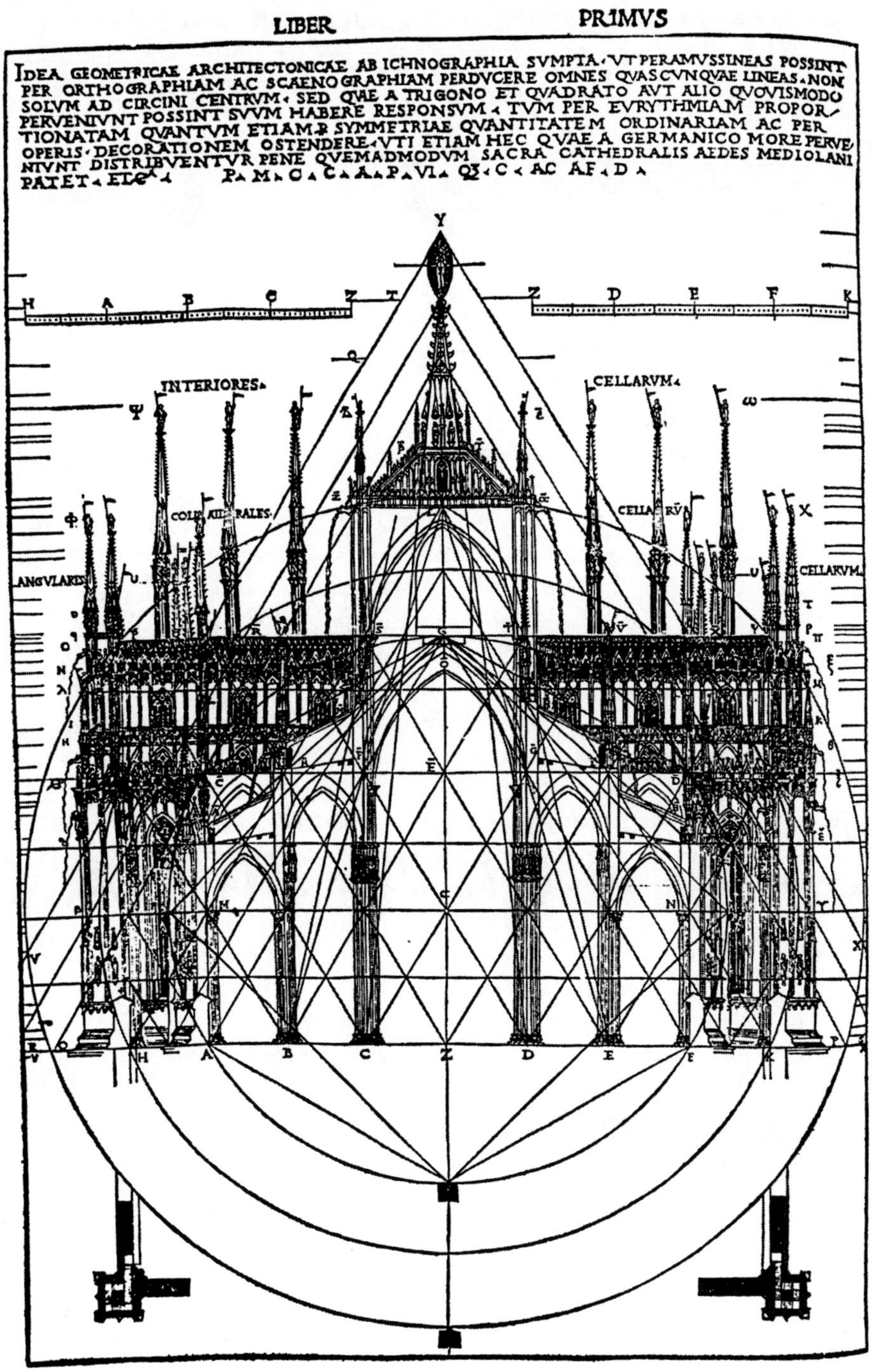

Fig. 62. Cesare Cesariano, Reconstruction of the Vitruvian geometric-design procedures called "*ad quadratum*," "*ad triangulum*," and "*ad circulum*" as illustrated in the Gothic fabric of the Cathedral of Milan (begun 1386). (Engraving published in Cesariano, *Vitruvius De Architectura*, 1521).

> 12. It can be divided by a bisecting [diagonal] line into two right-angled triangles with the catheti 16:12 *braccia*, or, in general terms, with the proportion 4:3. This is the Pythagorean triangle. This triangle, known and popular since ancient times, has the advantage of facilitating the marking of the right angle by means of sides in terms of whole numbers. For with the catheti 4 and 3 units long, the hypotenuse will be 5 units long. If the horizontal line is marked on the scaffolding with a string from the upper edge of one capital to that of the opposite capital, it is relatively easy to plot out exactly the height desired, because the hypotenuse of 5 units can be used as a control dimension.[7]

Much later, Vitruvius' authoritative description of "centering a design by string" (*ubi conlocentur cum calces*), and as employing the standard "Pythagorean triangle" motif, certainly became known to the Spanish architects, hence probably also to their usually literate site-foreman, an "artificer" (*alarife*). Vitruvius' wholly pragmatic explanation may be assumed to have been common knowledge among the Hispanic *alarifes*, at least after 1583, with the publication of a definitive translation into Spanish of the Roman architect's venerated treatise. This was prepared by a Toledan architect, Don Miguel de Urrea, and it was published as *M. Vitrvvio Pollión De Architectura* (Alcalá de Henares: Juan Gracián, 1583; the manuscript was, however, completed as early as 1569). The section from Vitruvius, as just quoted, is now subtitled "De la escuadra, o *cartabón*, que inventó Pythagoras por *la figura del orthógono, y del trigonio*." As Urrea's translation of this section begins: "También Pythagoras muestra haber hallado la escuadra [y] sin fábrica de artífices...."[8]

Being the work of a practiced builder, Urrea's translation — which is profusely illustrated throughout with woodcuts, many of which reconstruct the original illustrations alluded to in Vitruvius' manuscript — clarifies Vitruvius's often obscure Latin text. Urrea does so particularly by citing the use of those practical tools and instruments that were commonly employed by contemporary Spaniards, including the temporary wooden scaffolding (*el enmaderamiento*), which was supported by parallel, column-like posts (*los escapos*), with their intervals having been determined by various plumb-lines (*las plomadas*) and lead weights (*los lastres*) dropped from the temporary scaffolding. These devices are all used to determine the proportionately distanced footings of the columns (*los asentimientos de las columnas*), and such as all these locations were predetermined by use of a set-square (*la escuadra*), with this being used to produce the preliminary linear tracings over grids: "*cada cuadrado se traza*."

Even better, Urrea had reconstructed Vitruvius long-lost manuscript illustration of the Pythagorean theorem, that is, "the drawing of this ensemble [which was originally] attached [in Vitruvius' ms.]."[9] (Fig. 63) Urrea's woodcut shows us (in the far left) the respectively 3, 4, and 5 feet-long rulers which had yielded layouts of, respectively, 9, 16, and 25 square-feet. He also shows us (in the lower center) just how Vitruvius' staircase was inclined, and with a 3-unit module establishing its height; a 4-unit module provides the bottom length; and a 5-unit module fixes the essential diagonal. All along the 5-unit diagonal, plumb-lines have been dropped, creating thirteen, absolutely parallel lines in all; these fix the crucial intersections along the diagonal leg of the Pythagorean triangle, those obligatory points which

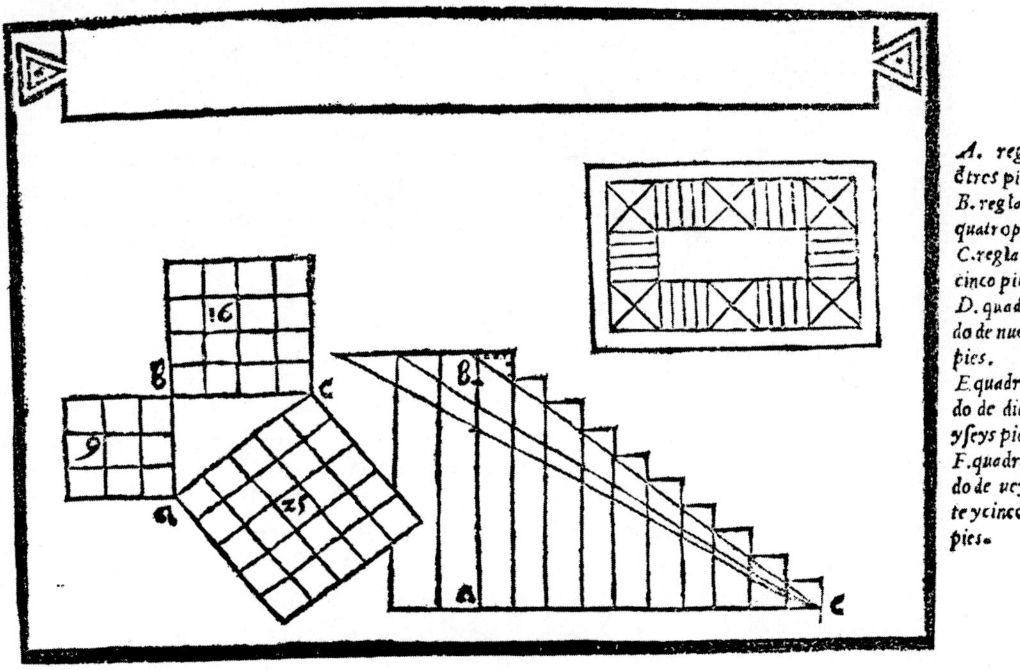

Fig. 63. Miguel de Urrea, Reconstruction of Vitruvius' drawing of "The Pythagorean Triangle," and showing its employment in the design of a four-story, right-angled staircase. (Woodcut published in Urrea, *M. Vitrvvio Pollión De Architectura*, 1583.)

finally establish the overall proportionate relationships, also the sub-divisions for each ascending step. Urrea's third vignette (in the upper-right-hand corner) shows the final ensemble: a staircase ascending around an inner patio; it proceeds upwards, with right-angle turns, and apparently reaches four stories in height.[10] Finally, his fourth vignette (appearing at the very top) represents a rectangular, metal ruler with notched handles. A combined "*regla y lastre plomado*," this instrumental motif appears in a similar location in many other woodcuts by Urrea clarifying the terminology of Vitruvius' text.

In any event, the decisive term in the sub-title to Urrea's translation of Vitruvius's *De architectura*—including its canonic description of the Muse-inspired *trouvaille* of a celebrated, simple (and celebrated as such) and standard, Pythagorean geometric procedure—is "*cartabón*," meaning a *template*, or imposed design module. This is the same simple tool which I believe was essential for the physical execution of the notoriously intricate, also inevitably proportionate, Arabic "interlace patterns" (*lacerías*), and I have already illustrated the actual manner of its functions within their design process (see fig. 7). This essential artificer's device was also described by Fray Laurencio de San Nicolás in the forty-seventh chapter of his *Arte y Uso de Architectura* (1639)—and, although he credits its operating principle to Pythagoras, he ascribes its initial, practical architectural application to Vitruvius. As he explained,

> The difference in wooden structural frameworks [*armaduras*] can be as diverse as they are needed to be employed by the artificer in his buildings. Since they only differ in being higher or lower, there can be many of these. Commonly, we only use two or three, but I will explain eight of these, showing the manner of drawing these, and from whence they take their names. And given that the *armaduras* take their names from their templates [*cartabones*], it will be useful to define a *cartabón*, its origins and its fashioning. According to Vitruvius (book 9, chapter 2 [and as he was quoted in Chapter 3]), the principle derives from Pythagoras, and from this he worked out the function of the square root [see fig. 63]. A template is a little wooden tablet [*tablilla*] shaped like the said [geometric] figure, and it serves to determine the angled joints of the wooden pieces composing the *armadura*. The manner of its fabrication follows. Above a horizontal line, AB, one places a half-circle; in the middle of AB, one established a compass-point, Y, and the top of the circle is labeled S. From Y to S, a [vertical] line is drawn. At the juncture of YB and YS a right angle is created. Next a [diagonal] line is drawn from S to B, and in this way you have created the triangle, or *cartabón de cuatro*. Since the derivation of this tool is clear, I need say no more about it.

He then goes on to describe the same procedure to be used for the generation of other *cartabones*: "de a cinco, de a seis, y siete, etc." All of these figures differ in having progressively more acute angles, and this is because the tops of these triangles, or templates, are located progressively distant from point S at the top of the hemisphere, and thus these become narrower as they move toward point B. Their common point of radiation is point A, and the author states that "line AB contains eighteen modules along its length, and from these markings we determine [respectively] the [height and length of the] four-, or five-, or six-unit templates." Using the same simple graphic illustration, San Nicolás then explains the working of the five-unit template, or *cartabón de cinco*, and this—which itself forms the often described "Pythagorean triangle"—was the triangular template most commonly employed by Hispanic artificers:

> First, you divide [horizontal] line YB into three equal parts; then from point M, which is one-third distant [from point Y], you draw a line parallel to [the vertical axis] YS. After that, you draw lines NA and NB [and line N rises vertically from point M]. Now, you will discover that [the shorter diagonal] NB measures one-fifth the circumference of the half-circle [between points S and B]. The point of contact of [the longer diagonal] NA with [the vertical] YS is at unit five, and this [triangular figure] forms the *cartabón de cinco*. Others derive this figure from the width, AB, and here is how that is done. They draw the semi-circumference [*cambija*], and placing the compass-point upon its center [at point S], they see that it touches the mid-point of line AB. There is scarce difference between one method and the other. However arrived at, it [the *cartabón de cinco*] makes for an *armadura* which provides excellent service for all manner of ceilings and, above all, it does so because not much woodworking is called for. Hence, this is the one you should employ in your designs.[11]

16

Diego López de Arenas: The Written Evidence for a Standard Application of Pythagorean Design-Modules by Hispanic Builders and Craftsmen

A late publication date (1633, with various later reprintings) of a popular artisan's handbook explaining those "*mudéjar*" (i.e., "neo–Islamic"), enduring Hispanic design principles—the kind characterized by geometric, non-figurative ornament arrayed in rigidly controlled, centrifugal patterns generated by compass and plumbline—documents the unprecedented persistence of such "Moslem" traits on a truly popular level in Christian Spain, even well into the Baroque period. (Fig. 64) Our *prima facie* evidence for the actual employment by Hispanic artisans of the ancient Pythagorean 3:4:5 design-modules is a practical, "how-to-do-it" manual, initially composed in 1619 by Diego López de Arenas, and entitled *Breve Compendio de la Carpintería de lo Blanco y Tratado de Alarifes* (but only later published in 1633, with reprintings in 1727, 1867, 1912, etc.).[1]

Although scarcely familiar in detail to most architectural historians, this is a unique document: it is the first publication describing in concrete detail both the design procedures—and the instrumental means for their practical implementation—belonging to a uniquely Hispanic *mudéjar* compositional mode which has been operational in Christian Spain since around 1175. Were it now to be published in translation in postmodernist America, an appropriate title would be "Neo-Mudéjar Architectural Design Principles for Dummies" (so corresponding to contemporary software manuals, the likes of "Windows 2003 for Dummies"). Another (less demeaning) title might just be "The 'Secret' of the Hispanic Builders."

According to its sub-title, this is a workshop manual (*tratado*) for "*alarifes*," these being either those master carpenters (*carpinteros*) or master masons (*albañiles*),

BREVE COMPENDIO

DE LA CAR·PINTERIA DE LO BLANCO,

Y TRATADO DE ALARIFES,

CON LA CONCLVSION

DE LA REGLA DE NICOLAS TARTAGLIA, Y OTRAS COSAS TOCANTES A LA IEOMETRIA, Y PVNTAS DEL COMPAS.

DEDICADO AL GLORIOSO PATRIARCHA SAN IOSEPH.

POR DIEGO LOPEZ DE ARENAS MAESTRO del dicho oficio, y Alcalde Alarife en el, natural de la Villa de Marchena, y vezino de la Ciudad de Sevilla.

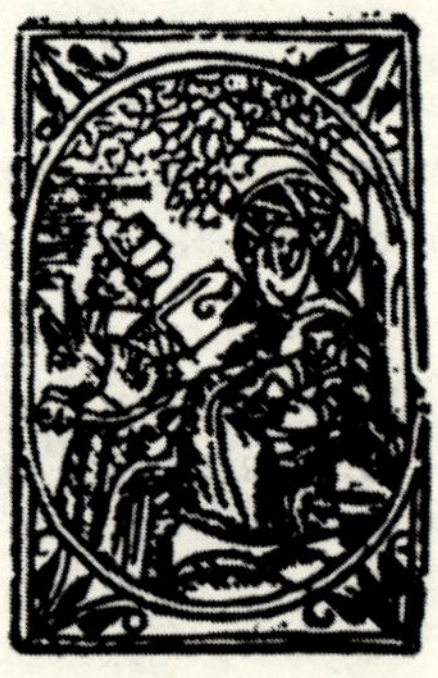

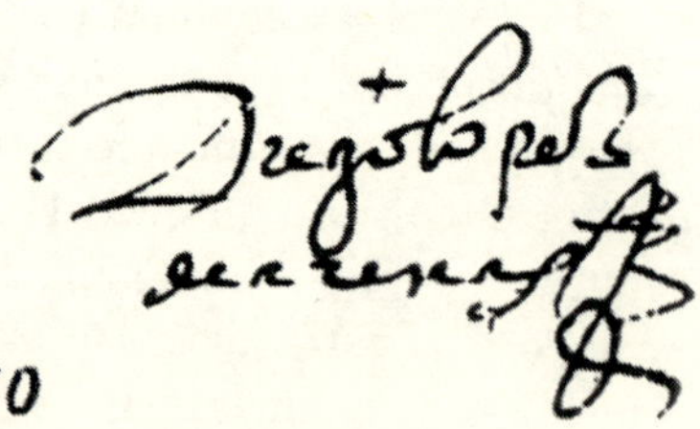

CON PRIVILEGIO

Impresso en Sevilla por Luis Estupiñan, en la calle de las Palmas.

Año de 1633.

Fig. 64. Diego López de Arenas, Title-page to *Breve Compendio de la Carpintería de lo Blanco y Tratado de Alarifes*, 1633 (with signature).

who acted as local foremen (*técnicos municipales*) for either public works (*obras públicas*), including churches, or as the supervisors for private commissions (*obras privadas*). The word *alarife* is itself wholly Arabic: *al-arif* meant "the expert," a "master" (*maestro*). Maestro Arenas' manual was wholly practical; specifically it was meant to provide a graduated, elementary to advanced, series of lessons for novice apprentices aspiring to become journeymen in the craftsmen's guild (*gremio de alarifes*) in Seville.

Now, in the context of the present inquiry, the text to López de Arenas' wholly practical, also profusely illustrated, handbook provides all the written documentation — the actual published "proof" — one had needed in order to corroborate one's initially empirically derived hypothesis of a given, universally applied, design-template. This was a model of supposedly "Pythagorean" inspiration, and this *aperçu* has already been repeatedly confirmed by — besides the *Arte y Uso de Architectura* published by Fray Laurencio de San Nicolás in 1639 — the *a priori* findings of a strictly post-modernist, cybernetic investigation (figs. 9, 15, 24–58). In sum, López de Arenas now provides us with the verbal — also graphic — evidence confirming the specific geometric hypothesis propelling this lengthy investigation.

As María Angeles Toajas explains in her invaluable, 60-page long, "Introducción" to the recent (1997) reprinting of this essential art-historical document, this is "a text of considerable difficulty for an uninformed [modern] reader," and it must be so, since it was written to serve as "an informative handbook for the use of professionals" (and likewise mostly incomprehensible to most of us are postmodernist software manuals). Fortunately, Toajas has appended a complete, 19-page, "Glossary of Technical Terminology" (*Léxico técnico*), so allowing the modern scholar to penetrate, even elucidate the arcane secrets of Arenas' *neo-mudéjar*, and now-obsolete, craft, here treated by Toajas as "architectural solutions" (*soluciones arquitectónicas*).

According to the Sevillan guild regulations set in 1635, all members had to be able to read, write, and to do simple arithmetic. Moreover, all journeymen *alarifes* had to prove that they were capable draftsmen (*debuxadores*); then they were allowed to design, then carve (*labrar*), decorated pillars (*pilares revestidos*), free-standing tombs and tabernacles (*tumbas, tabernáculos de gran arte*), altarpieces (*retablos*), and intricately carved choir-stalls (*y fazer coros de sillas ricas*). The most qualified maestros were those adept in solving geometric problems; they were called "geometers," *iumétricos*.

For Arenas however, the central problem is practical and narrowly focused: to provide working models for "frames"—*armaduras*— with these specifically being ensembles of joined beams braced together ("*vigas ensambladas y arriostradas entre sí*," according to Toajas). The *armadura* is made up of parallel, shaped-timber diagonals (*pares*) intersecting with horizontals, so fixing the proportionate distances between the parallel, vertical rows (*hileras*); these were invisibly braced by further diagonal boards (*nudillos*) attached behind the finished work. The primary use of such *armaduras* was to shape the highly decorated, geometrically arranged, wooden ceilings—*el artesonado*— the kind one still sees today throughout the Hispanic world.

Artesonado ceilings are characterized by *lacerías*, "interlace patterns" (as shown in figs. 8, 58), and each *lazo* (knot) making up the whole represented a geometry-derived design made by employing angular lines and, interlaced, they look like interwoven ribbon-bands made by conjoined segmental units. These elaborately decorative ceilings are composed from many different panels, each of which must be perfectly mated to the other components. Typically, the panels were narrow wedges: *armaduras de tijeras* (scissor-shaped frames); they were assembled as one might reintroduce slices cut from an apple pie. "*Carpintería de lo blanco*" signifies the sophisticated technique of assembling diverse pieces of carefully worked, wooden segments into a well-aligned and squared—*blanqueada*—conjoined whole ("*labores con maderas labradas en escuadría bien alisada, es decir, blanqueadas,*" as Toajas explains). The same overriding, "holistic" goal would also apply to decorative Hispanic stonework, including architectural motifs, even the sculptural (figurative) embellishments commonly applied to church façades.

Whatever the endeavor (or commission), the basic means to the end was the *template*, the *cartabón*. According to Toajas, these are "thin panels [*tablillas*] cut in the shape of right-angle triangles [*triángulos rectángulos*]." Their specific function was to determine the exact intersections, as fittings, of the diagonal beams (*pares*) and their angled joints (*limas*). For each job, there was made a special set of these *cartabones*, and each one was identified by the different acute angles (the *catheti*) assigned to them. These angles were empirically designed by using a simple protractor (*cambija*) to set the acute angles (*colas*); the right-angle (90°) corners remained, of course, invariable. As Arenas explains (in Chapter One), the minor angle of the template points upward diagonally, and the diagonal, the longest side, is laid along the horizontal base: "*Cathecus, a la cabeza; basis, la cola*" (as shown in fig. 7). These *cartabones* were named according to their angles (and one supposes that these names hark back to originally Arabic nomenclature). For instance, the *cartabón cuadrado* (square template) had two 45° acute angles, plus the inevitable right-angle.

One of the many drawings by Arenas, which were inserted into his holograph manuscript around 1618 (and which are far superior to the crude woodcuts eventually appearing in the published version of 1633), usefully illustrates both the desired end product and the instrumental means to achieve that end.[2] (Fig. 65) Above, we see the finished lattice-work that provides the intricate ceiling pattern; below are shown the three templates that were actually employed to determine the varying angles of intersection fixing the proper placement for all the individual, segmented pieces of wood. One of these pertains to the framework, the one labeled *armadura* a-b-c, and the other is an *albañecar*, with dimensions of b, c, and d, and that module was used to set up the all-encompassing *armadura*; the third is the *cos*, with that measure determining the diagonal angles belonging to the cut joints of each intersecting stick. The *albañecar*, which initially sets up the *armadura* fixing the structure for all the components to come, is a triangular template (*cartabón*) with 3:4:5 proportions. A compass-point had been added by Arenas into the center of the hexagonal opening at the top of the lattice-work; this is the generational axis of his composition, and so it functionally corresponds to the *rosetones* serving the same purpose in the two Mexican churches analyzed here (figs. 1, 24–28, 54–57).

As Toajas has discovered, "the most frequently used template was the *cartabón de cinco*." The five (*cinco*) in the name "5-unit template" refers to five, evenly spaced

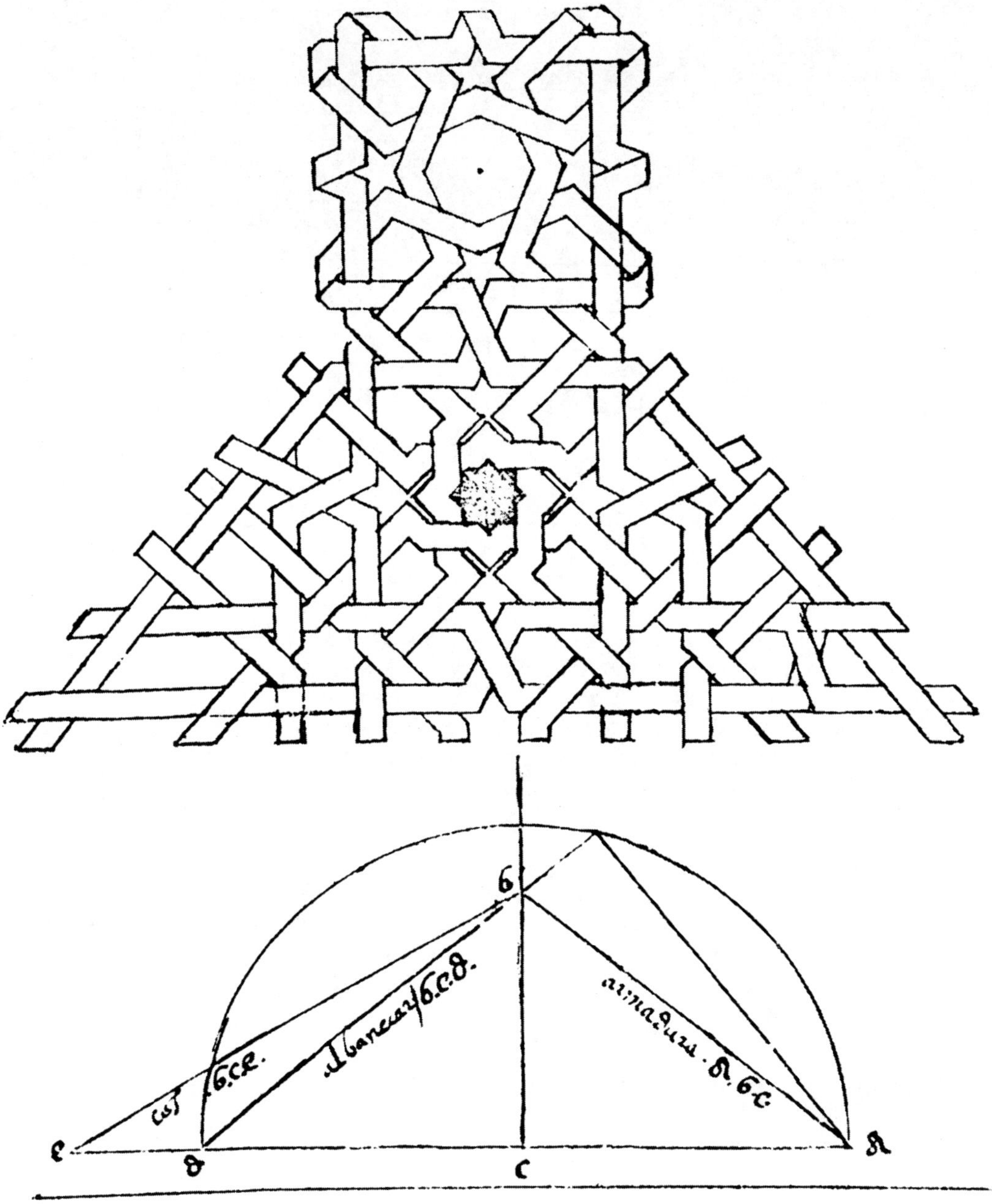

Fig. 65. Diego López de Arenas, Assembled ceiling-panels (***artesonado***) and the templates (***cartabones***) necessary for their design. (Drawing from "Ms. 1," ca. 1618, as published in M. Gómez-Moreno (ed.), ***Primera y segunda parte de las Reglas de la Carpintería, hecho por Diego López de Arena***, 1966.)

units that were ticked off along the diagonal; adjacent to the right-angle, there were three units marked on the shorter leg and four along the intermediary length. The result is, of course, a 3:4:5 triangle, meaning the one commonly called the "Pythagorean Triangle"—and just as this was explained by Fray Laurencio de San Nicolás in his *Arte y Uso de Architectura*. (Fig. 66) Helpfully, Toajas has prepared a

ARMADURA DE CINCO

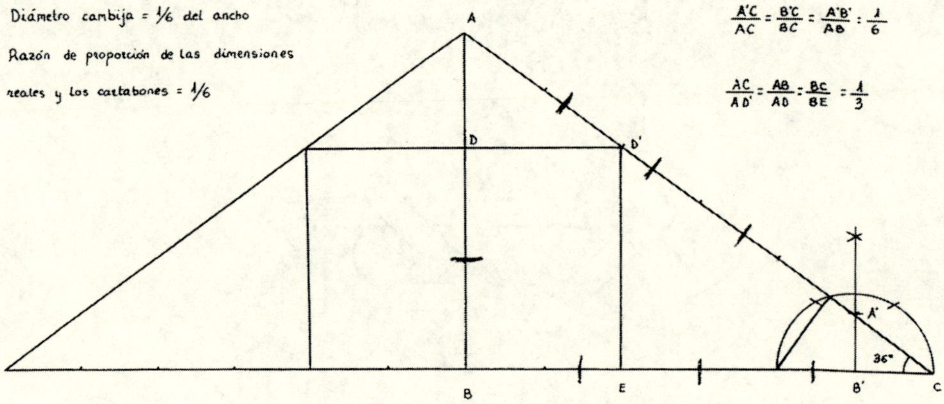

FIG. 4

Fig. 66. M. A. Toajas, Drawing reconstructing Diego López de Arenas's "Armadura de Cinco." (As published in her 1997 edition of Arenas' *Breve Compendio*.)

line drawing of Arenas' "*Armadura de Cinco*." The horizontal line below the apex is a *tirante*, or tie-rod, bracing the triangular framework. As she explains in an inscription, "the diameter fixed by the protractor equals one-sixth of the width. This establishes the proportion of the actual dimensions of both the real panel and that of the templates," with the latter establishing those "real" dimensions for the individual components of the lattice-work (and, even though this was not likely used by the artisan, she also includes the corresponding algebraic formulae in her "figura 4").

Next in popularity (or frequency of employment) was the "*cartabón de seis*"; given its six-unit diagonal, its proportions were a bit narrower. Nonetheless, it is a near match to the 5-unit template; the angle of the acute angle belonging to a given 6-unit template is 30°, while the angle (as measured by Toajas) of any given 5-unit template is 36°. The absolute mean between these two angles— 30° and 36°—is exactly 33°, and that is specifically the acute angle uniquely defining the celebrated *Pythagorean Triangle.* A corollary conclusion is that the design-module which I have called the "Pythagorean Triangle," and which I have found in so many Hispanic buildings (some regal and others merely rustic), is only generically correct.

To be more specific, one could now say that in all these examples, most likely, what actually determined the repeated 3:4:5 proportions revealed here (in figs. 10–15, 24–58) was instead the use of either the *cartabón de cinco* (in most cases) or, alternatively, the *cartabón de seis* (in a few other cases). But I will happily leave that time-consuming task of determining some possible 3-degree deviations from the 33° norm to other researchers. In any event, the main point has already been established: Hispanic buildings were certainly produced by means of "design-through-triangulation," and for these purposes a generic module—specifically a 3:4:5 triangle—was indeed the norm.

17

Fray Miguel de San Andrés: More Written Evidence for a Standard Application of Pythagorean Design-Modules by Hispanic Builders and Craftsmen

The techniques associated with "*carpintería de lo blanco*," an Islamic type of traditional woodworking that was always to be symbiotically associated with an architectural context, was not, however, restricted to Andalusia. It also became widespread throughout Latin America. As a point of great interest, the modern editor of Arenas' treatise, María Ángeles Toajas, presents

> the hypothesis of a possible movement of Arenas' text to Spanish America [*las Indias*], where his manual could have solidified the unquestionable diffusion of traditional mudéjar carpentry work in the New World, with examples of this perceptible as early as the sixteenth century, clearly so during the seventeenth century, and even extending well into the eighteenth century. Nonetheless, although this seems likely, especially taking into account the commerce and cultural interchanges between Seville and Latin America, it hasn't yet been possible to ascertain concrete facts relating to any movement of copies of Arenas' book.[1]

I will now provide that needed documentation, "concrete facts." In the library of the University of Texas at Austin there is an untitled manuscript of 170 folio pages (each measuring 41 x 27 cm); illustrated with 221 drawings and six groundplans, a part of the discursive text copies, even expands upon Arenas' book. Its composition has been dated between 1630 and 1642, and the author of record is Fray Andrés de San Miguel.[2] He was born in Spain in 1577, and was originally called Andrés de Segura de la Alcuña, after the name of his birthplace near Medina Sidonia in Andalusia. He came to Mexico in 1600, became a Carmelite lay-brother (*lego*), and died there in

1652. Since 1606, Friar Andrés was himself a practicing architect, executing commissions for his Carmelite order; these include the Hermitage at Desierto de los Leones, the Monastery of Salvatierra, and the Monastery of San Angel near Mexico City, built between 1615 and 1625.

San Miguel was a learned architectural writer, cut in the mold of Renaissance humanism. Besides Vitruvius, he was familiar with all the major Renaissance treatises; citing the prestigious writings of L. B. Alberti, Luca Paccioli, Andrea Palladio, Sebastiano Serlio and Jacopo Vignola (as published with Spanish texts), he also explicitly refers to Spanish writers, the likes of Diego de Sagredo and Juan de Arfe — also implicitly to Diego López de Arenas and Fray Laurencio de San Nicolás. As a model for contemporary Christian architecture, San Miguel lauds at length the largely mythical Temple of Solomon,[3] a legendary archetype which modern scholars have also shown to have provided the inspiration for the Escorial (figs. 42–52). In practice however, and not withstanding all of his classics-based erudition, Friar Andrés reverted to the traditional *mudéjar* models of his homeland, Andalusian Spain.

Besides showing nicely drawn examples of neo-mudéjar *lacería* (lacework) executed in wood, one of the folio-pages of San Miguel's manuscript includes carefully drawn reproductions of Arenas' *cartabones.*[4] (Fig. 67) The ones numbered "10" and "12" are the 3:4:5 proportioned templates called "*cartabón de cinco*" and "*cartabón de seis.*" Another template discussed in his text is the "*cartabón blanquillo*"; with acute angles of 33° 45' and 56° 15', it is a near match for the classic Pythagorean triangle.[5] Also of great interest is the Mexican friar's intricate drawing of an completed *artesonado* ceiling, for this particular *traza* does not directly copy any of Arenas' drawings (or woodcuts).[6] (Fig. 68) It is, however, inserted within what appears to be a brief (folios 72r–88v), but accurate summary of López de Arenas' manual.[7] Although this drawing admittedly looks very much "like" one of Arenas' many ink tracings (for instance, that one shown in fig. 65) — and it even includes a hexagonal, *rosetón*-like compositional point of departure — nonetheless, it is instead a wholly original composition, one created by carefully following the sort of neo-mudéjar procedures verbally described by Arenas, and which the friar concisely summarizes.

San Miguel's text succinctly tells that it is all done with preliminary ink drawings (*trazos de tinta*), and then the design is physically executed at the building-site. There, the artisans deftly wield plumb-lines ("*líneas a plombo*"), either vertically or diagonally, stretched cords (*cuerdas tiradas*), carpenter's squares and a compass (*reglas y compás*), triangular templates (*cartabones*), either flipped or rotated, and as aligned on to sequences of parallel lines (*líneas paralelas*). As he explains, specifically the *cartabones*

> are the basis for everything described here, because with these there are found all manner of joints imaginable, and this ability makes them common to all sorts of architecture as well —*con que se hacen comunes a toda suerte de arquitectura.* The *cartabones* are infinite in number, but they are all triangles—*siendo todos triangulares....* With these, one shapes all manner of angles, from right-angles to obtuse and acute angles, also all kinds of figures and bodies, and the *cartabones* unite, interlace and chain together all of these, one to another, and they do so with all the beauty, grace and perfection that human ingenuity can imagine.[8]

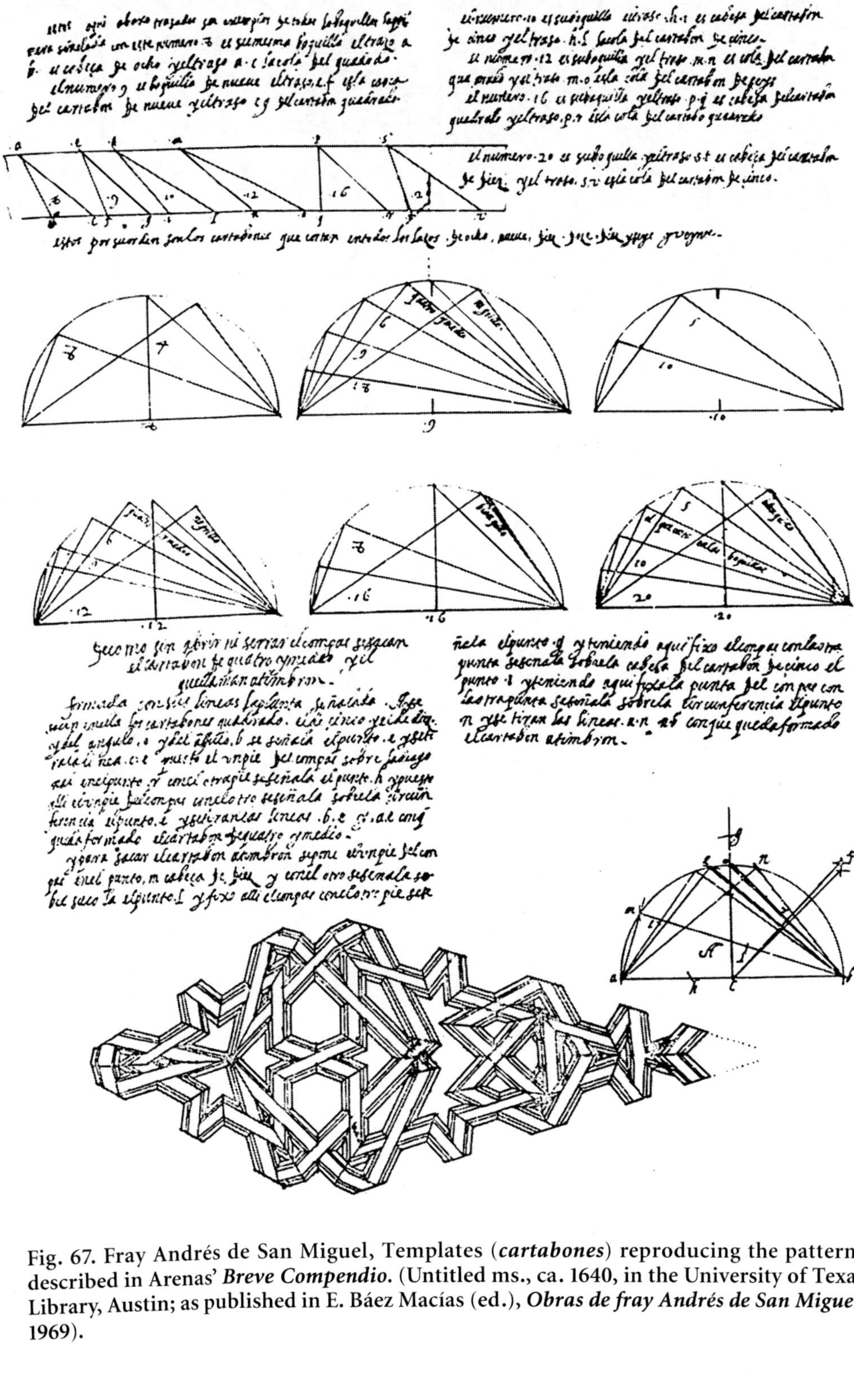

Fig. 67. Fray Andrés de San Miguel, Templates (*cartabones*) reproducing the patterns described in Arenas' ***Breve Compendio.*** (Untitled ms., ca. 1640, in the University of Texas Library, Austin; as published in E. Báez Macías (ed.), ***Obras de fray Andrés de San Miguel***, 1969).

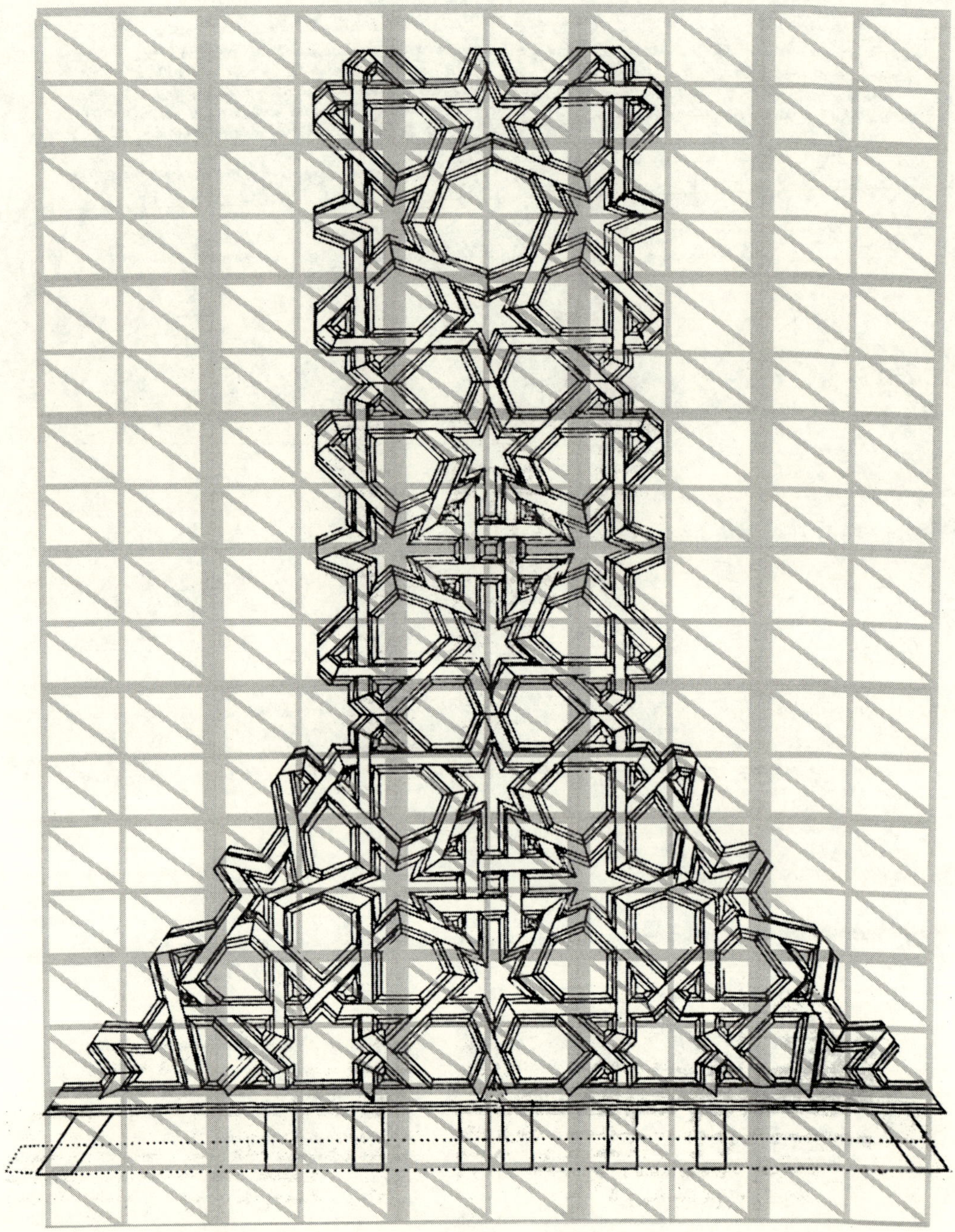

Fig. 68. Fray Andrés de San Miguel, Neo-Mudéjar ceiling (***artesonado***) recreating the patterns described in Arenas' ***Breve Compendio***. (Untitled ms., ca. 1640 in the University of Texas Library, Austin; as published in E. Báez Macías (ed.), ***Obras de fray Andrés de San Miguel***, 1969). The Mexican friar's drawing is here enhanced with the author's computer-generated grid-patterns of non-inscribed, 3:4:5 triangle-modules.)

Besides reproducing the handsome drawing by Fray Andrés de San Miguel (fig. 68), I have gone a step further, by again superimposing my 3:4:5 triangle grid-pattern over it. As before, we reach the same conclusions as those forthcoming when the overlay had been applied to extant examples of Latin-American architecture (figs. 24–28, 54–57): all the proportionate relationships establishing a parallel kinship between horizontal and vertical zones were fixed by a succession of horizontally-positioned diagonals generating inscribed rows of 33° diagonals in parallel ranks which intersect with parallel, vertical inscribed lines. In this case, and just as the friar explains, these alignments are produced by the manipulations of *cartabones* aligned on compass points: "*para sacar el cartabón, se pone la una punta del compás.*"

As has been pointed out, there is otherwise a paucity of architectural writings from the New World (the only other example, as quoted above, is that anonymous, also unillustrated and un-dated, late eighteenth-century manuscript, "Arquitectura mecánica conforme a la práctica de esta Ciudad de México"). Nonetheless, Andrés de San Miguel's now-published (in 1969) manuscript presently documents, and unquestionably so, that the subjects of the discussions by Fray Laurencio de San Nicolás and Diego de Arenas—especially their expositions of templates (*cartabones*) having the proportions uniquely belonging to a 3:4:5 triangle—represented common knowledge in Spanish America. Such "common knowledge" confirms use of this 3:4:5 triangle-motif as an essential design module in Latin-American architecture, a critical point initially verified by empirical means: the postmodernist employment of a cyber-generated grid overlay.

Given the graphic demonstration (fig. 67) that unquestionably Friar San Miguel also partook of this common knowledge, one might now suppose that the Carmelite architect even imposed the ubiquitous module upon his own buildings. (Fig. 69) In this case, the grid-pattern showing non-inscribed, 3:4:5 triangle-modules is layered over a modern drawing of the ground-plan of the entire Monastery of San Angel, designed by San Miguel around 1615 (meaning before the friar could have known of Arenas' manuscript, still unpublished in Seville). I take the domed crossing of the laterally placed basilica-church (built 1622–24) to represent the center-point of San Miguel's architectural complex, from which point the ratios for the locations of all other components comprising the monastery were drawn.[9]

According to the dictates of the Carmelites' "Original Regulations" (*Regla Primitiva*) relating to the layouts of monasteries, the church was to function as the initial module (*módulo*); so taken, it would ensure absolute proportional congruity for the entire monastic complex: "*y conforme a esta medida se guardará la proporción del arte para lo largo y lo alto.*"[10] Friar Andrés followed this injunction, noting that Carmelite churches usually had but a single nave, without side aisles, and that the width of the nave established the entire length, with the ideal ratio being 1:4.5 ("*cuatro anchos y medio y más lo que ocupen las pilastras*"). The same modularity determined the different heights of the interior features, so fixing "*la proporción de este género de templos.*" Working downwards, from the ceiling to the floor, he specifically states that the overall height of the nave, from the choir loft (*el coro*) to the church floor, constitutes ten *varas*; there will be six *varas* from the church floor to the base of the choir loft, and then four *varas* from the floor of the choir loft itself to the

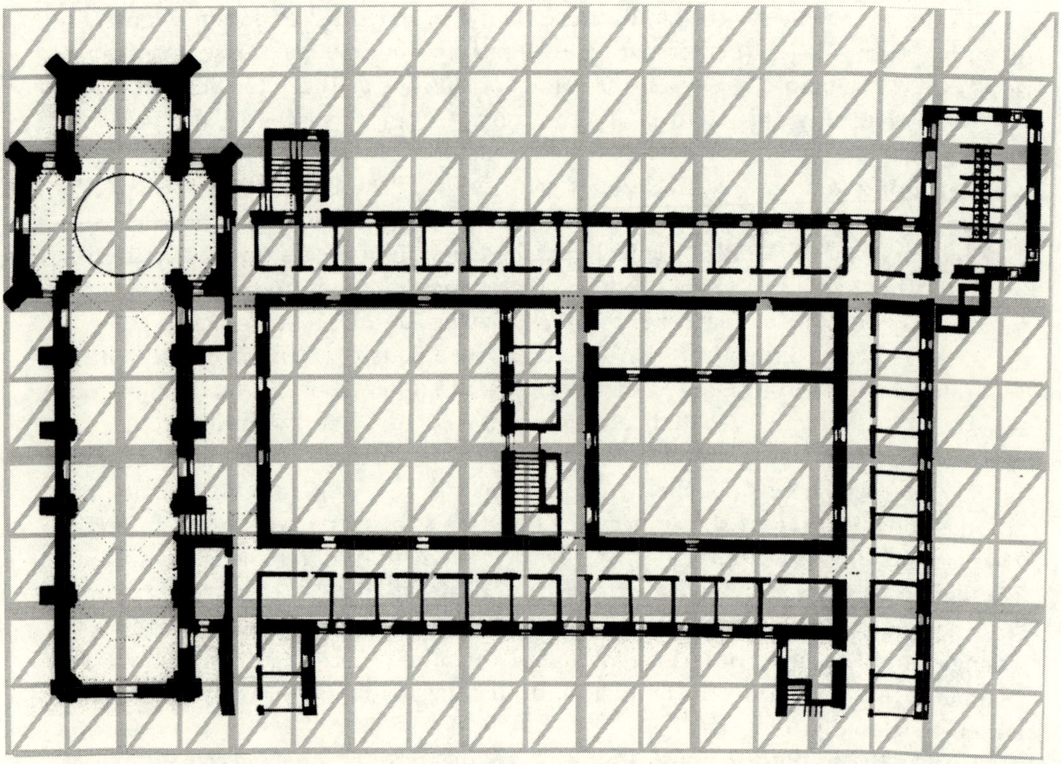

Fig. 69. Fray Andrés de San Miguel, Ground-plan of the Monastery of San Angel (Mexico), 1615–17. (Computer-generated grid-patterns of non-inscribed, 3:4:5 triangle-modules laid over a drawing given in Sebastián, *Arte Ibero-Americano*, 1985.)

springings of the arches. The desired result is that "*quedará la iglesia en buena proporción.*"[11]

The key to architectural proportionality is, of course, geometry, and Friar San Miguel quotes Plato to this effect: "*No entre él que ignore la Geometría.*"[12] Then he describes the ubiquitous motif of the 3:4:5 triangle at some length in a sub-section, "How to measure a squared quadrangle with right angles [*a escuadra*] and its corresponding opposite ends." His description of what he calls "the perfect triangle" sounds rather like that one provided much later in the U. S. Navy Seabee field manual (as quoted in Chapter 7); as he explained:

> One finds the size of a right-angle quadrangle in the same way as was done for a perfect square. This is shown by an example. In quadrangle *efgh*, the sides *eh* and *fg* are the longest, and each one of the longest sides measures 8 [or 4] *varas*, and each of the shorter sides has 6 [or 3] *varas*. 8 multiplied by 6 yields 48 [square *varas*], which is its interior size, and the square of 6 is 36, and 36 plus 64 equals 100, the square of which is 10 [or 5]; this is also the length of the diagonal. This is the line which starts from angle *e* and goes to angle *g*, so dividing the quadrangle into two triangles with the same dimensions. [...]
>
> By employing other means, one can find the number of *varas* on the surface of this triangle by using the same line, the one starting from one of the angles and

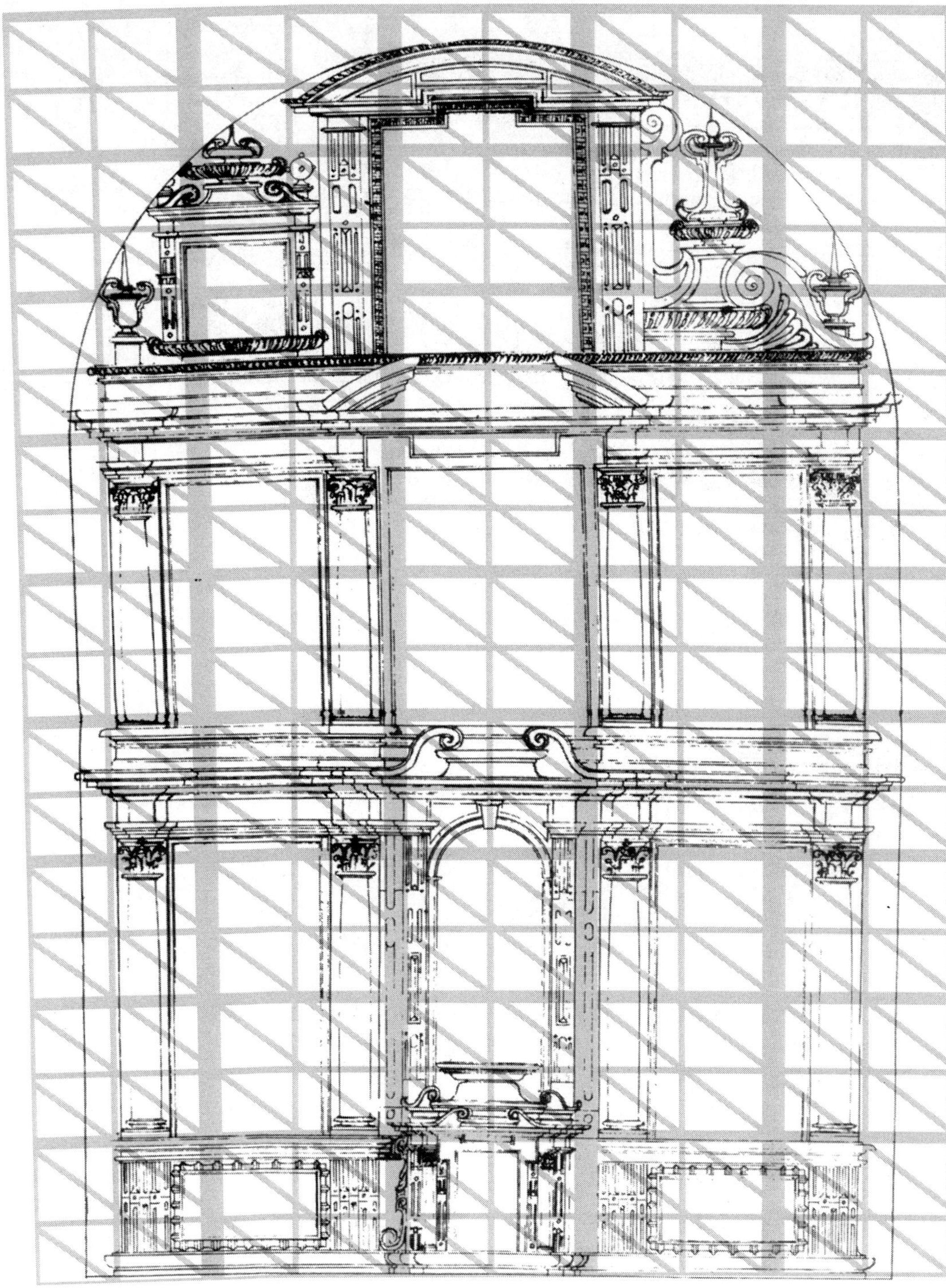

Fig. 70. Fray Andrés de San Miguel, Elevation of a side altar. (Untitled ms., ca. 1640 in the University of Texas Library, Austin; as published in E. Báez Macías [ed.], *Obras de fray Andrés de San Miguel*, 1969. The Mexican friar's drawing is here enhnaced with computer-generated grid-patterns of non-inscribed, 3:4:5 triangle-modules.)

> ending in the middle of the base on the opposite side, so dividing the triangle into two equal parts. This plumb-line [*línea a plomo*], or perpendicular, is found by in this manner. Multiply one of the equal sides by 13 and divide the results by 15, and one of the sums produced will represent the number of *varas* which the plumb-line contains. To learn by means of this diagonal line what the surface of the triangle might be, simply multiply the length of this line by half of one of the sides of the triangle, and the result represents the amount of that ground-plan [*planta*], or the area of the triangle [and he gives various examples and their numerical results].
>
> However, without any recourse to the plumb-line, one can measure any other kind of triangles in another manner. Contract into one all the sides of the triangle which you wish to measure and take a half of what results; with this half, the sides of the triangle may be taken separately, and then half of the sum is multiplied by any of the sides ... and the results are again multiplied by one of the other sides, and what results is again multiplied by one of the other sides and [eventually] you get the square root, and that will be the number of the varas contained in the proposed triangle. It doesn't matter in this multiplication process where one commences because, wherever one starts, a number always results [and he gives further examples].[13]

Now that we have seen graphic proof for San Miguel's commitment to *neo-mudéjar* design principles and procedures (fig. 67), we may now see how he adapted these to an architectural ensemble employing strictly classical, or non-arabesque, motifs. (Fig. 70) In this case, we have the friar's design, a typical architect's projection, for a side-altar seen in elevation ("*alzado para [un] retablo*").[14] Whereas the architectural style represents a typical example of seventeenth-century classicism, and very much in the contemporary European "Baroque" mode, nonetheless the imposition of the grid with the 3:4:5 triangle patterns shows significant departures from its European counterparts. A description of the role played by the intersections of inscribed, 33° diagonals meeting with similarly inscribed, vertical lines exactly matches the operations of similar patterns which we first encountered in the Islamic portals of the Great Mosque in Córdoba (figs. 9–15).

In brief, a diagonal anchored on the lower right-hand corner of the side-altar ascends at a 33° angle; its juncture to the upper left fixed the base-line for the footings of the pilasters in the second story. Another diagonal, drawn from the base-line of the pilasters in the second story, establishes a horizontal line setting the bases of the Corinthian capitals in the same story. Passing through the vertical axis-line in the center of the composition, the same diagonal line fixes the point of maximum extension of the projecting cornice separating the second and third stories. Another diagonal, drawn from the upper end of the second-story cornice, fixed both the height of the left-hand niche in the second story and the center of an ornamental vase on the opposite side; rising further (and outside the composition), it establishes the base of the flattened arch crowning the entire ensemble. Another diagonal, traced just below it, sets the horizontal line for the height of ornamental vases in the third story. In a smaller, but no less significant, detail, another diagonal, drawn from the bottom of the altar, establishes both the height and width of the altar table. And so it goes, and throughout the composition.

18

A Sixteenth-Century "Mexican Mosque"

In the preceding parts of this study the graphic, also written, evidence attesting to a consistent employment in Hispanic architecture of a distinctive system of inscribed design modules arising from Islamic building practices has been presented. The present task is to reveal the existence of other, complementary kinds of neo–Islamic building practices operating in architectural projects fashioned by Spanish supervisors in Latin America. These revelations of a calculated revival of Moslem typologies establish the broader historical context encouraging repeated employment of the particular compositional topos we have exhumed here: the Pythagorean triangle, with this usually serving as the essential architectural module determining the proportionate placement of the various design components.

In his exhaustive and still fundamental study of *The Open-Air Churches of Sixteenth-Century Mexico* (1969), John McAndrew has cited numerous instances of the employment of *mudéjar* motifs in ecclesiastical structures erected all over *Nueva España* under the direction of European-born Augustinian, Dominican, and Franciscan monks. However, those who actually built these often imposing structures employing the obviously imported neo–Islamic devices were Native Americans, *los indios.* Among other early chroniclers, Father Jerónimo Mendieta stated (ca. 1580) that "the Spaniards, masters of many crafts, rarely do more than give the work to the Indians, and they tell them how they want it done. And the Indians do it so well that no one could do it better."[1] Since the first known professional architect, Toribio de Alcaraz, did not arrive in Mexico until 1544, one must suppose that the first, Renaissance-style architecture erected in the New World had been designed by autodidact friars. As may be further inferred from identical schemes appearing in distant towns, it appears that they were also working from master drawings circulating from one monastery to another; alas, none of these *trazas* or *planos* has survived.[2] Another qualified Spanish architect, Claudio de Arciniega (1526–ca.1590), a native of Burgos in Castilla la Vieja, who came to Mexico around 1545, is documented as working in Mexico City in 1559 on San José de los Naturales, an open-air chapel, which is the major focus of this chapter.

The materials used by the first builders of Christian architecture in Latin America were exactly those employed by mudéjar *alarifes* in Spain: bricks, rubble, mortar, stucco, wood and tile (*ladrillos, la mampostería, la argamasa, el yeso, la madera y la cerámica*). As cited by McAndrew, the individual, specifically *mudéjar*, architectural elements, which may still be observed in dozens of sixteenth-century religious buildings anonymously erected throughout Mexico, include *alfarjes*, polychromed ceilings assembled from fitted wooden components to make repetitive interlace patterns; the ubiquitous *alfiz*, or foot-less, rectangular moldings placed upon façades; *artesonado*, painted and coffered, wooden ceilings (which may also be observed in the *haram* of the Great Mosque of Córdoba); and *merlones*, series of squashed, prismatic pinnacles looking like battlements, likewise to be seen on the roof of the Great Mosque of Córdoba (see figs. 9–11). This neo–Islamic legacy makes perfect sense; McAndrew notes that it had long been the custom in Spain that "captured mosques had been promptly and easily turned into churches, with minimal alterations. The practice had not stopped, for when Charles V took Tunis in 1535 the old mosques became churches almost overnight."[3] It may be also mentioned that when Hernán Cortés mentioned Aztec temples in his dispatches to Charles V he often referred to them as *mezquitas*, that is, mosques.

Another link to the sixteenth-century outburst of Mexican *mudéjarismo* is the native, pre–Columbian style of architectural decoration, *tequitqui*. McAndrew observes that "*tequitqui* is the American cousin of *mudéjar*. The word is the Náhuatl term for 'one who pays tribute,' or a 'subject person,' while *mudéjar* is an hispanization of the Arab *mudayyam*, which means more or less the same." As in *tequitqui*, so too in *mudéjar*:

> The whole [decorative *tequitqui* ensemble] holds together chiefly by an all-over surface tension, and not by the subordinating power of a few dominant forms or currents of movement. Coherence comes not from any organic or sequential interrelation of parts, but from their insistent repetition in tight and seeming endless patterns, and from the incessant interaction of their outlines. Almost everything seems equally important because almost nothing submits to subordination.... Zones of flattened relief stand out in front of a neutral flat ground which has been cut back to an even depth behind them, so that everything lies as strictly on the two levels as if the ornament had been neatly punched out with a cookie-cutter and laid on a flat surface. There is little or no modeling.
>
> Everywhere in [post-conquest] tequitqui one sees European forms pressed out flat and presented in sharp silhouette, raised evenly above a half-seen neutral background [with] dense patterns, flat patterns which often made strange tensions by contrast with the inert bulk whose surface they enlivened.... The half-abstracted forms are repeated and multiplied by *horror vacui*, until they make the typically dense, never-ending, unaccented patterns which are antagonistic to the phrased spacing with which [non–Hispanic] Europeans had habitually deployed these forms. In western Europe, only in Spain was this kind of difference in rhythm understood in the sixteenth century, because *only in Spain was an important non–European heritage still alive: a heritage from the exotic art of Islam.*[4]

Although referring to sixteenth-century Mexican architectural decoration, this formal description exactly fits scores of medieval mudéjar ensembles which are still to be seen throughout Spain. Although he observes how in both styles, *tequitqui* and *mudéjar*, "detail would be assembled in zones of dense and carpet-like patterns," and how "simple modular relationships with simple accordant proportions" were consistently employed,[5] McAndrew never tells us just how such intricate decorative ensembles were actually designed in order to have their various components assembled in an obviously well proportioned manner. The material evidence suggesting the most likely answer, the common employment of a triangular design module, has been presented in the preceding chapters.

Another issue which needs to be addressed is an architectural format seemingly unique to the New World. Having no *Christian* parallels in Spain, an ecclesiastical forecourt — the "*patio*," a vast, tree-planted atrium with an open chapel facing it — was invented in Mexico to serve new demands, the mass conversion of a non-European native population. In 1564, Doctor Vasco de Puga observed that "the first thing [the Spaniards] do on the land they are settling is to lay out the monastery, with its great expanse of gardens and open yards [*patios*] ... and when it is all laid out, they begin the church and monastery buildings, which they put up very quickly and very well."[6] Padre Mendieta also made reference to an architectural innovation implied to be unique to *Nueva España*:

> All the monasteries here in New Spain have a large walled *patio* placed in front of the church.... and usually they are adorned with trees [usually *naranjos*, or orange-trees] set in orderly rows.... The patios were made to be used mainly on holy days, so that when all the [indigenous] townspeople are gathered together, they can hear Mass, and be preached to in the patio, since [due to their numbers] they will not fit inside the body of the church.[7]

In effect, the enclosed but roofless *patio*, later also called an *atrio*, substituted for the nave of the parish church, with the distinct advantage that thousands of catechumens could be ministered to simultaneously.

Granted land for its building in 1527, the first permanent church to be erected on the American mainland belonged to the Franciscan monastery of San Francisco set in the center of Mexico City (formerly Tenochtitlán). Named San José de los Naturales, according to McAndrew's extensive research, "it was probably this early example which established the precedent for thousands of [open] church forecourts laid out during the next four centuries."[8] As he also recognized, the specific architectural model for this seemingly unprecedented, but obviously essential, innovation were most likely "the patios of the mosques abandoned by the Moslems in their retreat through Spain."[9] These mosque *patios*, called a *sahn* in Arabic, were walled, contained a well for ablutions, and were usually planted with rows of trees.

Today, the Spanish archetype for the Mexican *atrio* can be viewed in Seville at the walled and gated, "Patio de los Naranjos" (or orangery) — and just as it had originally been designed by Islamic builders in the late twelfth century. This famed garden *patio* is attached to the north side of the Cathedral of Seville, which itself exactly conforms to the site of an enormous *masjid alhama*, an enormous hypostyle hall with seventeen naves constructed by the Almohades between 1172 and 1176. The Patio de los Naranjos originally served as the forecourt (*sahn*) to the old mosque. Still attached to it is the original minaret (*alminar*), now called La Giralda (fig 30). Another functional parallel to the Mexican *patios* is set by the Islamic *mushalla*, a detached (mosque-less), plain, gated and walled courtyard with a *mihrab*, or prayer niche, oriented to the east, the direction of Mecca. Erected as a place for collective prayer upon special occasions (for instance, to end a drought), *mushallas* were once to be seen in al-Andalus at Almería, Córdoba, Granada, Málaga, Totosa, and Valencia.[10]

In Mexican towns, to attend mass Spanish families would go to one of their own parish churches and these were, naturally, built in the traditional, meaning roofed, format. To the contrary, the far more numerous Indians in the same quarter (*barrio*) would instead go to their own *capilla de indios*, inevitably an open-air church or *patio*. The first of these, so setting the formal precedent for all the rest, was San José de los Naturales, and as erected within the grounds of San Francisco in Mexico City. The idea for its establishment seems to have been conceived by Fray Pedro de Gante, a Fleming born around 1480 in Ghent and then named Peeter van der Moere. Educated as a humanist, he had good connections; he was either an uncle or a great-uncle of the Emperor Charles V. Arriving in New Spain in 1522, he became the first European to teach the natives in their own language, Náhuatl. In 1526, he established what he called "San José de Belén de los Naturales" (Saint Joseph of Bethlehem, belonging to the natives).

As McAndrew recognizes, this building was "of enormous import, for it established the open chapel in the Mexican [architectural] repertory as its most original item, and it set an example soon to be followed many times in many different towns." Unfortunately, this highly influential building — an *Urbau*, the kind spawning innumerable replications, like the Temple of Solomon — has long since disappeared; the site is now incongruously occupied by a 43-story skyscraper erected alongside the High Life Department Store and Cucaracha Tourist Cocktail Club. Thanks, however, to McAndrew's diligent research, its original appearance — which is what he calls that of a "Mexican mosque"—can now be retrieved.[11]

The first vault of the church of San Francisco was raised in 1525, and beside it, to the north (as is the case with the Sevillian Patio de Naranjos), Fray Pedro had the Indians erect the original, *patio* version of San José de los Naturales. Although the historical evidence is scarce, it is logical to assume that the initial construction was simple, no more than a small, plain, thatched shed, or *xacal*, facing an open court. A year or two after, it was recorded that a grand Christmas *fiesta* was celebrated there with Indian participants. A walled *patio*, with a two-hundred foot high cross carved from wood erected in its center, is known to have been in use by 1532. This is confirmed by Bishop Zumárraga writing in 1534, then mentioning the "patio where Mass is said for the crowds of Indians on Sundays and feast days," so showing the existence of a large, open or unroofed, chapel. In 1539, the Bishop stated that the walled *capilla* had already become the "chief sight of the country, with its covered gallery and pulpit correctly arranged—*con el corredor y oratorio en forma*." At the same time Zumárraga praised a "*capilla*," presumably San José, to which he planned to add a fifth story to its tower, which, already reaching four stories, was then the highest structure in the city.[12] As one deduces from these comments, San José then had the appearance of a one-bay deep, high portico, with an apse set in the middle of its rear or eastern wall, a tower at its north end, and a capacious forecourt extending westwards.

Soon after, in 1547 and 1551, disastrous earthquakes struck Mexico City, and in 1552 Fray Pedro wrote the Emperor that San José "has been done over again, good and well built, so that the Holy Offices can [again] be celebrated there impressively." In 1554 Doctor Francisco Cervantes de Salazar described San José as including a

> little chapel, which is [placed] behind a wooden trellis, yet it is wholly open and plainly visible from the front. Its roof is set high above the ground and is carried by tall tapering columns of wood. The material is ennobled by its workmanship.... It is arranged in such a way that the crowd of Indians, big as it is, who flock here from all around on feast days, can see [the Host] without hindrance and can hear the priest as he performs the Holy Service.[13]

This suggests a wholly new, post 1551, design for the covered gallery facing the forecourt, and most likely on a much greater scale, especially since the First Council of the Mexican Church was convened there in 1555. In 1558, Fray Pedro now announced San José to be "the first church made in the land," and also claimed, perhaps with some exaggeration, that it could hold 50,000 worshippers within its *patio*.[14] The importance of San José was once again clearly attested by its designation, in 1559, as the official site for the elaborate funeral obsequies of Charles V. Claudio de Arciniega was then appointed the architect in charge of the extensive remodeling, which included the addition of soaring pillars, perhaps reaching a height of fifty feet, and an arched, stone portico. The roofed part of the building was then further extended to the west, now acquiring a depth of six bays (around 85 feet) and a width of seven bays (around 100 feet), so providing space (around 8,500 square feet) for around 2,000 people, probably all Spaniards. McAndrew estimates that the area embraced by the open forecourt would have been 17,000 square feet, so suggesting a crowd of perhaps 5,000 standing *indios*. The new appearance was recorded by Dr. Cervantes de Salazar in 1560:

> On the left [or north of San Francisco] there is a [covered] chapel called San José, to which one ascends by two steps. It is very large and it is supported by many columns which make seven naves [or bays] which have been painted to look like marble.... It is a sight to see because it is so ingeniously roofed with wood [most likely as *artesonado*] laid over many columns. In front it has a set of stone arches [the portico]. It is very well lit because the chapel has a high ceiling and is open in the front [to face the patio]. The stone arches [in the portico] are low, and serve more for ornament than for shelter or support.... Because there was a set of fourteen very well made arches which hindered the view of the *Túmulo* [the Emperor's catafalque, as placed in the center of the roofed chapel], he [Arciniega] had ordered them to be taken down and [later they were] moved to the front [of the chapel, so extending its depth].[15]

What is especially singular about the building of 1560—particularly including its extension to the front, the open *patio*—is that it then structurally becomes what McAndrew calls a "Mexican mosque." As he emphatically affirms, "No matter when or how it got its final form, and no matter how it may best be classified, there can be little doubt that the seven-naved San José was patterned on a mosque and its forecourt." His arguments may be reiterated as follows.[16] The repeated feature of a forecourt —*sahn*— with its multiple quincunxes (intersecting five-part star-patterns) of trees was a Spanish-Moslem architectural specialty, with this format initially established first in the eighth century at the Mosque of Córdoba. The typical placement of a minaret (*al-minar*) at a corner of the mosque near the *sahn* (as at Córdoba, see fig. 73) also corresponds to the placing of the soaring bell tower at San José.

From Dr. Cervantes' description of the Capilla de Indios in Mexico City we discover that its roofing system must have resembled that of an *apadana*, or "Hall of Columns" used by ancient Persian royalty. This multi-colonnaded, or hypostyle, scheme was later adapted by Moslem builders as the preferred format for a mosque. A typical example, the Great Mosque at Kufa in Iraq, was described in 1184 by Ibn Jubayr, who spoke of

> a vast mosque, the *qibla* side has [a width of] five aisles, whereas the rest have [a depth of] two only. The aisles are supported by columns like masts, composed of hard blocks of stone superimposed piece by piece, bedded upon lead, and they are not surmounted by arches. Extremely high, they go up to the ceiling of the mosque. I have nowhere seen a mosque of which the columns are so tall or the ceiling so elevated.[17]

However, being Flemish, clearly Pedro de Gante cannot have been sufficiently familiar with the architecture of mosques to have had the ability to recreate a mosque in Mexico. In fact, he would probably have had no idea of what one looked like. Therefore, the idea would have more probably come from one of his Spanish companions intimately familiar with Andalusia, where a few mosque buildings of this type were still among the arresting sights. One of the original Twelve Friars was from Córdoba, and another came from Seville, and there certainly must have been other Andalusians in the next groups of Franciscans who arrived before the mosque-like design of the chapel was actually produced around 1560.

Another feature of San José which specifically echoes a corresponding device at the Mosque of Córdoba is the low stone arcade running across the entire façade facing the patio. Since the portico was placed well below the roof, as Doctor Cervantes had noticed, and essentially without having anything substantial to support, its arches must have been free-flying; perhaps (as McAndrew suggests) they once looked like the multiplied, pink-and-cream-striped, free-flying arches still to be seen inside the Great Mosque of Córdoba. If this arcade was easily moved, as it seems to have been, then it was probably as unstructural in its original location as it was after it had been moved by Arciniega some four bays forward, and finally set up as the façade now intruding into the patio. However, if the classicizing Claudio de Arciniega was willing to use these arches, even in a subordinate position, then it is unlikely that they would have thereafter retained any explicitly Moorish details.

Nowhere in Spain, however, and not even in mudéjar Andalusia around 1560, were there churches explicitly designed in the guise of mosques. Moreover, ever since the elegant architecture of the Mozarabic refugees from Andalusia had gone out of fashion six centuries before in northern Spain, churches had not there been designed with a mosque format, either consciously or unconsciously; San José is the exception, but it spawned a number of Mexican replications. Nonetheless, there was a highly visible Moorish character distinguishing later mediaeval architecture in most parts of Spain. It reached its climax in the splendidly decorative mudéjar art of the late fourteenth century, and particularly as revived in Isabelline architecture in the late fifteenth century; clearly, architectural *mudéjarismo* was far from extinct in the early sixteenth century. This taste did not, however, make mosque-like churches, and naturally so, for the plans and general schemes of mudéjar builders were employed to serve orthodox Christian purposes: mudéjarismo delighted not in Moslem plans but in Moslem trimming. Although mudéjar taste occasionally imitated Moslem structural elements, their structural nature was subsumed into decorative patterns, as in the typical and dazzlingly intricate *alfarje* ceilings. Disinterested in complicated spatial effects, this is an art of the highly patterned plane, for which proportionate systems are required for clarity.

Mudéjar customs are occasionally documented in Latin America. The Conquest had coincided in the homeland with persecutions of the unbaptized Mudéjares. In Spain, the last of their mosques were closed in 1525, and the last Islamic stragglers were forcibly baptized within a year (and thus made *moriscos*); at the same time, the first large groups of *indios* were being baptized in Mexico. Despite the fact that migration of Moriscos was banned, records listing the names of settlers show that a few with Arabic names—for instance, Alonso de Mora, Diego de Azamir, Diego de la Mezquita ("mosque"), Hernando Tarifa, Francisco de Vendabal (or Ben-Abal), Pedro Morrón, Velázquez Mudarra, Alonso Zambrano, etc.—had soon slipped into New Spain among the supposedly pureblooded Spanish immigrants. Moreover, a special edict sent from Spain in 1543 again prohibited the entry of Moriscos into Mexico, but this time "because of the trouble with those who have already come [*los que han pasado*]."[18] It was not only these migrant Moriscos who perpetrated the late mudéjar phase of Spanish taste; it is even recorded that some pure-blooded *encomenderos* (plantation owners) liked to sport turbans and engage in Arabian-style tournaments; as Dr. Francisco Cervantes de Salazar wrote in 1558, "*Jineta y cañas son contagio moro.*" Moorish influence appeared now and then, capriciously here and there, but it did not reach its height in Latin America until early in the seventeenth century.

In 1545, another, far more positive, mention of these emigrant Spanish Moslems comes from Santo Domingo on the Caribbean island of La Española (now the Dominican Republic). Here the Mudéjares are referred to as "Barbary slaves" (*esclavos berberiscos*), adding that there were some one hundred households of them (*cien piezas de ellos*), and that even more were to be found in the hinterland (*tierra de adentro*). However, other Moriscos are cited who had been granted royal license, since "they are married and have children"; those *berberiscos* who had been designated to be "free citizens" (*personas libres*) had been specifically granted this royal pardon because they were "professional builders and carpenters [*oficiales de albañiles y carpinteros*] and practitioners of other trades deemed most advantageous for the population of the land." Another royal decree approved of this sanction in the case of "*los berberiscos con oficios que está bien,*" meaning those with useful mudéjar building skills.[19]

Under the peculiar building conditions of sixteenth-century Mexico, mudéjar forms were usually simplified, and restricted to a few ornamental features on buildings not otherwise explicitly mudéjar. As previously noted, the most telling motifs include pyramidal and stepped merlons, octagonal piers, *alfarje* ceilings, and especially the *alfiz* (a rectangle of molding enframing an arch), and as often accompanied by scalloped, polylobed, or mixtilinear arches. These anomalous elements originally go back through Spanish mudéjar to the late decorative phases of Moslem work in Spain, beginning around 1200. However, some other characteristic forms of earlier Moslem work were not revived in New Spain any more than they were in Spain. For instance, even though there is not a single colonial-period *arco de herradura* (horseshoe arch; see figs. 9–15) to be seen in Mexico, there had been a fine set of these was included in the sixteenth-century church at Santiago de los Caballeros in Santo Domingo (alas, it was destroyed by an earthquake in 1842).[20] However, many more examples were not to be expected, since that particular motif had disappeared from the standard Spanish architectural repertory long before the time Tenochtitlán fell. Ironically, although the horseshoe arch is often taken to be a strictly Islamic device, its source is actually Christian, for it was a standard feature in Visigothic architecture, and the Moslem builders of al-Andalus acquired it from that local, Spanish source.[21]

Uniquely, but including its dozens of replications, San José de los Naturales is akin not so much to mudéjar artifice as it is to actual Moslem building. Specifically, it is related not so much to recent Christian work as it is to the classic, hypostyle Hispano-Islamic mosques of past centuries. So doing, it becomes a building as strange historically as it must have been visually. If more concrete facts were known about later Spanish mosques, such as those built by the Nazaríes in Granada during the fourteenth and fifteenth centuries, perhaps a convincing single prototype for San José might be found. Nonetheless, even though this is unlikely, because no really large examples are recorded, San José is clearly related to a large mosque, but we do now know exactly which one. Since morphologically it *is* a mosque, and of a type built in Spain only slightly before 1200, I will later suggest that its most likely model was the ninth-century Sevillan mosque, with significant twelfth-century modifications, now known as El Salvador.

A true Islamic mosque was used very differently than is a Catholic church, and not just in the meaning but also in the specific choreography of its rites. Although there might sometimes be crowds of the faithful assembled in the garden-like *sahn* leading into the mosque, they were there only to wash ritually before entering the mosque proper, and not to witness any service emanating from an architectural feature equivalent to an open chapel. The *mihrab,* in form an apse-like, prayer niche (see figs. 12–15), was sunk in the back, usually eastern, wall of the typical mosque; if one stood in alignment with the main portal, and with one's vision aimed from the *sahn* toward the east and through the rows of columns erected inside the multi-bayed *haram* (nave), the mihrab was made visible from the courtyard. Sometimes the *iman* (director of prayers) might channel collective concentration by standing in front of the mihrab, often used to serve as a dignified place to keep the Koran, there giving instructive readings from the Holy Book. The prayer niche was usually set inside the central point of the *qibla,* a bare wall at the back of the mosque and marking the compass direction of the Ka'aba at Mecca.

Unlike a Christian church, the Islamic mosque was not held to be the house of

God (*Beth-El* in Hebrew). Instead, it was only presented as a gathering-place for collective prayer; thus its function was more like that of a synagogue, which had also influenced its form. Between one part and another of the mosque, there was little differentiation in the meaning or prestige accorded to the building's components. Except for the *mihrab*, there were no areas specially shaped for distinct functions. Since Islam had no symbolic ritual equivalent to the Mass, it had no need for any specialty consecrated place in its *masjid*, a prayer-hall. In the traditional mosque there was no channeled nave, sanctuary, or major axis, or any need for one; the mosque lacks the spiritual focal point — the Eucharist — obligatory in any Catholic church. The column-supported roof of the *masjid* covered an unfocused area without directional or localized emphasis; consequently, its large interior might bewilder a Christian visitor by its undirected repetitiveness.

A notable exception is the Great Mosque of Córdoba, for the late tenth-century reformations of al-Hakam II had given it a directional, ceremonial axis focused upon the *maqsura*, a towering, three-bay grid set before the *mihrab* (see fig. 73, nos. 14, 23). Jerrilynn Dobbs observes that "this particular plan of three privileged aisles is unknown before this time in any congregational mosque," and the result was "an architectural setting that reproduces in many ways that of a [contemporary Spanish] church."[22] Certainly, many of the conquistadores coming from Andalusia to New Spain in the sixteenth century were familiar with this building and its unique axial focus upon an apse-like *mihrab*, and Dodds also specifically points out the Christian-Visigothic architectural sources of this symbolic niche.

When the faithful gathered at the only required time — Friday noon — it was for communal praying and to listen to readings taken from the Koran. They stood in rows to face the arabesque-decorated *mihrab*, hence they oriented themselves toward Mecca. Special merit was accorded the front row; to enable as many as possible to enjoy this visual delight the mosque was planned, at variance with the church, for expansive width much more than for focused depth. This change in orientation — from depth to breadth — occurred in Spain during the period of the domination of the Almohades, who occupied Seville in 1147. The same visual priority determined the plan of San José de los Naturales, where the focal point now became the elevated Host. Although there was considerable formal similarity between the architectural format of the Hispano-Islamic mosque and its later derivation — the later Mexican chapel-plus-atrium scheme — there was, and naturally so, given a wholly different liturgical purpose and procedure, little functional similarity. Only on Fridays, when all of Moslem Córdoba, Granada, or Seville then went to the Great Mosque, and when there might be overflow in the orange-shaded *sahn*, was there any significant resemblance to what was later to take place on Sunday inside the *patio* of San José.

Besides functioning as a place of ablutions, what was the larger meaning attributed to the Islamic *sahn-patio*? Likely, it had the same significance as that one later to be attributed to the cloister-gardens placed alongside rural Romanesque monasteries, and these were explicitly called a *paradisus*, and following the Edenic model given in the Book of Genesis (2: 8–14). When compared with their garden-less European counterparts, Spanish metropolitan cathedrals oddly incorporate paradise-gardens; notable examples include — besides the Patio de Narajos flanking the Cathedral of Seville — the *paradisus claustri* attached to the Cathedrals of Santiago de Compostela, Barcelona, Tarragona, Burgos, and Toledo, among others. This Spanish peculiarity must be explained by its unique Islamic heritage. As explained by John D. Hoag, a specialist in Islamic architecture,

> Throughout the Koran, when Mohammed speaks of paradise, he constantly uses the phrase, "gardens beneath which rivers flow." Once he says, "For them are upper chambers, and upper chambers are built beneath which rivers flow." There the elect were to be clothed in green silk and to be entertained by large-eyed maidens beside the four rivers of paradise. In the Koran, the word *firdaws* is used several times for paradise. It comes from the Persian *faradis*, derived from a word meaning a place walled in, or the finest and highest part of a garden. Such walled courts ... seem deliberately intended by their builders to provide a setting for the anticipation upon earth of the pleasures of the hereafter.[23]

No one in Mexico would have known this in 1560, for there had been no public practice of the Moslem religion in a large hypostyle mosque in Spain — nor anywhere else in Western Europe — for a generation, and before that, until 1492, such gatherings only happened in the beleaguered Naṣrid Kingdom of Granada. More likely, San José must have been planned by someone who had lived in a Spanish city where a big mosque had been recently expropriated, more or less intact, later to be quickly resuscitated as a big church — and this was prominently the case in, to cite the best known examples, Córdoba, Granada, and Seville. The Spanish designer of San José probably had himself worshipped in one of these converted mosques. Thereafter he could have applied what he might remember of it, decades later, to the solving of a church-building program in Mexico that was as much novel as it was urgent.

For example, along with all the other mosques in Granada, the largest and most important *masjid* had been exorcised and consecrated as a church soon after the city fell in 1492. Part of the Granadine mosque was pulled down when the *Capilla Real* was built as a burial place for Ferdinand and Isabella. However, most of it still stood intact for some decades after 1528, when Diego Siloé began his work on the Renaissance-style Cathedral which would rise in time to replace it; in fact, part of the mosque was still in use as a church as late as 1561. A large eleventh-century mosque, with nine by eleven aisles divided by marble columns, the one at Granada must have resembled the mosque–Cathedral of Córdoba, especially since the stone colonnades continued outside of its *haram* and directly into the *sahn*, then in the form of aligned quincunxes of lemon trees.

Another candidate, to my mind the most likely, which could have served as the formal model for San José was the first metropolitan mosque of Seville; founded in 829 by Abd al-Rahman II, it was expanded in 1195. After the reconquest of Seville in 1248, the old mosque became known as El Salvador and it remained the premier Christian temple until the completion of the present Cathedral in the late fifteenth century. Most significantly, its *haram* and *sahn* remained intact until 1671; thereafter, it was drastically reformed in the then-popular Churrigueresque manner.[24] The result of this architectural mutilation is an important gap in the architectural record which is almost as great as the lacuna stemming from the loss of the Almohad *masjid alhama* which, beginning in 1402, was replaced by the Cathedral of Seville. However, since Seville was granted in 1503 the monopoly of trade with the New World, with all Spanish ships debarking for the Americas being required to pass through its port, the original appearance of this old mosque would have been known to the Spanish conquistadores— including Claudio de Arciniega, coming from Burgos around 1545.

Obviously, the designer of the mosque-like *patio-capilla* of San José de los Naturales must have known one or more of these newly christianized mosques, most likely El Salvador in Seville, or perhaps elsewhere, in either Córdoba or Granada. The Mexican designer chose to adopt a building type common in Spain for centuries, the hypostyle *haram* with a *sahn* extension. Although a structural type long dead ideologically in the homeland, it was now again made useful in solving a new ideological problem in the New World. So doing, the designer of the mosque-like San José, most likely Claudio de Arciniega, is revealed as an independent-minded and ingenious amateur, free from both the chronological and sectarian imperatives limiting contemporary architectural practice.

The earliest, thatch-roofed version of San José was evidently nothing more than a rude *xacal*, so it was much less a "mosque" than was its famous successor; perhaps it was more like some indigenous, pre-conquest monuments in its structural details. Although we can only guess at its appearance, it is known that Aztec shrines, or *teocallis*, commonly had thatched roofs, but Islamic mosques never did. The Indian workmen erecting the initial structure could have appropriated a familiar model, the *xacal*, with more speed and dependability and less supervision than an exotic, also more architecturally complex one, the *mezquita*, that is, provided that the friars did not think the finished work was too provocatively heathen. But there were not yet many friars to do the supervising, and those available had more pressing duties.

If any lesson from Aztec precedents was heeded in the earliest version of San José, and if any of these were retained in its far more architecturally ambitious, mosque-like successor, then, and with wonderful historical appropriateness, there had been put to work ideas drawn from both the religions which the Spaniards had fanatically set about to extirpate from their realms. In an apparently opportunistic fashion, the two pagan faiths were made to work together for the Catholic faith imposed in their place. As various anecdotal records indicate, the result was so different from anything seen in Spain that it had to be explained to newly arrived Spaniards. Although novel, the mosque-like format obviously suited both their needs and their taste.

A much better documented case in point is the *Capilla Real* (royal chapel) erected in the town of Cholula over the site of a pagan temple dedicated to Quetzalcóatl. As for its historical importance, McAndrews succinctly observes that "the Capilla Real at Cholula is an out-and-out copy of San José de los Naturales."[25] The first version, likely built around 1565, was nine bays wide and seven deep, and it then measured 190 by 170 feet. It suffered an untimely fate; according to the *Relación de Cholula*, written in 1581:

> There is in this city a monastery of the Order of St. Francis; the monastery building, cloister, and church are very sumptuous and well made.... As the great

> confluence of natives could not fit within this church, next to it, and within the same circuit of walls, they made a big *cassilla* [meaning a portico facing on to a *patio*], almost square and supported by many arches, and with towers at its ends. After it had been vaulted, in order to celebrate a feast in it impressively, the centering of the arches and vaults was taken down. However, because the mortar had not fully set, that night all the vaulting fell to the ground, leaving nothing standing but the walls. It was a miracle that God made it fall at night, for had it happened the day before, there would have been great havoc for there had been more than four thousand people inside the building. It has since lain in ruins because, since the Indians are now [due to a plague in 1576] so many less, they have not been able to take up the task of rebuilding. This structure had been the most sumptuous that the Indians have built hereabouts.[26]

Around 1590, Luis de Arciniega (1537–ca. 1600), a younger brother of the architect of San José de los Naturales, was appointed to supervise the reconstruction of the Capilla Real at Cholula. The work went slowly; in fact, the final version, presumably still closely hewing to Arciniega's original plan, was only completed in 1731. The Capilla Real is now famed for its mosque-like, quincunxes of tiled and aligned domes, with these presently presenting the densest constellation of domes to be seen in the Western world. (Fig. 71) Like San José, its inspirational source is, once again, "eastern," specifically meaning Islamic. Before it was covered up, the façade of the

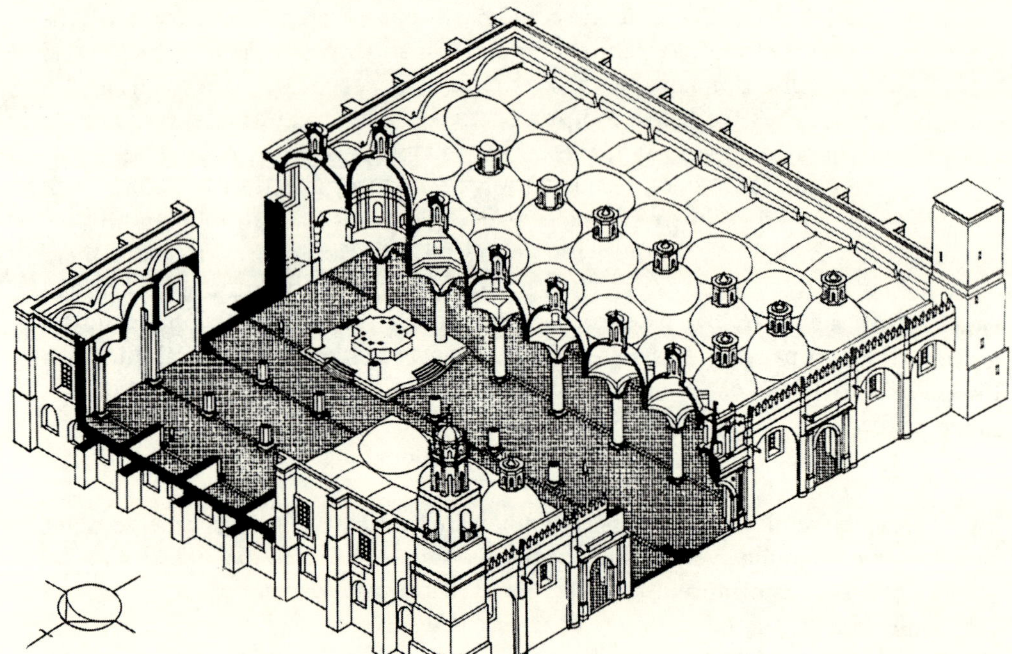

Fig. 71. Luis de Arciniega (attr.), the Capilla Real attached to the Monastery of San Gabriel de Cholula, as designed ca. 1590. Isometric reconstruction (as given in Sebastián, *Arte Ibero-Americano*, 1985.)

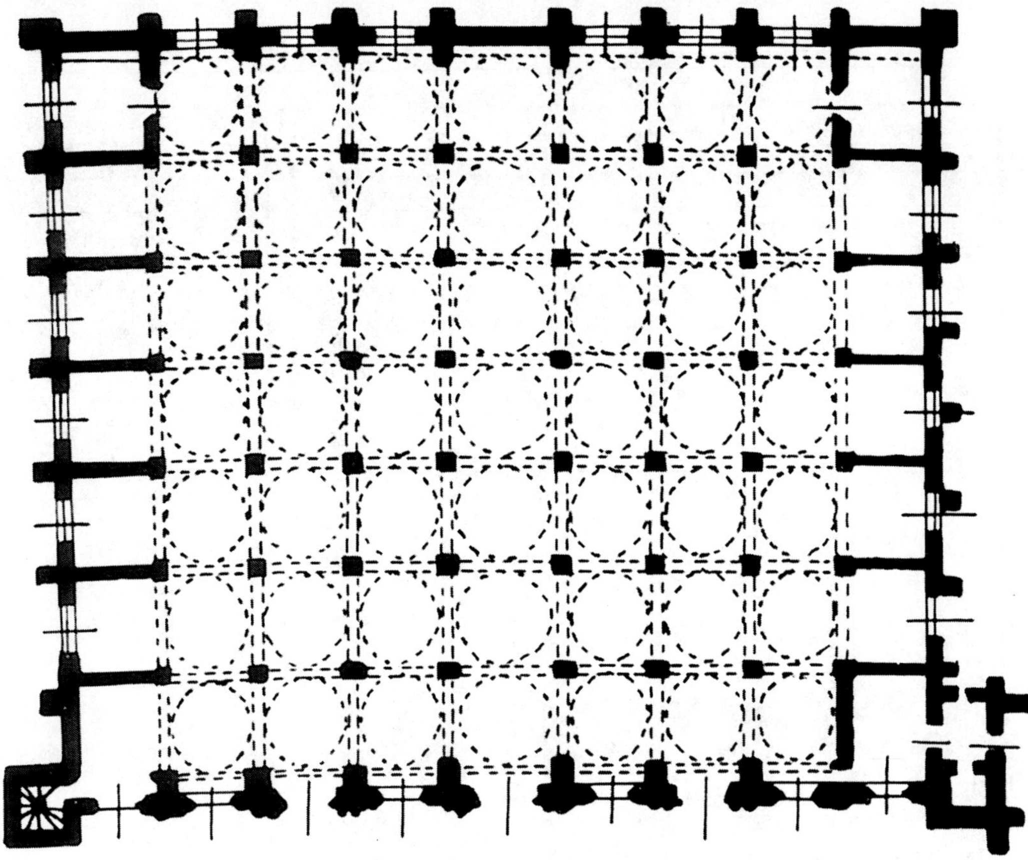

Fig. 72. Luis de Arciniega (attr.), ground-plan of the Capilla Real attached to the Monastery of San Gabriel de Cholula, as designed ca. 1590 (as given in Sebastián, *Arte Ibero-Americano*, 1985). Note: East is at the top of the plan.

Capilla Real must have looked very much like that one facing the *sahn* at the Great Mosque of Córdoba. For this assertion we have the testimony of Father Bernabé de Cobo; writing in 1630, he succinctly recognized the Capilla Real at Cholula as having been clearly built "in the style of the Mosque of Córdoba."[27]

Corroboration for its Islamic derivation is found in the still existing ground-plan of the Capilla Real. (Fig. 72) Presumably, its hypostyle-hall format faithfully replicates the lost model of San José de los Naturales. The arcades (as shown in fig. 71) run perpendicular to the altar wall. This decidedly atypical scheme exactly echoes the arcades running perpendicular to the *qibla* wall containing the *mihrab* in the Great Mosque of Córdoba, and such as these were initially laid out in 787. (Fig. 73) In fact, although built on a reduced scale, the overall scheme of the square-plan, multiple-bay system operating at the Capilla Real at Cholula, including its open-air atrium attached to the north, generically replicated the archetypal format of the Great Mosque of Córdoba, and such as this was known to Spaniards in the sixteenth century. In its turn, the Capilla Real spawned other replications in Mexico.[28]

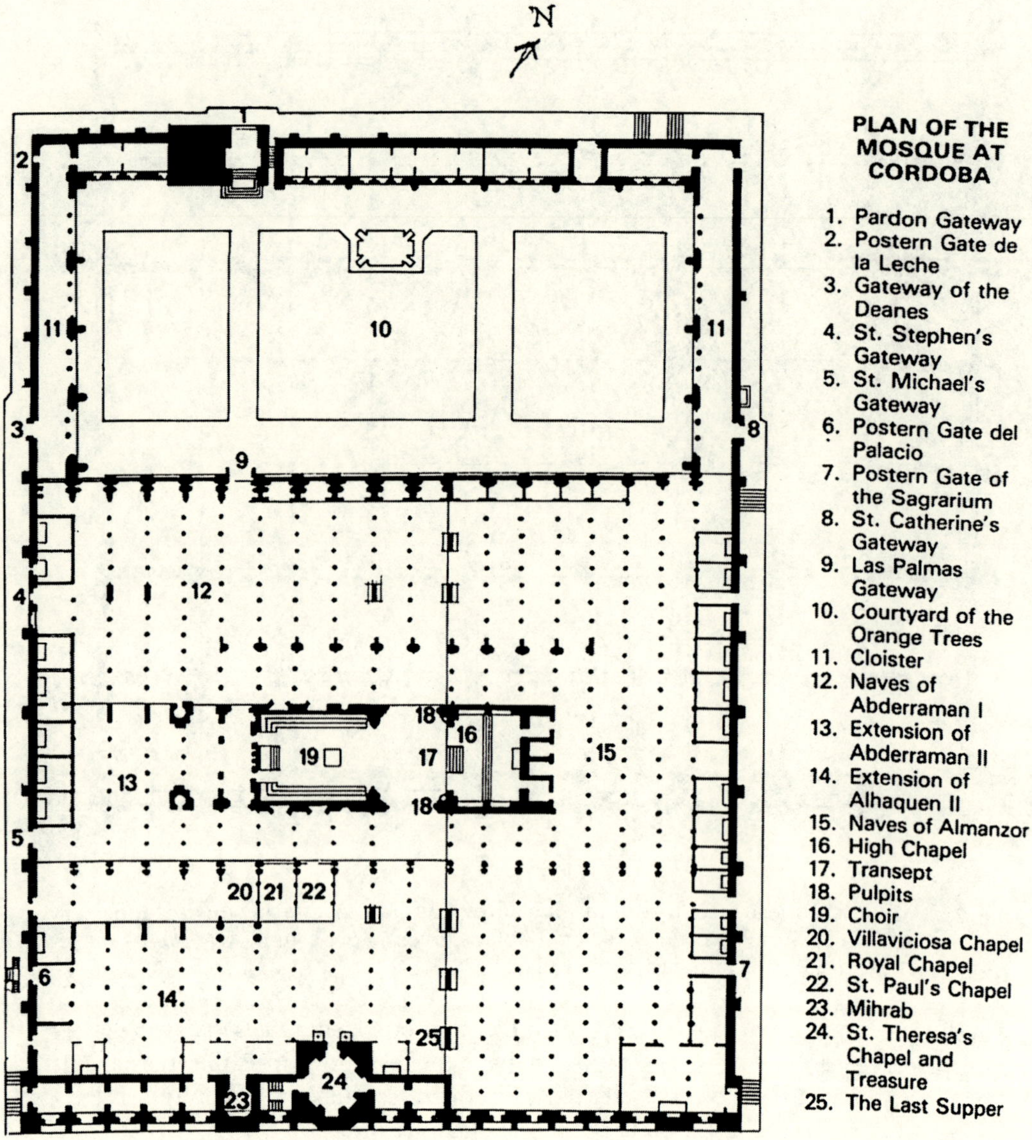

Fig. 73. Plan of the Great Mosque of Córdoba, Spain, including the Cathedral installed in 1523.

Since we are fortunate to have a measured architect's drawing (fig. 72), in this case we can further demonstrate that the rudimentary proportionate ratios determining the *plano* of the Capilla Real were themselves arrived at by means of a simple application of an equilateral-triangle design module. Put simply, the interior of the church is bisected by a central passage, a processional walkway, with this running from west to east, towards the high altar. Flanking this corridor are two matching naves built to accommodate the crowds of eager worshippers, and each of these colonnaded halls measures three bays wide by eight bays deep.

To create the interior of the Capilla Real, the architect first drew a square. Bisecting this elementary figure, he obtained two diagonals. Along each of these diagonals, he placed eight, equally spaced tick marks. Running diagonally from one corner of the square to the other, these points established the footings for all of the columns defining the aisles. By further dividing each tick mark by one-half, the designer then found a compass-point for each dome. The operation was repeated by drawing another diagonal exactly parallel to the first one, with this second line being placed one tick mark distant from the first. This process was to be repeated six more times. The accuracy of the proportionate scheme would be checked by drawing parallel diagonal lines in reverse, that is, at right angles to the first set. Once the criss-crossed pattern for the nave was fixed, then it was a simple matter to extend the pattern by one unit, so marking out the side chapels.

Another rudimentary geometric scheme may be deduced. First subtracting the transverse aisle running south to north, one then divides each of the flanking halls in two, so finding that each half then becomes a symmetrical figure of 3:3. For instance, when initially anchored in the center of the compound pier placed at the near edge of the opening of the bisecting passage-way, the bottom line of this diagonal inscribed line touches the outer wall at the fourth pillar; the same effect results if the initial point fixed (or "flipped" with a template, as shown in fig. 7) at the eastern end of the of the passage. When diagonal ("inscribed") lines are repeatedly drawn by hand over the exact centers of all the supporting columns defining the parallel aisles, the result is a web of eight, exactly parallel diagonal lines, and the points of intersection define the width of each bay.

Ironically, while the various successive versions of San José de los Naturales were being built, especially those following the influential mosque solution evidently datable to 1559, with this later spawning the imaginative replications at Cholula and elsewhere, the closest Spanish prototypes were being destroyed. The Great Mosques at Seville and Granada were eventually wholly obliterated by the Christian cathedrals laid over them. While the Great Mosque of Córdoba has survived intact for the most part, its interior was then being mutilated by Christian efforts to make it into a more usable church, in fact, the metropolitan Cathedral, inserted there in 1523 (nos. 16–19 in fig. 73). What he perceived to be fumbling, insensitive efforts prompted even the Emperor Charles V to exclaim to the Cathedral Chapter, "If I had known what you had wished to do [to the Great Mosque], you would not have done it, because what you are carrying out there is to be found everywhere, and what you had there formerly does not exist anywhere else in the world."[29] Oddly, even the political leader of Christendom saw the architectural merits of the world of Islam. As this study has attempted to demonstrate using other kinds of historical evidence, the same sentiments must have been shared by the humble artisans of the trans–Atlantic Hispanic world, those ingenious but mostly anonymous *alarifes* and *albañiles*.

This affirmation brings to mind another historical irony. In fact, the Great Mosque of Córdoba had itself begun as a Christian church, the basilica of San Vicente, as constructed by the former overlords of post–Roman Hispania, the Visigoths. The story is best told by an Arab chronicler, Ibn Idari:

When the Moslems conquered al-Andalus [they initially decided] to share the churches with the Christians. They had already followed this practice in the church at Damascus and at other places. Thus, the Moslems shared with the Christians their Great Church of Córdoba [then named San Vicente] which was in the center of the city. The Moslems constructed their *masjid alhama* [prayer hall] in one part and they left the other half to the Christians; all the rest of the churches [in the city] were then destroyed. As the Moslem population steadily grew in al-Andalus, the Arab chieftains and their armies installed themselves in Córdoba. The [shared] mosque became too small, and even though some galleries were added on, the crowd suffered much discomfort due to its narrowness.

When Abd al-Rahman I took over al-Andalus and made Córdoba his capital, he decided to expand the mosque and to make it perfect. He summoned the Christians of Córdoba and asked them to sell him the part of the church which still remained to them. To comply with the treaty which they signed, as compensation he allowed them to rebuild the churches which had been destroyed during the conquest. So they left their church, and Abd al-Rahman then transformed it into a mosque. Both the dismantling of the church and the construction of the mosque began in the year 785. In 786, the walls were raised and the pavement was laid down. All this happened in one year, and it is said that Abd al-Rahman spent eighty thousand dinars [on the mosque].[30]

19

Some Triangular Conclusions

On the one hand, the strictly practical and mechanical procedures employed in the execution of designs for the façades at Tepotzotlán and Ocotlán (figs. 1, 24–28, 54–57) are "European," meaning that they were based on certain workshop operations common since (at least) the early Gothic era in central Europe. On the other hand, since it is a typically "Hispanic" edifice, the underlying general formal/stylistic design characteristics shown to have been operative at both Tepotzotlán and Ocotlán (among other Hispanic examples treated here: figs. 3, 23, 29, 31, 33, 36–53) additionally embrace a scheme of vertically ascending, planarized enframements, the arrangement of which is beholden to a rudimentary system of equally proportionate and modular, sequential additions.

These typically "Trans-Hispanic" traits are, as we have also revealed, unquestionably "Neo-Islamic" in their broad conception. Needless to say, none of the people involved with, for instance, Tepotzotlán — neither the Jesuit fathers, nor its unknown (Spanish?) designer, nor the humble (Mexican-Indian) artisans working under his direction — would have admitted, or even would have been aware of the now-evident fact of the specifically "*neo-islámico*" basis of their building. Nonetheless, and particularly as is shown by the mosque formats adopted for San José de los Naturales and the Capilla Real at Cholula (figs. 71, 72), wholly Islamic architectural solutions had been eagerly embraced in Latin America since the sixteenth century.

Unfortunately, the actual historical origins of the simple design technique employed in Moslem Andalusia (figs. 9–15), and later throughout Latin America (figs. 1, 24–28, 54–57, 71, 72), had (and have) been long since conveniently forgotten.[1] Nonetheless, it now becomes fairly obvious that a 3:4:5 triangular format, with this determining the basic proportionate relationships found in so many Spanish and Latin American façades, had long since become a traditional workshop procedure in the Hispanic architectural repertoire. As our graphic evidence shows (figs. 24 ff.), the procedure was common to generally literate *alarifes*, also to the generally unlettered *albañiles* and *carpinteros* working under their direction, all of whom used these simple but infallible means to design many Christian churches in Iberia and Latin America. The solid (even if sparse) documentary evidence to support that claim has also been laid out here.

As to the ultimate sources for all this, a final question may be posed, and then logically answered. Not only did Vitruvius (IX, proemio, v–viii) laud the figure we call the "Pythagorean figure," or 3:4:5 triangle (fig. 63), but he even went so far as to claim that "when Pythagoras discovered this, in the belief that the Muses had inspired him in the discovery, he is said to have thanked them and sacrificed victims to them." Use of the 3:4:5 regulating line-pattern was common centuries afterwards, for example, such as it was illustrated by Cesariano's schematic overlays of Vitruvian figures imposed upon the Gothic fabric of the Cathedral of Milan (figs. 16, 62). Nevertheless, neither does Vitruvius' praise of the Pythagorean figure, nor does its use by Gothic builders much later, account for its initial appearance in an important Hispano-Islamic construction like the Great Mosque of Córdoba (figs. 9–15, 73). If not from Greco-Roman architecture, then where did the Moslem builders derive their continuing interest in this figure?

Another Roman writer, Plutarch, provides the answer (as is also visually demonstrated in the Appendix). As he tells it, the so-called "Pythagorean triangle" had been popular in Egypt since time immemorial (and, from other sources, we know that its prestige probably arose from the Pharaonic ceremony called "The Stretching of the Cord"). Therefore, Islamicized Egypt must be the specific geographical source for the geometrical figure that was to be reiterated in Córdoba and many other mosques in Moorish *al-Andalus* as a unifying compositional leit-motif distinctive of Hispano-Islamic architecture. As Plutarch put it in his *Treatise on Isis and Osiris* (ca. 110 AD),

> And there are good grounds for supposing that the Egyptians intended to compare the nature of the universe to the triangle, which is the most beautiful of all figures, and which Plato seems to make use of to the same effect in his books on the *Republic*, as composing a nuptial figure; and this triangle is such that the side which makes the right angle is three units in length, the base four, and the third [diagonal] line, which is called the hypotenuse, is five, whose power equals the powers of the two others that form the right angle. Thus [according to Plato] we should compare the line that falls perpendicularly on this base to the male, the base to the female, and the hypotenuse to that which is born of the two.[2]

Whether these kinds of grandly cosmic (or more narrowly connubial) considerations ever actually entered into the motivations of the technically proficient *alarifes* of Tepotzotlán and Ocotlán (figs. 1, 24–28, 54–57) is, of course, debatable (also rather unlikely). Nevertheless, recognition of the ubiquitous figure of a 3:4:5 triangular figure ordering the façade of an obscure Jesuit church in the Mexican highlands, at Tepotzotlán, does unquestionably betray its compositional kinship with the heterodox beliefs of the Islamic builders and designers of the Great Mosque of Córdoba in *al-Andalus* centuries before (figs. 9–15). Centuries later, one's own nimble digital pursuits with strictly postmodernist computer software reveal that the venerable neo–Islamic design procedure is, even today, to be found throughout Latin America (figs. 36–39, 72), also (as one should expect) in Spain itself (figs. 29–31, 40–53)—and even in famous monuments where its distinctive presence had never before been suspected.

There is, admittedly, a limit to the applicability of this procedure. Put simply, this is because most buildings are not (often can not be) photographed exactly perpendicular to the viewer; additionally, there will often be distance distortion, especially noticeable at the top of tall buildings. As a result, there will typically be parallax distortions in most published photographs (for instance, as in fig. 1, which I have mostly corrected using the Photoshop "transform" procedures). Even when the cyber-researcher is working with "transformed" photographs, in which some distortion inevitably remains, nonetheless the *consistency* of such geometrical alignments, particularly those typically found at the parts closest to the photographer, reveals the presence of the hypothesized geometrical operation.

However, were they available, I would have greatly preferred to work with architect's scaled drawings of elevations and floor plans[3] and, especially, with modern, careful and accurately measured, highly detailed drawings of the combined sculptural-architectural ensembles placed on Hispanic façades. Unfortunately, the latter are rare indeed, but the reason for this lacuna is obvious: photographers are cheaper than are skilled draftsmen, the kind who work in the obsolete manual, labor-intensive, mode. Nevertheless, since my analytical tool — digitized grid overlays (used as "layers") — is very easily fabricated, then you can make your own and, with these in hand (or in your hard drive), most likely you will find other usable pictorial examples to corroborate my findings.[4]

Again, the consistency of the alignments discovered by my overlay operations, especially in those few cases where I have had an accurately measured drawing of an elevation to work with (figs. 3, 10, 12, 23, 28, 30, 31, 36, 43, 44, 46, 47, 49, 51, 70), shows that, indeed, a predetermined system of geometrical and inscribed patterns was essential to the initial design procedure. In effect, we are dealing with a once-ubiquitous, Hispanic architectural *topos*, one hidden to our eyes until now. The name for the tool then commonly employed to achieve such consistently harmonious results was the "*cartabón de cinco*" of the Sevillan *alarifes*. Another tool was the "*parilla*" or "*cuadrícula*," a grid, the kind drawn either on paper or with strings attached to a rectangular frame.

In his exhaustive 1953 monograph on the use of modules and geometry-derived proportionality in Córdoban architecture, Emilio Camps Cazorla also presumed the consistent use of *cuadrículas* with "inscribed figures" (*figuras inscritas*), and he illustrated the only piece of surviving graphic evidence supporting this logical assumption. This is the fragment of an equilateral triangle drawn (ca. 950) upon a wall at Medina az-Zahra (near Córdoba) in order to determine the placement of the voussoirs in a horse-shoe arch; evidently, radiating diagonal lines were drawn from the bottom *catheti* by means of templates.[5]

But that tantalizing, but only fragmentary bit of graphic evidence still leaves open the larger question: what would the architect's sketch for the entire elevation have looked like? As I believe, the appearance of the working drawing actually produced by the unknown Córdoban designer of the Portal of al-Hakam II to guide his gang of *al-banna* workers would have looked very much like fig. 10: a linear schematic of the projected design, over which a pattern of diagonal and vertical intersections was drawn. These schematic drawings then showed the craftsmen on the work site where to fix the necessary horizontal and vertical alignments upon the wall gradually ascending behind the wooden scaffolding. Once the wall reached its full height, then the individual components made in nearby workshops—decorative panels or sculptures—would be attached to their predetermined sites upon the vertical plane.

In the case of a much later Renaissance application of the same procedure, the designer's schematic, including "inscribed figures," probably looked like figs. 16–22. These are actual working drawings known to have been used by Renaissance architects and artisans. Therefore, in no way can these (see also figs. 59–61) be confused with those modern, hence speculative, recreations produced *a posteriori* by inquisitive scholars, some of which are quite elaborate. Typically, these modern reconstructions-after-the-fact are *too* elaborate, especially given the fact that a rustic artisan most likely could *not* have made heads or tails of most of them.[6]

To the contrary, and as based on the surviving visual evidence (figs. 4, 5, 16–22, 59–61), the initial step — the "*mapa*" — would have been a quick thumbnail sketch, one looking like fig. 23, and showing just the intersection of two massive, inscribed 3:4:5 triangles, with these determining, besides the vertical axis of the central façade, both the width of the entire building and the height of the flanking bell-towers. Only then did the architect, Juan de Herrara in this instance, prepare his second "layer" (as in fig. 40), the fine network of parallel diagonal lines ascending with 57° angles. These "secondary interections" in his grid determined the height of all the horizontal subdivisions, also the lateral spacings of all attached columns and window openings.

Due to a recent, almost archaeological, analysis of both the physical and the textual evidence, we are beginning to know a great deal about the actual means of production in Italy of pictorial artworks during the Renaissance, including the (occasional) melding of routine workshop practices to contemporary design theory.[7] Unfortunately, much (and similarly archaeological) research is still needed in the study of Renaissance architecture, even in Italy. As James Ackerman pointed out back in 1954, even in Rome during the High Renaissance, the practice of architecture was "not strong in theory," and "generally the solution of structural problems had to be left to masons and carpenters." Moreover, High Renaissance architects actually did "little designing." Although the clearly proportionate character of their buildings strongly suggests the use of preliminary sketches with inscribed designs, the physical evidence for this practice has since mostly disappeared; the rare exception are the drawings (and texts) by Francesco di Giorgio (figs. 17, 18).

Speaking of those now-lost schematic designs, Ackerman assumes that "very few of them were intended to be used in constructing a building, or even to be seen by anyone other than the architect. They are nearly all rapidly sketched studies of tentative ideas." An altogether different sort of graphic material were the highly finished "presentation drawings," but these, too, "are rare, and they would not have been much use for construction because they almost never include measurements or a scale." A second type of drawn architectural guide "was intended for use in construction, but it is limited to details—a window, an entablature — and was intended only to guide masons and carvers."[8]

Such graphic evidence attesting to actual design practice is even rarer in Spain — and practically non-existent in Latin America. Particularly in the case of the distinctive Hispanic architectural phenomenon, the big gap in our knowledge is not its "theory," but instead the routine craft practices and procedures that actually produced it, particularly the ubiquitous *fachada retablo*. These routine workshop procedures I have attempted to exhume here. As they are now revealed to have operated in the New World during the seventeenth and eighteenth centuries, these humble artisan techniques are to be further understood as *anachronisms*. Given the known dates of their usage, they must be so called, for they are as much "medieval" and they are demonstrably "Islamic."

Initially, this study began as strictly visual research, pure formal analysis. When a consistent pattern eventually emerged (following, naturally, many dead-end leads), then the strictly scholarly work began, the search for historical documents providing verbal confirmation for what was initially discovered just by pursuing the visual clues. Either way, as visual analysis or as archival research, it all boils down to "detective work" (and I cheerfully confess to being a diehard *aficionado* of the television drama called "Law and Order"). Detectives only pursue *facts*; doing so, naturally they must also follow many dead-end leads. To the contrary, when one is blinded by *theory*, some *a priori* notion, then what is in plain sight remains unsought, thus cannot be explained. Most "theories," for being typically unsupported by mundane *facts*, would never hold up in a court of law (and to see how those courts actually operate, again see "Law and Order").

Since the same results (or revelations) should be forthcoming from any interested person adept with a either PC or a Mac, its scanner and the software appropriate to it, said results (conclusions) are *repeatable*. And one really should check my results by employing *alternative* grid-solutions, for example, the equilateral triangle, or "lozenge-grid," variant (as shown operating in figs. 8, 58).[9] Or, for that matter, one may challenge my findings by coming up with any *other* explanation (and, unquestionably, an explanation is called for). Whatever that *other* explanation may be, *it must, in the end, be wholly compatible with the needs of simple artisans*, the ones who actually did the work at the building site. That is the bottom line, the minimum requirement for credibility. Put another way, my reconstruction of those long-lost workshop procedures would now allow nearly anyone (me, too) to design the same kind of buildings.

As we may again recall in closing, "repeatable results" is another way of saying "scientific proof." Among other discoveries, we have provided the graphic — also textual — evidence attesting to a hitherto unsuspected Islamic presence, one which is now made visible throughout Latin America in its most characteristic architectural monuments. *Quod erat demonstrandum: ars sine scientia nihil est.*

Appendix: Viollet-le-Duc Explains Architectural Proportionality and the Design Function of the Pythagorean Triangle

Eugène-Emmanuel Viollet-le-Duc (1814–1879) exercised a major influence on nineteenth-century architecture, especially by his comprehensive discussions of theories of rational design. Especially interested in the practice of architectural restoration, Viollet-le-Duc was initially trained in the design techniques of medieval structures in a wholly practical manner during the renovation of Saint-Germain-l'Auxerrois (1838). He later worked on the restoration of such important Gothic structures as, among others, Sainte-Chapelle in Paris, Nôtre-Dame de Paris, the Abbey-Church of Saint-Denis and the Cathedral of Amiens. First published in 1872, his illustrated architectural treatise, *Entretiens sur l'architecture*, profoundly influenced the development of modern "organic" construction techniques and forthcoming "functional" concepts of design. For our purposes, his most important contribution is an extended discussion of the fundamental importance of architectural proportionality, also including detailed analyses of the specific role played in that process by the Pythagorean-triangle motif.[1]

The specific technical issue analyzed here has been *architectural proportionality.* As shown in our examples—no matter whether found in either Spain itself or in Latin America—typically, Hispanic façades had been profusely decorated, at times to the point of *horror vacui.* Nevertheless, and has been noted by all students of architectural design, the overall effect consistently remains one of a finely-tuned, proportionate distribution of all the parts, from the largest to the smallest components. Unfortunately, to date no scholar has succeeded in showing the actual workings of a simple mechanical process employed to ensure the generation all that intricately reiterated proportionality which is so distinctive of Hispanic architectural design. Nonetheless, although never making references to any Spanish buildings, Viollet-le-

Duc did explain the workings of all the technical issues which we have already examined here in their specifically Hispanic — also Islamic — context.

As he explained, when we find ourselves discussing the use of proportionality in ancient buildings, in effect we are dealing with a *mystery*, for the task at hand is the recovery of once commonplace, but now wholly forgotten, technical procedures:

> In order to show the real position which is offered to the architect in the middle of the nineteenth century, we shall have to examine the nature and extent of the acquirements which the architect of the present day ought to possess. These requirements are of two kinds, some theoretical, the rest purely practical. [...] No one disputes the fact that an extensive knowledge of geometry is the ground work of all architectural labors. Now the study of archaeology proves to us the application that had been made of geometry to [historical] styles of architecture which are very different in appearance. [...] The architect ought not only to possess a large acquaintance with descriptive geometry, but also to be sufficiently familiar with perspective to be able to draw a design, or parts of a design, in every aspect. Perspective should be a practical science to him, and to such a degree that, in designing his geometrical projections, he can realize in thought the effect which the prominent or elevated parts will produce at different planes of distance, and according to the character of the site, the declivity of the roofing, the thickness of the walls, etc. When the horizontal plan has been determined, the architects of former days were accustomed to prepare a set of perspective elevations, and in this they proceeded wisely and spared themselves deceptions. [...]
>
> We all know what has been said and written respecting architectural proportions since the time of Vitruvius and, perhaps, before him. But all that may be summed up in this principle, namely, that there are proportions which were regarded as beautiful in ancient times, and that we cannot do better than recognize them as such now. We certainly observe the proportions utilized by the Athenians for the architectural orders, and those were proportions independent of their [actual] dimensions. [...] The Greek artists, who were so highly gifted, appear to me to have established a harmonic system, but not a formula, as did, later on, the Romans, who were veritable engineers. Going further back, and examining the monuments of Ancient Egypt, we also recognize the influence of a harmonic method. [...]
>
> Incontestably during the Middle Ages, and probably also in Classical Antiquity, certain methods were adopted, according to which the proportions of buildings were pre-determined. We have [today] but little information on the subject; the loss of traditions and the utter decay of official instruction have allowed the clue, with which the professional guilds were formerly so familiar, to slip from our hands. [We have lost those workshop techniques] which formerly guided architects through the labyrinth of "mysterious" branches of knowledge, those architectural methods employed by our predecessors in their art and which enabled them to produce masterpieces. [...] We are [presently] apt to revenge ourselves on knowledge which we do not possess by despising it, but in the nineteenth century contempt is not demonstration. For [now-forgotten] geometrical methods, which had been elaborated and consecrated by long experience, we have substituted empirical formulas whose origin and reason it is absolutely impossible to assign, precisely because these formulas have been transmitted to us only at second or third hand. [...]
>
> In his *Nouvelle théorie déduite du texte même de Vitruve* (Nîmes, 1862), M. Aurès has lately published some very singular results respecting the relative proportions of the [classical] orders. This author proves with demonstrations that the Greeks took

their *modulus* at about the center of the column—and not at the base, as was formerly suspected. He thus succeeds in proving mathematically the exactness of the measures given by Vitruvius. Whereas it is useful to know the actual proportions which the ancients gave to the orders, it is probably even more useful to discover what were the generating proportions adopted throughout Classic architecture, also in the Middle Ages, and even in Renaissance architecture—although the latter often loses sight of the pure traditions, so giving itself up to caprice. It would be self-deception to imagine that proportions in architecture are suggested by instinct. There are absolute rules and geometrical principles, and if these principles accord with ocular instinct, it is because sight is a sense resembling hearing, which cannot encounter a discord without being offended, however little one may be acquainted with music.

It would certainly be strange if architecture, which is the offspring of geometry, could not demonstrate geometrically how it is that the eye is annoyed by a defect in the proportions of a building. We must therefore consider the question from a higher and, especially, a more demonstrative point of view. The *triangle* was regarded by the Egyptians, who were very good geometricians, as a perfect figure. The equilateral triangle is particularly that figure which completely satisfies the eye. It presents three equal angles, three equal sides, and a division of the [encompassing] circle into three parts; a perpendicular dropped from the vertex divides the base into two equal parts, and also permits the formation of a hexagon inscribed within the circle which divides it into six equal parts. No geometrical figure affords more satisfaction to the mind, and none fulfills better those conditions that please the eye, that is, regularity and stability.

Now we can see that the equilateral triangle was employed by the Egyptians to give such a proportion to important parts of architecture as should please the eye. If they had to construct pillars destined to carry a lintel, leaving (as is often the case in their oldest edifices) as many voids as solids, the proportions of the height of these pillars in relation to their width and to the voids was often determined by a series of equilateral triangles (see drawing A). [Fig. 74] These piers are so divided that either the [vertical] axis of each of them meets the apex of the [inscribed] triangle, as is seen at points *a–a*. Or, if more slender dimensions have been intended, then the base of the equilateral triangle will never go beyond one of the sides of the pillar, just as the drawing indicates at point *b*. Thus they satisfied a requirement of the eye, which demands that the part borne and that which bears it shall not go beyond the angles of an equilateral triangle. If we run counter to this principle, as indicated [below] at point B, we then abandon the conditions of a good proportion. The eye will then remark a want of stability, for it requires that the axis—placed at point *c*—should find laterally two solid points of support, at an equal distance at the base of the triangle, as found at points *d* and *d'*.

For example, if similarly inscribed within an equilateral triangle, the façade of [Christian] basilica, which is a building consisting of a nave and two side-aisles, will give a satisfactory proportion in its ensemble. [Fig. 75] If openings have to be made in the gable of the church, and if these are arranged so as to be inscribed within equilateral triangles, these voids will be pleasingly proportioned in reference to the façade. The eye will instinctively draw the [inscribed] lines—either *a–a'–b*, or *c–a'–d'*, or *a–c'–d*, or *e'–a'–f*—without finding any of them traversed by voids; in fact, to the contrary, the placement of those openings are to be fixed by them. And this requirement of the eye will be in harmony, as it always is, with the rules of stability.

The Greeks were not unacquainted with this simple principle [of proportional-

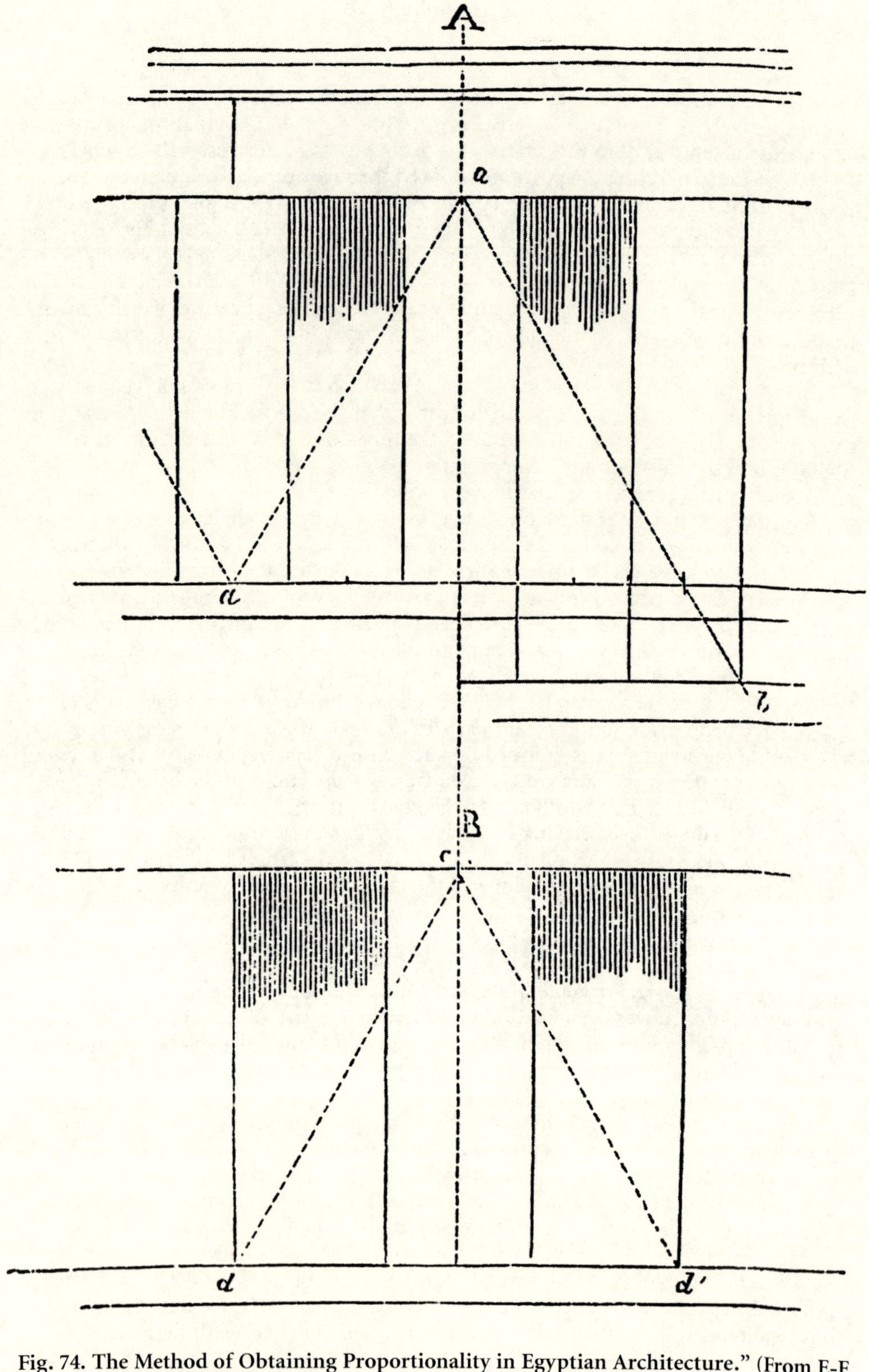

Fig. 74. The Method of Obtaining Proportionality in Egyptian Architecture." (From E-E. Viollet-le-Duc, *Entretiens sur l'architecture*, 1872, "Lecture IX," fig. 1.)

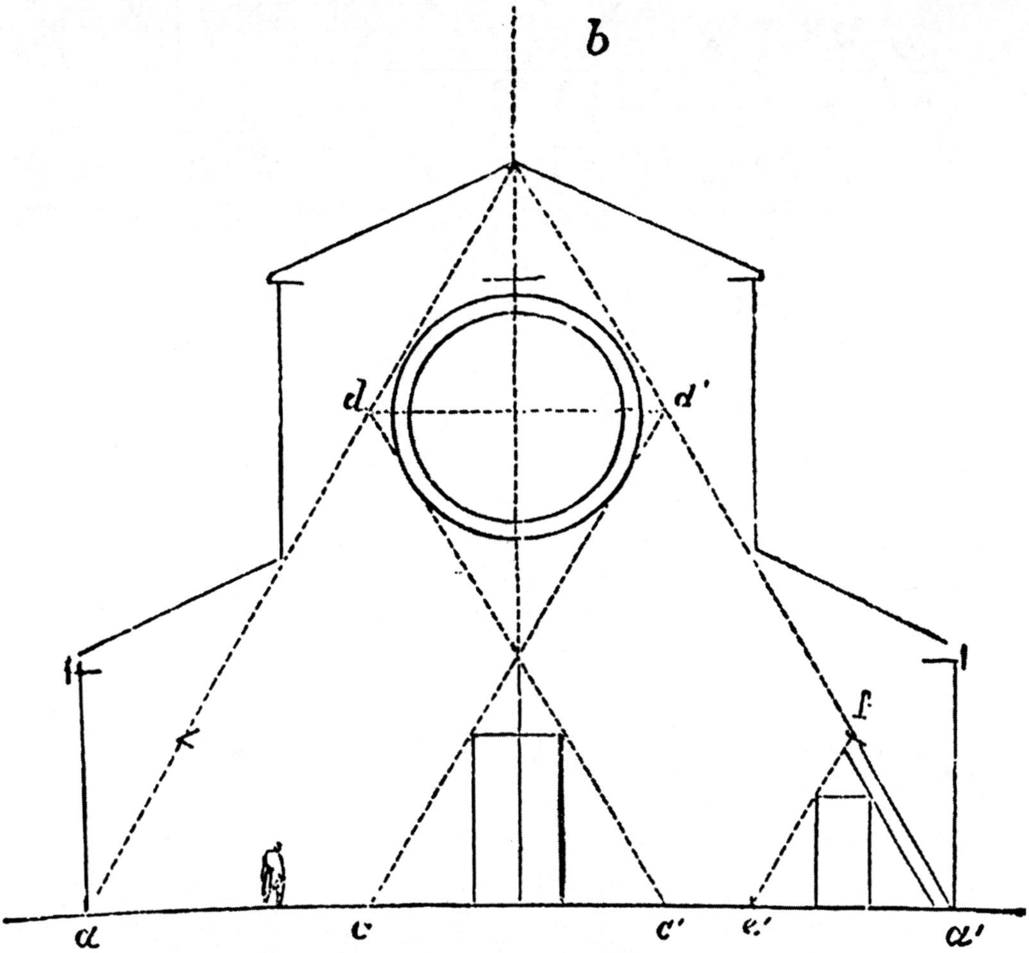

Fig. 75. "A Method for Obtaining Proportionality in a Christian Basilica." (E-E. Viollet-le-Duc, *Entretiens sur l'architecture*, 1872, "Lecture IX," fig. 2.)

ity]. Thus, if we analyze the [Doric] order of the Temple of Corinth (see drawing A [in Fig. 76]), we find that the [inscribed] equilateral triangle, whose vertex is placed under the abacus [the top of the capital] located at point *a*, has for its two other angles the centers of two adjacent columns erected above the base-axis, line *b–j*. When the Dorians wished to secure a more slender proportion, as they did in the order of the Temple of Concord at Agrigentum (see drawing B [in Fig. 76]), they allowed the angles of the equilateral triangle to project to the exterior edge of the shaft of the column. Or, if they decided to give a greater width to the inter-columniations, as they did in the order of the Temple of Aegina (see drawing D [in Fig. 76]), they selected the upper (rather than the lower) side of the abacus as the vertex of the equilateral triangle. The equilateral triangle is not the only figure that has been adopted as the generator of proportions. The pyramid with a square base—whose vertical section from the vertex, parallel to one of the sides of the base, gives an equilateral triangle—had also been used [by the Egyptians]. The section of a pyramid through one of the diagonals of its base will give the triangle *c–d–e* (see

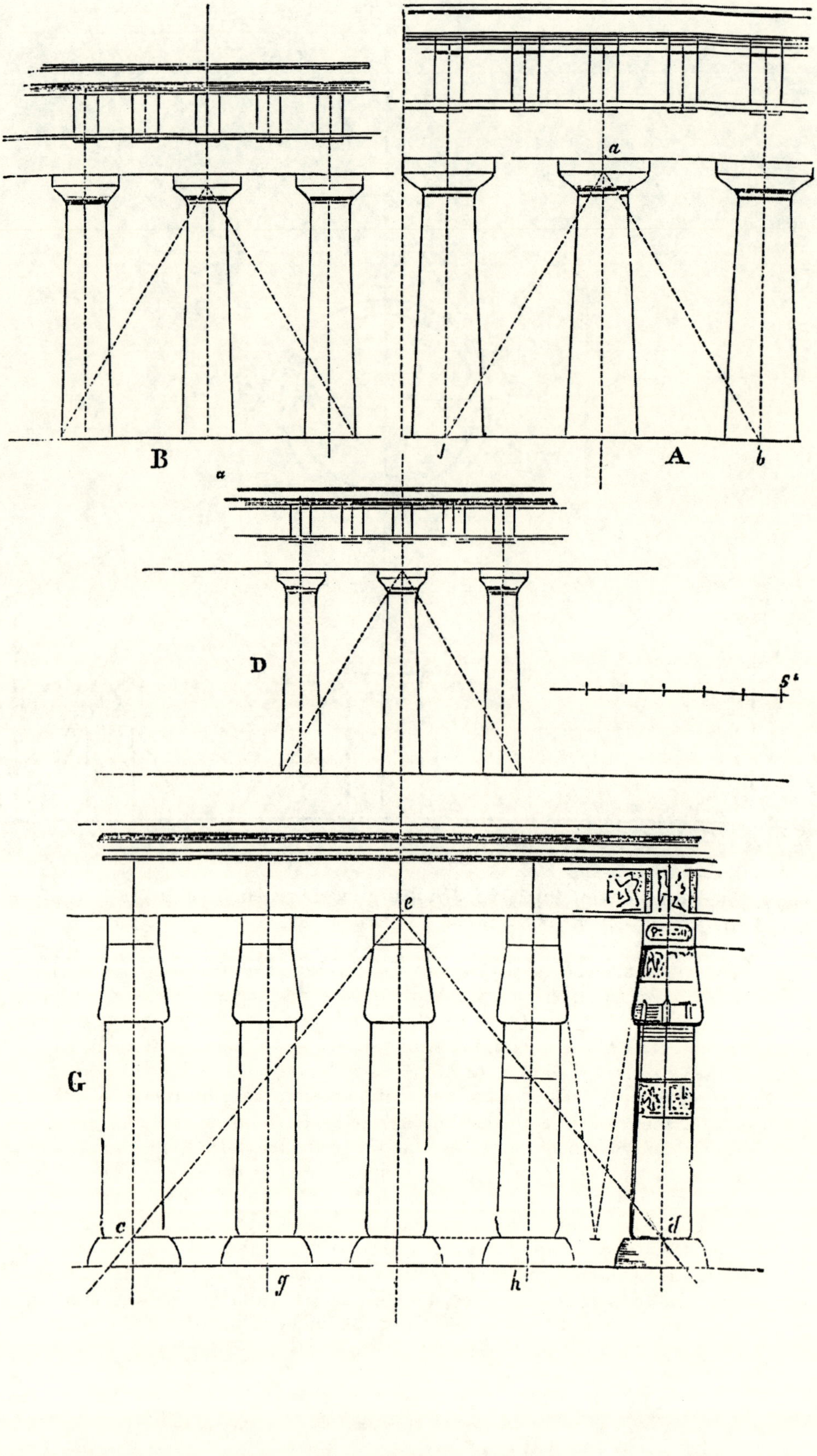
B
A
D
G

drawing G [in Fig. 76]), and this triangle was adopted in some colonnades in the ancient monuments of Egypt, and especially in the portico of the Temple of Khons at Karnak (20th dynasty).

This equilateral triangle — drawn on the diagonal of the square base of a pyramid whose vertical projection, parallel to one of the sides of the square — appeared to give the inclined lines that were most pleasing to the eye. [...] If we apply this equilateral triangle to the Parthenon, for example, we are led to curious observations. [Fig. 77] We see that this [inscribed] triangle fits exactly between two vertical lines falling from the middle of the exterior line of the angle columns and the extreme apex of the pediment. The point where the [diagonal] sides of this triangle meet the lower line of the architrave — along point B — fix the two [vertical] axes of the third columns, to the right and left [of the central axis]. If we then divide the [interior] interval *a*–to–*b* into three equal parts, and then mark off one of these divisions on the right and on the left, we shall have then determined the axes for the six central columns. The [foot of] angle A of the triangle gives the foot of plumb-line B falling from the architrave. The horizontal line C–D, as drawn at the intersection of the [diagonal] side of the triangle with the axis of the second column, gives the height which has served to fix the relative proportions of the edifice — in a word, its *modulus*. [...]

Of all architectural monuments, those which are isolated, and whose character is not so much that of utility as of art, in and for itself — for example, the triumphal arches built in such numbers by the Romans — should most endeavor to present a perfect harmony, the result of an attentive study of proportions. [...] The Arch of Titus at Rome, small as it is, presents an agreeably proportioned mass, one which completely satisfies the eyes. [Fig. 78] Let us then inquire into the principle of proportion adopted in the design of this structure. Here the equilateral triangle was the generator. The keystone of the arch is placed at the vertex of an equilateral triangle, whose base *a–b* is determined by the distance which separates the two axes of the two piers. The space of the opening *c–d*, up to the springing of the arch, at *e–f*, forms a perfect square. The lower lip of the cornice passes over the vertex of another [inscribed] equilateral triangle, of which the diameter *e–f* forms the base. The lower bed of the attic-cornice passes over the vertex of an equilateral triangle. whose base *g–h* extends over the entire width of the structure on the upper level of the bases of the columns, and whose profile, continued around the pier, presents a well-marked horizontal. The square niches placed between the columns attached to the face of the piers have their lintels placed upon the [intersecting] vertices of [other, inscribed] equilateral triangles; extending downwards, their diagonals establish the base for the inter-columniation. The dedicatory tablets placed above these niches do not rise above the apex of the equilateral triangles, whose base is determined by the width of the piers making up the compound-columns. [...]

Take, for [another] example, the Basilica of Constantine at Rome. [Fig. 79] On the transverse section of this edifice we place the [inscribed] triangle ABE; we see that the diagonals AB and AC fix the axes of the outer walls, B and C; they also establish the height of the cornice of the great order by their intersection — at points D and F — with the face of the walls pierced with great arches forming the aisles. These sides — AB and AC — also give the springings, G and H, belonging to the

Opposite: **Fig. 76. "Proportion by Triangulation in Greek and Egyptian Architecture: A) The Temple of Corinth; B) The Temple of Agrigentum; D) The Temple of Aegina; G) The Temple of Khons at Karnak." (E-E. Viollet-le-Duc, *Entretiens sur l'architecture*, 1872, "Lecture IX," fig. 3.)**

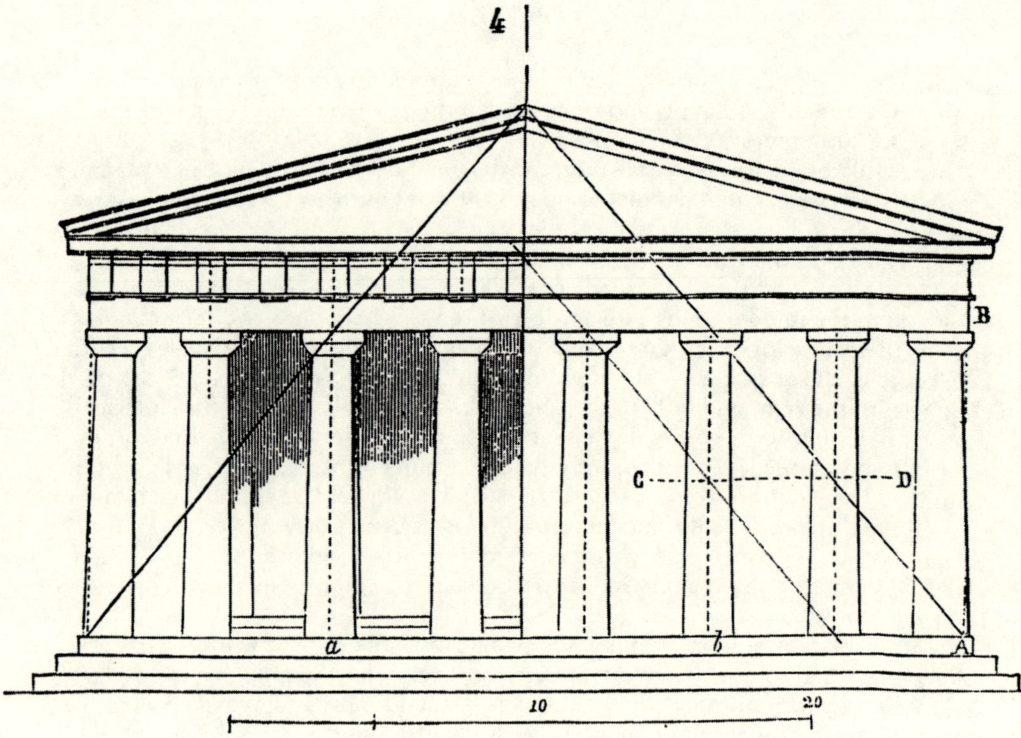

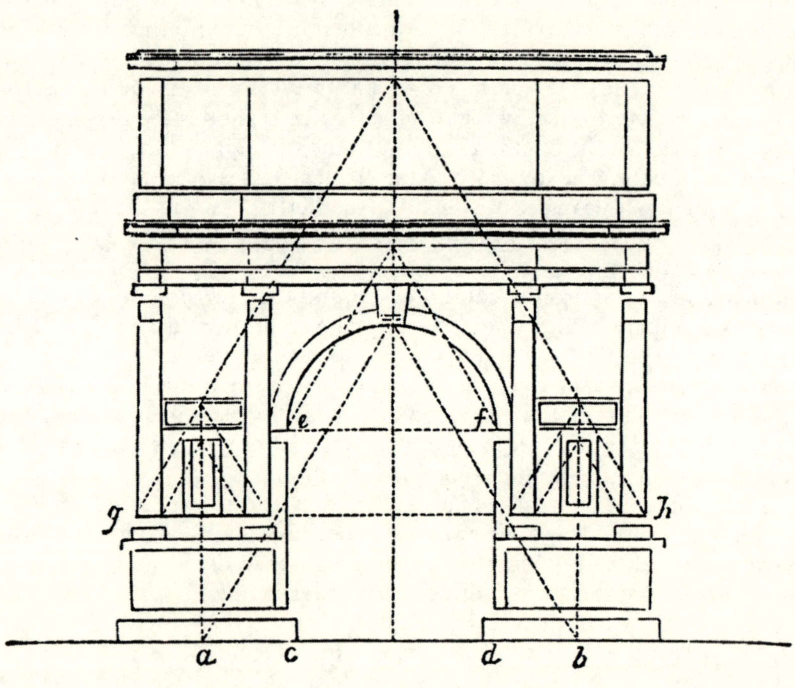

Top: Fig. 77. "Triangulated Proportionality in the Parthenon." (E-E. Viollet-le-Duc, ***Entretiens sur l'architecture***, 1872, "Lecture IX," fig. 4.) ***Bottom:*** Fig. 78. "Triangulated Proportionality in the Arch of Titus." (E-E. Viollet-le-Duc, ***Entretiens sur l'architecture***, 1872, "Lecture IX," fig. 5.)

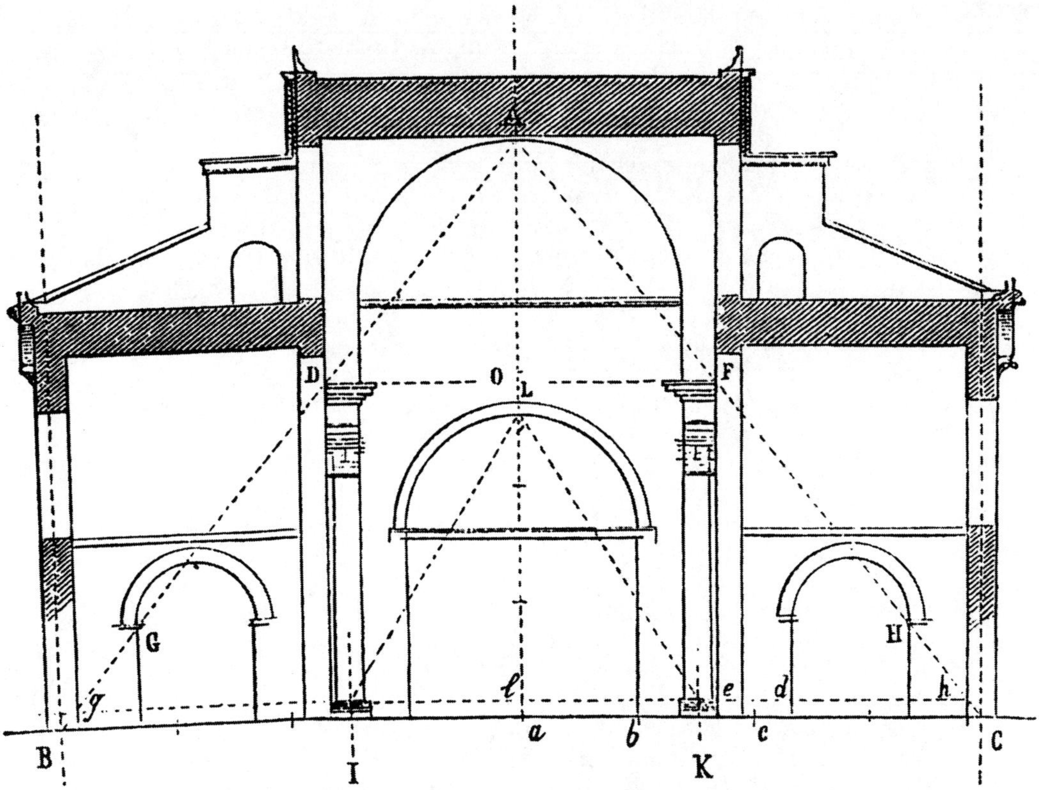

Fig. 79. "Triangulated Proportionality in the Basilica of Constantine." (E-E. Viollet-le-Duc, *Entretiens sur l'architecture*, 1872, "Lecture IX," fig. 8.)

smaller arches of the side-aisles. Two giant columns, I and K, are placed to rise from above the bases of the axis of these columns, so forming an equilateral triangle, I–K–L; thus we find the height of the *intrados* of the keystone belonging to the arch of the tribune-arch. Taking the half, *a–b*, of one of the four divisions of the base of the [greater] triangle ABC, we find the bases for the arch. Point *c*, which divides one of these sections at the quarter, yields the axis of the pier *e–d*. According to this drawing, the width between the two wall-faces found at the [upper] level—DF is the same as the height—OA, just as is the width of the wall-faces *g–h*, as taken from above the level of the base, is the same as the height *l*–A. Here, as in the Parthenon and the Arch of Titus [Figs. 77, 78], the eye finds points of stability at [for instance] points A and D and G and at points *g*, which are so many landmarks taken on the [diagonal] side of a triangle which the ancients regarded as perfect.

As Viollet-le-Duc finds, *all* deliberated applications of architectural proportionality did begin with the ancient Egyptians. Moreover, not only was the basis of their proportionality a *triangle*, but he adds that the most desirable *modulus* of all was—specifically—the one which we have called the "Pythagorean triangle." Here is how Viollet-le-Duc explains the exact workings of all the technical issues involving the same 3-4-5 modular figure which we have previously examined here in its specifically Hispanic—also Islamic—applications.

The importance of the subject [i.e., the ancient sources of architectural proportionality] justifies reference to Plutarch, who says [in his *Treatise on Isis and Osiris*]: "And there are good grounds for supposing that the Egyptians intended to compare the nature of the universe to the triangle, for this is the most beautiful form of all. And, composing a nuptial figure, Plato seems to make use of the same effect. This triangle is so composed that *the side which makes the right angle is three lengths in height; the base is four, and the third line, which is called the hypotenuse, is five in length*; its power equals the powers of the two other lines that form the right angle. Therefore, we should compare the line that falls perpendicularly on to the base to the male, and the base to the female, and the hypotenuse to that which is born of the two." It is evident that the right-angled ["Pythagorean"] triangle — whose base is 4 units, and the square of which is 16, and having the side of 3 units, whose square is 9 — has for a hypotenuse a line of 5 units, whose square is 25 (16 + 9). We shall presently return to this [triangular] demonstration, whose importance [in architectural design] is considerable, as shall be seen. [...]

Let us now return to the text of Plutarch, as quoted above. The Great Pyramid of Cheops at Giza was designed according to the method given by this author. It will be necessary to give [a graphic] demonstration here. [Fig. 80] At the lower extremity, point B, of the line AB, which is divided into four parts, there is erected a perpendicular line, BC, which is equal to three of the divisions of the [horizontal] base-line, AB. To this, we join [diagonal line] AC, which forms the [new] hypotenuse; it contains five such parts (the length of the base plus one-quarter of this length). This is the [3–4–5] triangle *par excellence* of the Egyptians, and as described by Plutarch. From point D, which is the center of the base-line AB, we erect a perpendicular, and we give it a length equal to half the hypotenuse AC; this line, DE, will be the half of five parts, that is, two parts and a half.

Joining the points AE and BE by two [diagonal] lines, we have the [equilateral] triangle formed by the Great Pyramid of Cheops; the line DE is its height, and AB is one of the [four] sides of its square base. A line perpendicular to the hypotenuse, as drawn [upwards] from the angle B, will likewise give the height of this pyramid, and the [new, diagonal] line AF is equal to one of the [original] sides, either AE or BE. By continuing the diagonal BF up to the [inscribed] circle, within which the [complementary] triangle ABC is drawn, we get the [extended] line HB. Next, by letting fall a perpendicular from point F, as placed on the side BC of the [invisible] triangle, we get a [new, horizontal] length, FK. If we divide each of the four parts of the base, AB, into two lengths, and if we then divide each of these subdivisions into six parts, we get 48 units. Dividing the perpendicular BC in the same way, we get 36. Dividing the two and a half parts of the height DE, we get 30. If we proceed to divide the hypotenuse in the same manner, we get 60.

Now, 60 = 5 × 12; and 30 = 2 × 12 + 6 (the half of 12); and 36 = 3 × 12; and 4 and 8 = 4 v 12. We have, therefore, dimensions measurable by 4, by 3, and by 5 (also by 2½). If we divide each of the parts of the base-line AB into 100, we get 400; and by cutting the vertical BC in the same way, we get 300. In vertical-line DE, and operating in the same way, we get 250 units; and, for the chord BH, the number is 480; similarly, the part AF of the hypotenuse yields 320, with line FC giving 180, and the perpendicular FK gives 144 (or 12 × 12). Thus, by means of this [geometrical] figure, we obtain both decimal and duodecimal divisions. In such a system of proportions, the duodecimal system has the advantage of being easily divided into halves, quarters, and thirds, and the combination of the two systems applied to our figure gives useful results. Thus, the base-line AB, as divided by the duodecimal system into 48 units, has a proportional relation to the chord BH as divided by the decimal sys-

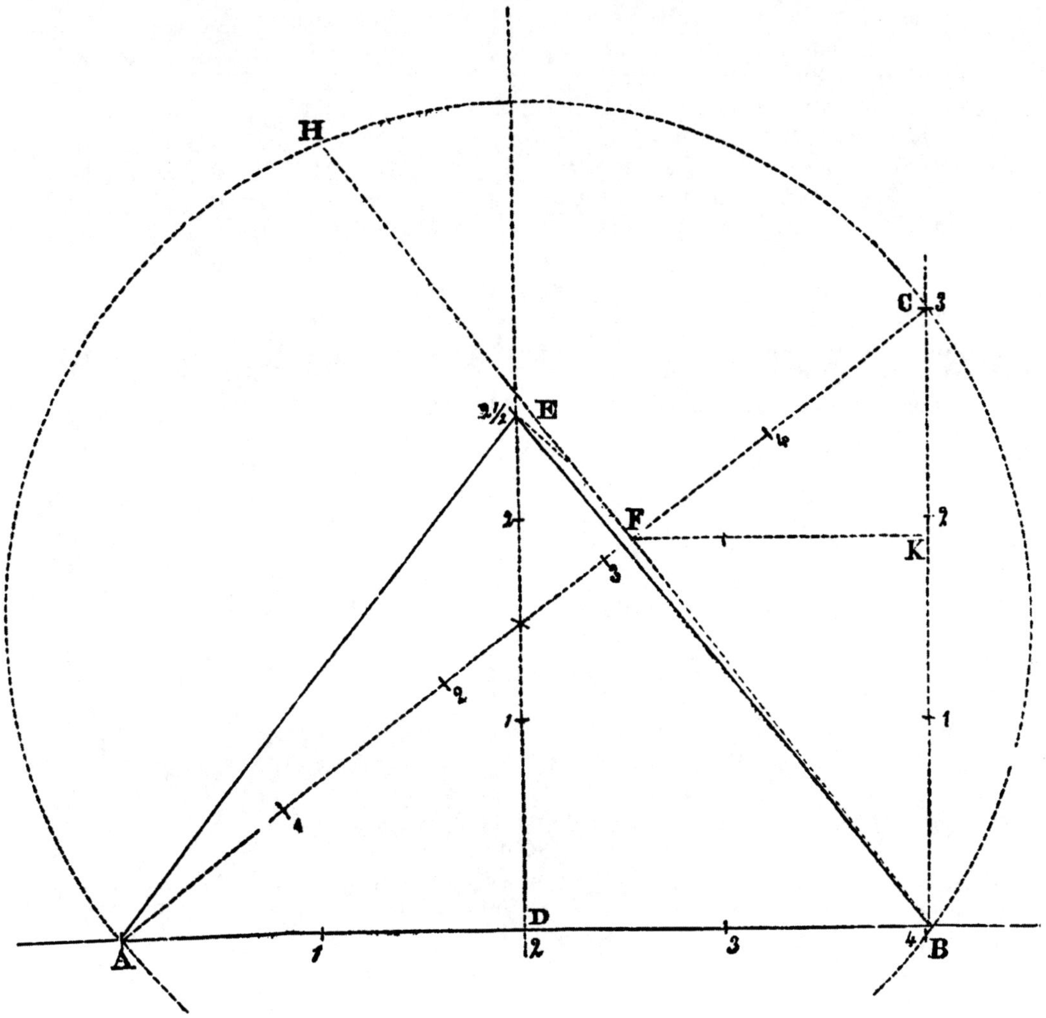

Fig. 80. "The Pyramid of Cheops with Inscribed Egyptian Triangles." (E-E. Viollet-le-Duc, ***Entretiens sur l'architecture***, 1872, "Lecture IX," fig. 7.)

tems, which gives 480 (or 48). Perhaps the architects of the [Hellenic] ancient world made use of this [Egyptian] figure; it is certain, as we shall see presently, that the masters of the Middle Ages made it [the 3–4–5 triangle-module] the generator of some of the great [medieval] buildings. [...]

One of the conditions to be fulfilled when we wish to give agreeable proportions to a range of architectural features is to avoid similarities, particularly equal lengths or surfaces in direct relations to each other. The artist [responsible for the Basilica of Constantine: see Fig. 79] had taken good care not to divide the diagonal line *g* to A into two equal parts. The ratio of diagonal g–D to diagonal D–A is 29 to 21, and [in this case] the Egyptian [or Pythagorean] triangle, taken as a generator of proportions, has the advantage — with its base being to its height as 4 is to 2½— of avoiding in its very principle those [identical] ratios of widths and weight, for those are the most displeasing of all. In fact, the eye comprehends dimensions only by contrasts.

Thus, the nave of a [Gothic] church appears lofty because it is narrow in proportion to its height, or [conversely] it appears wide because it presents greater width than height. But if you get a perfect ratio of width to height, you then will have the key to the proportions that should be applied to your building. The Egyptian [or Pythagorean] triangle was considered by the Ancients as perfect, and it must often have served as a standard of proportions. By its divisions it facilitated architectural design, as was shown above. Unfortunately, there are so few ancient Classic buildings complete in all their parts that it is difficult to apply this method with certainty to a great number of them in order to determine their [original] proportions [in detail]. Happily, this is not so with regard to the Middle Ages, during the period when art was passing from the cloisters into the hands of secular artists. At that time, whether because the architects were acquainted with certain ancient writings, or because certain workshop traditions were preserved in their original purity among them [as] a principle (or mystery) known only to the initiated or to adepts, the [geometric-modular] method was retained among them. It is certain that we find in their system of proportions, as applied by them to architecture, laws evidently derived from certain principles of Classical Antiquity, even though these [medieval] artists had no idea of imitating [literally] the forms of ancient architecture. The very starting-point of their system of construction was, as I believe I have demonstrated above, absolutely foreign to the system adopted by the Greeks or Romans.

Let us analyze one of the oldest buildings of the secular school [of architecture] that is really French, and in which Gothic form appears as early as the twelfth century. [Fig. 81] Let us take [as an example] a transverse section drawn through the nave of the Cathedral of Paris. The total width of the church being established, let [base-line] AB represent half that entire width; it is then divided into four, equal parts. Starting from the [vertical] axis-line A, the first [lateral] division gives the facing wall of the great nave, starting above the columns. The second division [unit 2] gives the interior face of the [centered] columns of the double aisle. The third unit [signalled at point P] gives the axis of the outer wall, which rises up to the sills of the windows. The fourth unit [at point B] gives the outer face of the first course of the buttresses rising upwards from the level of the interior floor. Taking as a base of operation the upper part of the bases of the piers in the nave, on this horizontal — at point A — we erect the perpendicular AC; on this vertical line we mark five divisions which are equal to each of those [four] divisions placed on the [horizontal] base-line AB; this [length] gives us the total height of the nave [at point D]. Next, we connect the upper end — point D — of this [four-unit] line to point B [at the far end of the base-line], and the ratio to the half of the base-line [of this diagonal] is that of 5 to 4. We have, therefore, the bottom half — base-line ABD — of the Egyptian triangle [as shown in Fig. 80].

The intersection [in Nôtre Dame de Paris] of this [diagonal] side, BD [with *five* units], with the perpendicular, AD [with *four* units], as erected on the first part of the [horizontal] base-line, BA [with *three* units], fixes, at point E, the springing [of the arches] of the upper vault. By intersecting with the diagonal BD, a vertical rising from unit 2 upon the base-line BA marks the level of the sills of the gallery-windows at point F, and the intersection of diagonal BD with a vertical rising from unit 3 fixes the top of the windows in the outer side-aisle at point G. The first division, point H, of the perpendicular A–D places the level for the springing of the arches in all the side-aisles, whose center points are raised twelve inches above the capitals. The third division, found at point I and also fixed upon the vertical line AD, provides the level of the crowning of the triforium-gallery vaults. The hypotenuse KI [traversing AD] gives the inclinations of the slope of the flying-buttresses, and their

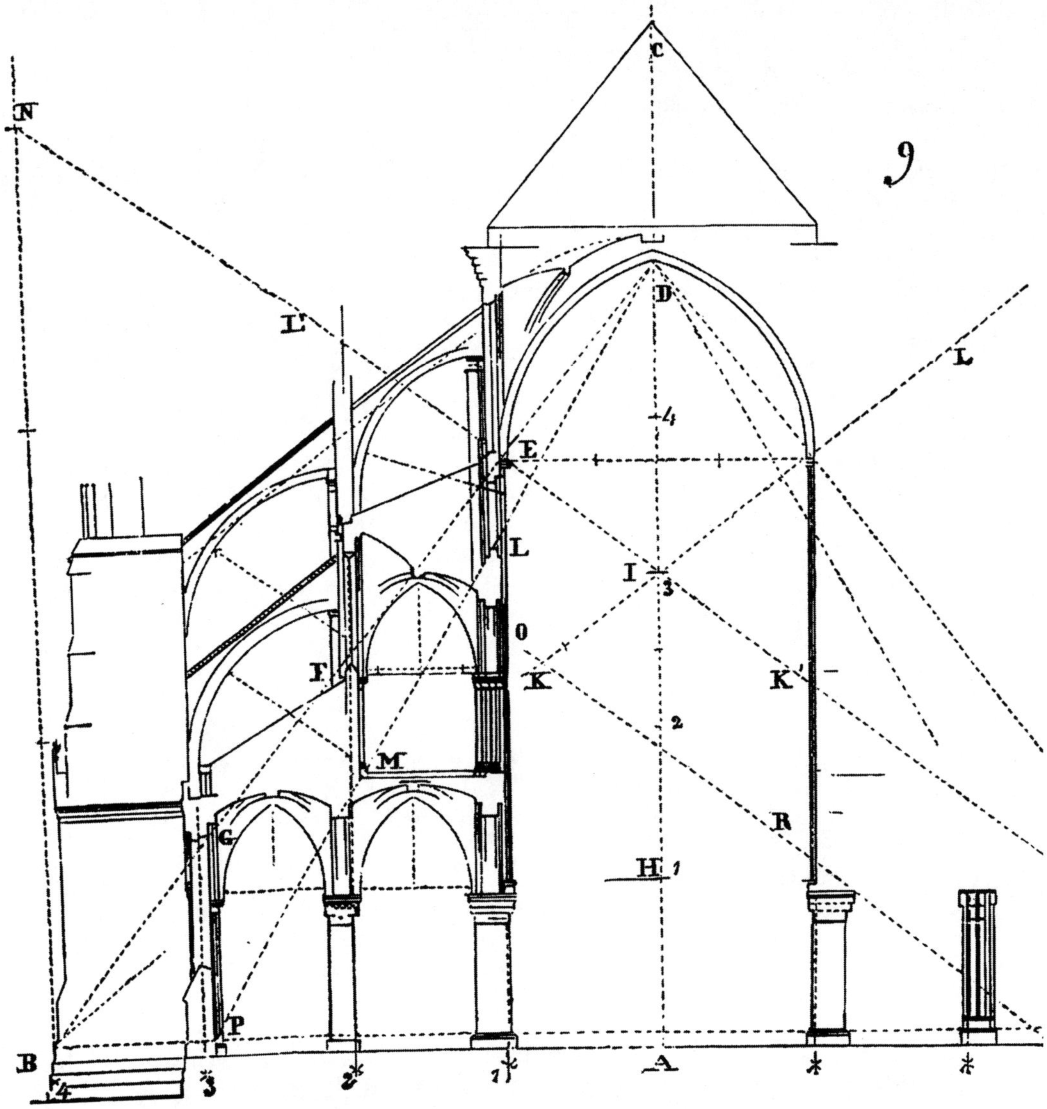

Fig. 81. "Triangulated Proportionality in Nôtre Dame de Paris." (E-E. Viollet-le-Duc, *Entretiens sur l'architecture*, 1872, "Lecture IX," fig. 9.)

centers are placed on the axes 1 and 2 [their base is also 3 to 4]. If we now take [the vertical axis] AD as the height of an equilateral triangle, we see that the [diagonal] side of this equilateral triangle, PD, which cuts the section lines of the construction partly determined, provides, at point L, the lower sills of the rose-windows. At point M, it fixes the floor of the triforium-gallery, and we place the interior face of the outer wall at point P [at ground-level]. Finally, a [diagonal] line, RO, is drawn [beneath and] parallel to the hypotenuse, L'K'; in crossing the level of the crowns of the gallery vaults, it gives the original inclination of the outer triangles of these vaults. The gables and roofs all exactly conform to a triangle, one like the Egyptian triangle [Fig. 80], whose half is ABD. [...]

Let us now analyze a transverse section [similarly] drawn through the nave of the Cathedral of Amiens. [Fig. 82] It would be a mark of genius to give harmonious proportions to a building whose height was double the width of another building whose proportions already presented a perfect whole. This is what the architect of Nôtre Dame d'Amiens accomplished with rare skill. All who enter this cathedral are struck with the apparent grandeur of the whole and by the perfection of the respective proportions. Entering the vast interior, the eye is satisfied at once; it understands without effort a conception originating in a single impulse of a superior mind. Is this harmony the result of a series of tentative efforts? Is it just the effect of happily hitting off the details in harmony with the general design? I have no great faith in luck, especially in architectural design; nor am I more inclined to believe in happy hits, of their coming from mere instinct. If a work is good, it is because it springs from a sound principle, one methodically carried out. Through analysis of a transverse section of the nave of the Cathedral of Amiens, we find a harmony of proportions, in this case one obtained by means of the superimposition of two Egyptian [or Pythagorean] triangles.

The base-line, AB, of the lower triangle ABC runs through the top of the bases of the nave piers. Here again we find the same foundation level being used for the entire design's geometrical scaffolding. This base-line AB has for its length the distance which separates the two exterior faces of the outer walls of the side-aisles. Its [inscribed] apex, located at point C, fixes the height of the bottom of the string-course [running along the facing nave walls], as marked at point D; made conspicuous by a wide band of ornamentation, this string-course runs without interruption all around the interior of the edifice. Once divided into four equal units, the base-line AB gives, at points 1 and 3, the exterior faces of the great piers (in the inset-drawing of PIER P, this contact point—*a*—is given by the tangent *gh*).

These four basic divisions are next bisected at points O, 1', 2', and 3'. By means of a line drawn parallel to the diagonal AC, subdivision 1' determines the site of point E. The intersection of point E with the diagonal BC, sets the level of the astragals [convex moldings] of the capitals belonging to the engaged columns XXX [as shown in the inset-drawing of PIER P]. Another diagonal parallel to AC, as drawn from point 1, then determines, at point F, the crown of the ribbing of the vaulting in the side-aisles. Another parallel diagonal, as similarly drawn from point 3', will then give the inclination of the window sills in the side-aisles. And yet another parallel diagonal, as drawn from point 2, determines by intersecting a horizontal line drawn from point F, the edge of the keystone of wall-rib *f*. Another parallel, as drawn from point 2', determines, at point *e*, the inner face of the engaged pier; due to its point of meeting with the horizontal drawn from point E, the thickness of the pier (of which the detail is shown in inset-drawing P) had been previously decided by the proportion of the weight supported by this compound pier.

The vertical axis of the entire nave, GC, is next divided into five units; each of these divisions, and in accordance with the drawing of the Egyptian triangle [Fig. 80], then becomes equal to one of the eight divisions in the base-line AB. The first [vertical] division—point 1—gives the level of the window sills in the side-aisles, and division 3 determines the bottom of the capitals on the central cylindrical part of the piers, as is shown in inset-drawing P. If the capitals of the engaged demi-shafts belonging to this compound pier—X–X—and those of the central pier have the same abacus, then the astragals of the great cylinder are placed lower than those

Opposite: **Fig. 82. "Triangulated Proportionality in the Cathedral of Amiens."** (E-E. Viollet-le-Duc, ***Entretiens sur l'architecture***, **1872, "Lecture IX," fig. 10.**)

10

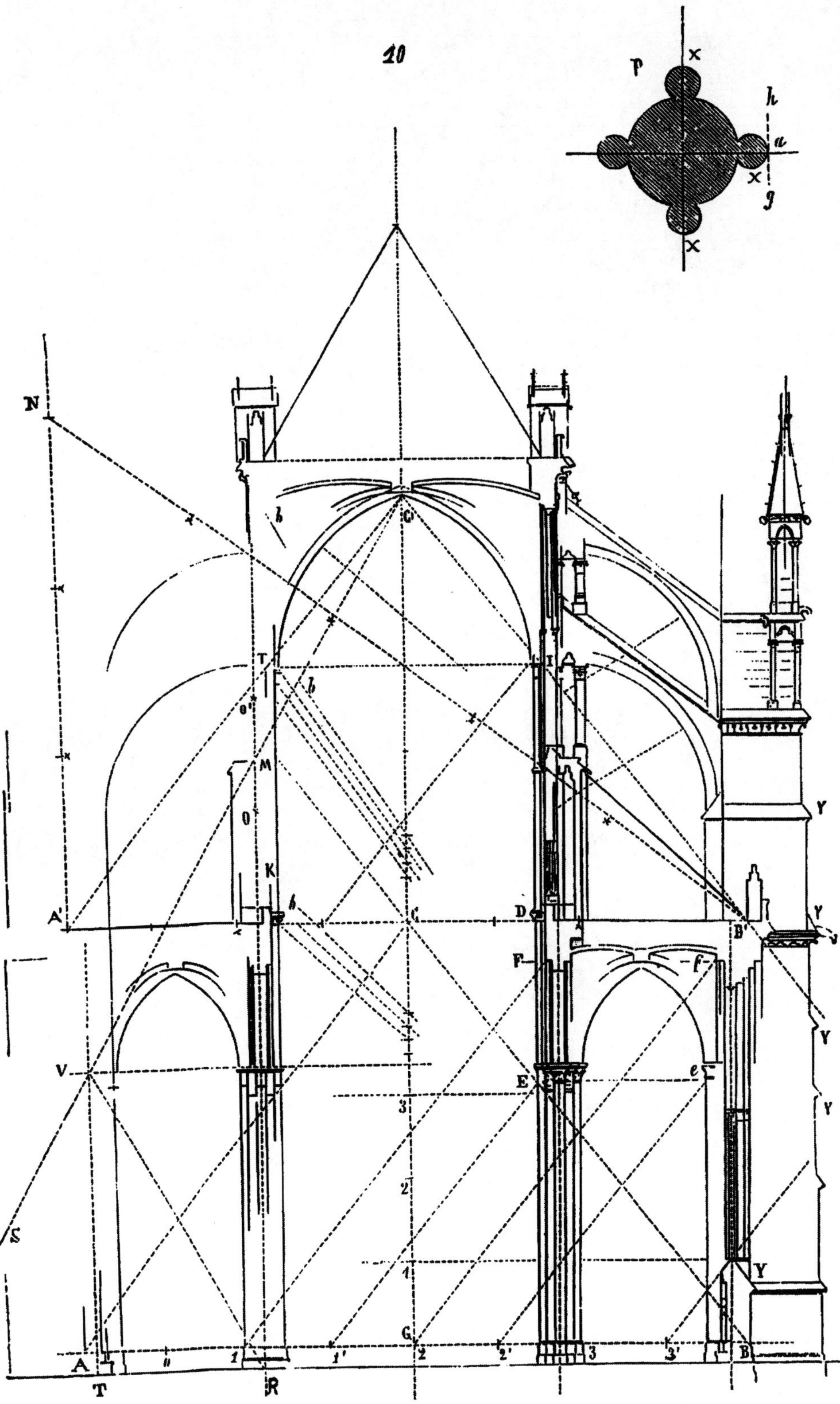

of the demi-shafts. On the next floor-level A'–B', which terminates the first ordinance belonging to the side-aisles—the architect resumed the lower base-line—AB—and marked it by the internal jambs of the openings which pass through the buttresses. On this new base-line, A'–B', he raised a second Egyptian [or 3–4–5] triangle, the apex of which is point C'. The [diagonal] sides of this [Pythagorean] triangle—A'–C' and B'–C'—meet with the walls or spandrels along the vertical line K, so giving—at point I—the height of the springing of the arches, and point C' then fixes their crowns.

In drawing these arcs he has followed the same method that was adopted in Notre-Dame de Paris [as shown in Fig. 81]. In Amiens, the axis of the nave-piers—R—has been extended, and the centers of the flying buttresses—O + O'—have been marked on these axes, with [apex-point] C' now being taken as the vertex of an equilateral triangle whose [left] side now becomes [the diagonal line] C'–S. The intersection of diagonal C'–S with the [vertical] axis R determines, at point M, the height of the upper passage of the triforium, and the intersection of C'–S with the [vertical] axis T determines the outer wall of the side-aisle, as found at point V, so also establishing [a horizontal line for] the springing of the arches. By its intersection with axis R, another diagonal, VR (also the side of equilateral triangle), determines both the floor-level of the side-aisle and the height of the compound-columns placed there.

The slant of the roof is likewise that of an equilateral triangle. Here the hypotenuse B'–N [as intersecting the vertical axis G–C'] fixes the height of the nave [where the arches spring, at points T–I]. All the profiles [of the parallel cornices] drawn at point *b* [below base-line T–I] are so arranged as to perfectly developed, and they do so by taking point B [at ground-level] as the axis of vision. Even the diagonal lines of the capstones in the flying buttresses are drawn parallel to the hypotenuse B'–N. This Egyptian [or 3–4–5] triangle has so completely served the purpose being the generator for the whole design of the building that even the sloping of the outer weatherings (Y) are all cut parallel to the hypotenuses AC and BC; it even seems that, in order to determine their angle of inclination, the designer had worked by running a set-square over the various parts of his diagram and by starting from a general base-line.

It is certain that, even if many processes of detail may have escaped our observation in the drawing of this section, nonetheless all the divisions, even the most minute, had clearly been obtained by means of the intersections of vertical lines with lines drawn parallel to the [diagonal] hypotenuses. If the fact of this [working] principle is not questioned, then it may still be asked why these geometrical methods give such satisfactory proportions. It is simply because they establish a constant and harmonious relation between the lengths and the breadths. As did the Egyptians, we must also recognize that the [Pythagorean] triangle, module generated by sides with 3 to 4 to 5 proportions—and where the sides 3 and 4 form a right angle—is *a perfect standard.* [...] The determination of proportions by [Pythagorean] triangles is advantageous because it furnishes points of articulation which direct the eye instinctively to the general system, even if it does not [fully] comprehend the [geometric] method followed. There can be no proportion without unity, and there is no unity without plurality, for pluralities imply not "similars" but differences. [...]

Next we may examine the manner in which Viollet-le-Duc explains the exact workings of all the technical operations involving the same 3–4–5 modular figure, now analyzing its functions within a specifically Islamic context. His discussion of the Arabic appropriation of the Pythagorean design motif may begin with an analysis of a primary, also most characteristic, component of the Great Mosque of Córdoba, its "horseshoe arches" (or *arcos de herradura*: Figs. 9–15). The Islamic horseshoe arch has its French counterpart (perhaps even offspring) in the *ogive* arch, a recurrent feature in Gothic architecture. Preceding his shrewd analysis of the essence of Islamic design, the geometric methods employed to design ogive arches are initially explained by Viollet-le-Duc as follows:

> No one thought [after the collapse of the Roman Empire] about abandoning the semicircle. The semicircle remained the generator of the arch generally, also barrel vaulting, groin vaulting, and the cupola. We consider one of the three triangles described above — namely, 1) the equilateral triangle; 2) the triangle taken on the diagonal of a pyramid with a square base, whose vertical section, from the summit parallel to one of the sides of the base, forms an equilateral triangle; and 3) the triangle which we call Egyptian [or Pythagorean] — as the three generators of proportions. However, because all three triangles give an angle at the summit of less than 90ffl, we are obliged to give any arch [derived from them] a form other than that of the [simple] semicircle.
>
> Suppose, for example, we take the Egyptian [or 3–4–5] triangle as the generator of proportions for a pointed arch. [Fig. 83] Its [horizontal] base is AB; this line provides both the diameter and functions as the base-line for the springing of the arch. If we wish the apex of this arch to reach the summit, C, and making use of compasses to draw the arch, we must find on line AB not just one center [or compass-point], as one does for drawing a round arch, but instead two centers; the two compass-points are the two points of intersection of [diagonal] perpendiculars drawn from the middle points of lines AC and BC to join with the base-line AB. We shall thereby draw a curve having two [opposed] segments; these segments are composed of the arcs of circles derived from the system of proportions imposed by the [inscribed] triangle ABC. This figure is what is now called, but very improperly, the *ogive* (or pointed arch). It is quite certain that the post-sixth-century Byzantine architectural schools adopted this arch, one which the French architects of the twelfth century also employed admirably, and they made the ogive the starting point of a new system of architecture. [...]
>
> With the advent of Islamic architecture [in the seventh century], the [Byzantine] remnants of the Greek school found themselves in the presence of other barbarians, but they left them great liberty, for the Moslems [then largely lacking any architectural traditions of their own] manifested no taste or preference in matters of art. Love for elaboration of form and a delicate observation of proportions was regained, and [Islamic sensibility] endeavored to associate the arch with the column, and sought curves other than that of the semicircle. We shall now see how these essays frequently produced architectural schemes singularly beautiful in their lines and proportions.
>
> The Mosque of Amrou was being erected at Cairo in the year 21 of the Hegira (AD 641). [Fig. 84] Certainly the Mosque of Amrou at Cairo was designed by none other than the Greek [Byzantine] artists who had then taken refuge in Egypt, and it was built with marble columns [that is, *spolia*] taken from great Roman buildings. This mosque consists merely of immense porticos erected around a courtyard. We shall examine two bays of a portico, and we see that the Greek architects (for no one will suppose that Omar had [Arabic] architects with him) employed columns despoiled from edifices built under the Roman Empire. They then raised arches upon these capitals that were novel in form, for these arches had complex curves, but with lines

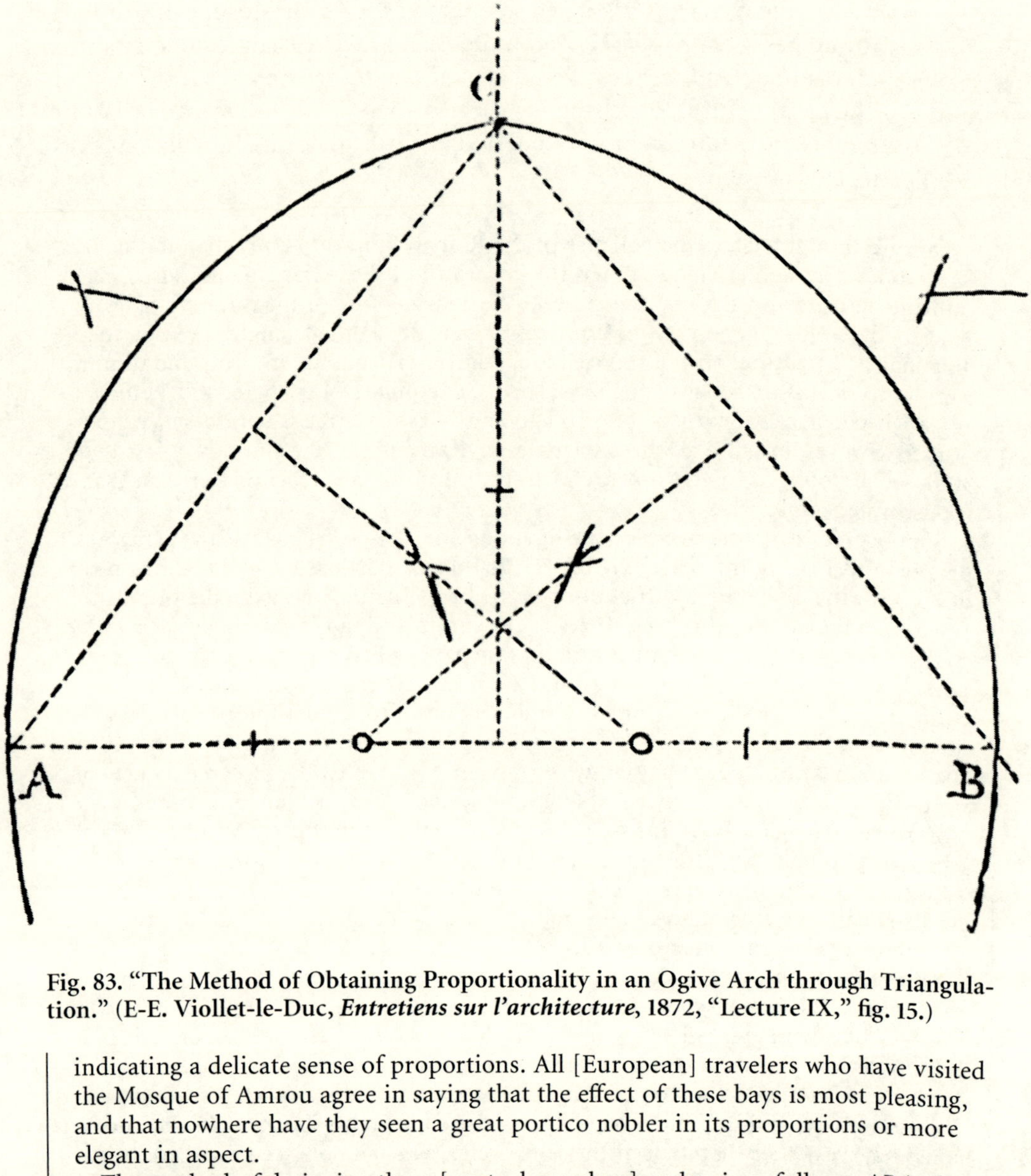

Fig. 83. "The Method of Obtaining Proportionality in an Ogive Arch through Triangulation." (E-E. Viollet-le-Duc, ***Entretiens sur l'architecture,*** 1872, "Lecture IX," fig. 15.)

indicating a delicate sense of proportions. All [European] travelers who have visited the Mosque of Amrou agree in saying that the effect of these bays is most pleasing, and that nowhere have they seen a great portico nobler in its proportions or more elegant in aspect.

The method of designing these [proto-horseshoe] arches is as follows. AB is the [inscribed] horizontal line where the compass-points are placed, and figure ABC is the Egyptian triangle [but one containing two, conjoined back-to-back, Pythagorean, 3–4–5, figures]. The arch was designed according to the method previously given in our drawing of the ogive arch [see Fig. 83]. The apex-point, C [in Fig. 84], is taken as the vertex of another equilateral triangle, DEC, and the base of that [larger] triangle gives the upper level of the abacus of the capitals. The Egyptian [Pythagorean] triangle established both the centers of the arch and its diameter in relation to its height. The equilateral triangle also gives the general proportion of the arch above the capitals. In order to avoid the weak line that would be produced by a simple vertical line, if drawn from F to A, instead the architect continued the curve of the arch past point A, to point G. In this manner, he obtained a stop above the parallelopiped [an inscribed, six-sided prism] which surmounts the capital; in its execution, this stop lends an appearance of particular firmness to these arches. [...]

The square and compass became supreme [in Islamic architecture], and yet with these means, apparently so restricted and with so dry a theme, the artists whom we call "Arabs" succeeded in producing marvels. It was natural, however, that the study of proportions should then become one of the most efficacious means of giving a pleasing appearance to buildings. In fact, in this architecture of the Caliphs, *proportion is everything*, because nothing masks its defects. Ornamentation lends its aid in rendering the harmony of that proportion manifest, but it is effective only in the mass. Decoration engages the attention only as embroidery on a [well-proportioned] tissue; it pleases without distracting. [...] The Greeks had been the instructors of the Romans, and they became the instructors of the barbarian hordes of Arabia. This last effort of their genius [for proportions] has also an influence that survived even to the fifteenth century, and made itself felt on the confines of the West. [...]

The Arabs did not change the Roman system of structure, they only modified its

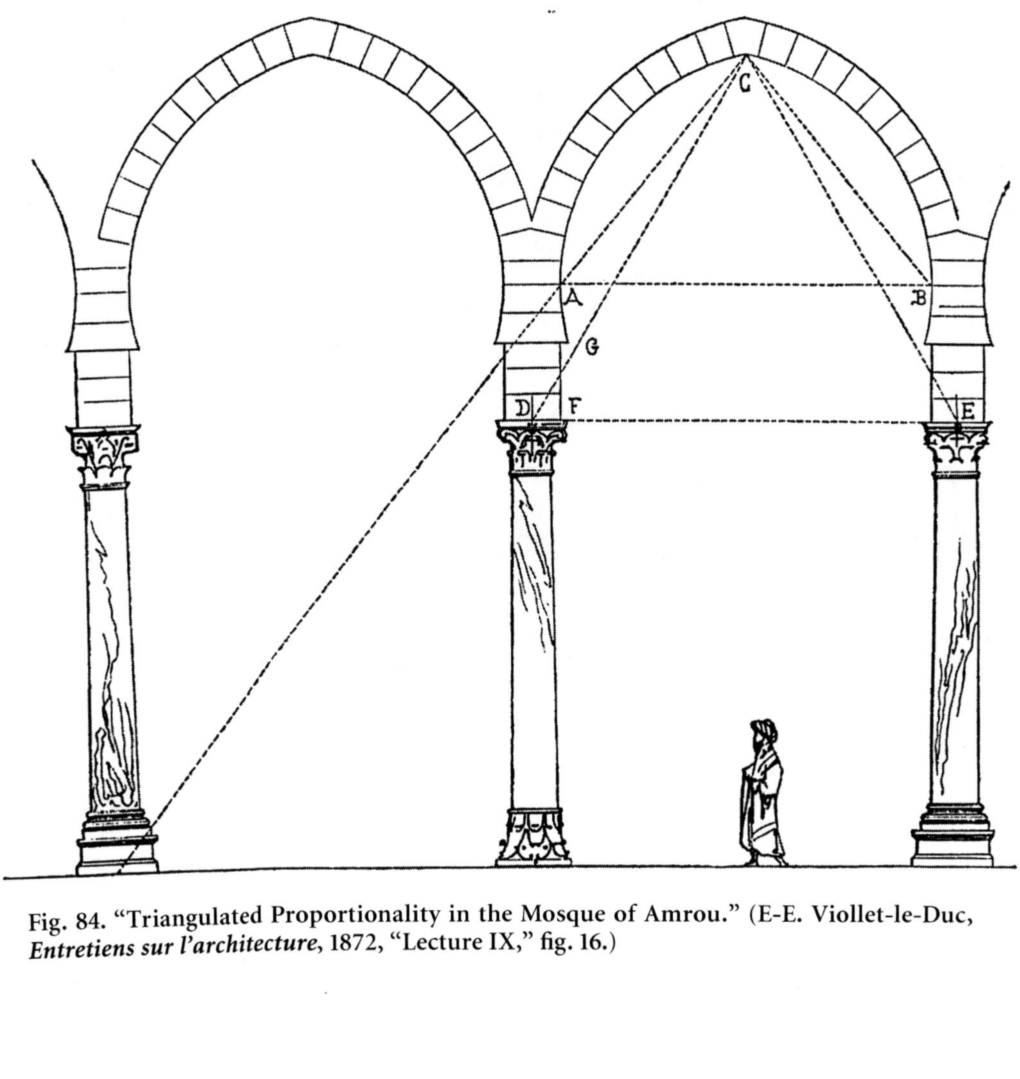

Fig. 84. "Triangulated Proportionality in the Mosque of Amrou." (E-E. Viollet-le-Duc, *Entretiens sur l'architecture*, 1872, "Lecture IX," fig. 16.)

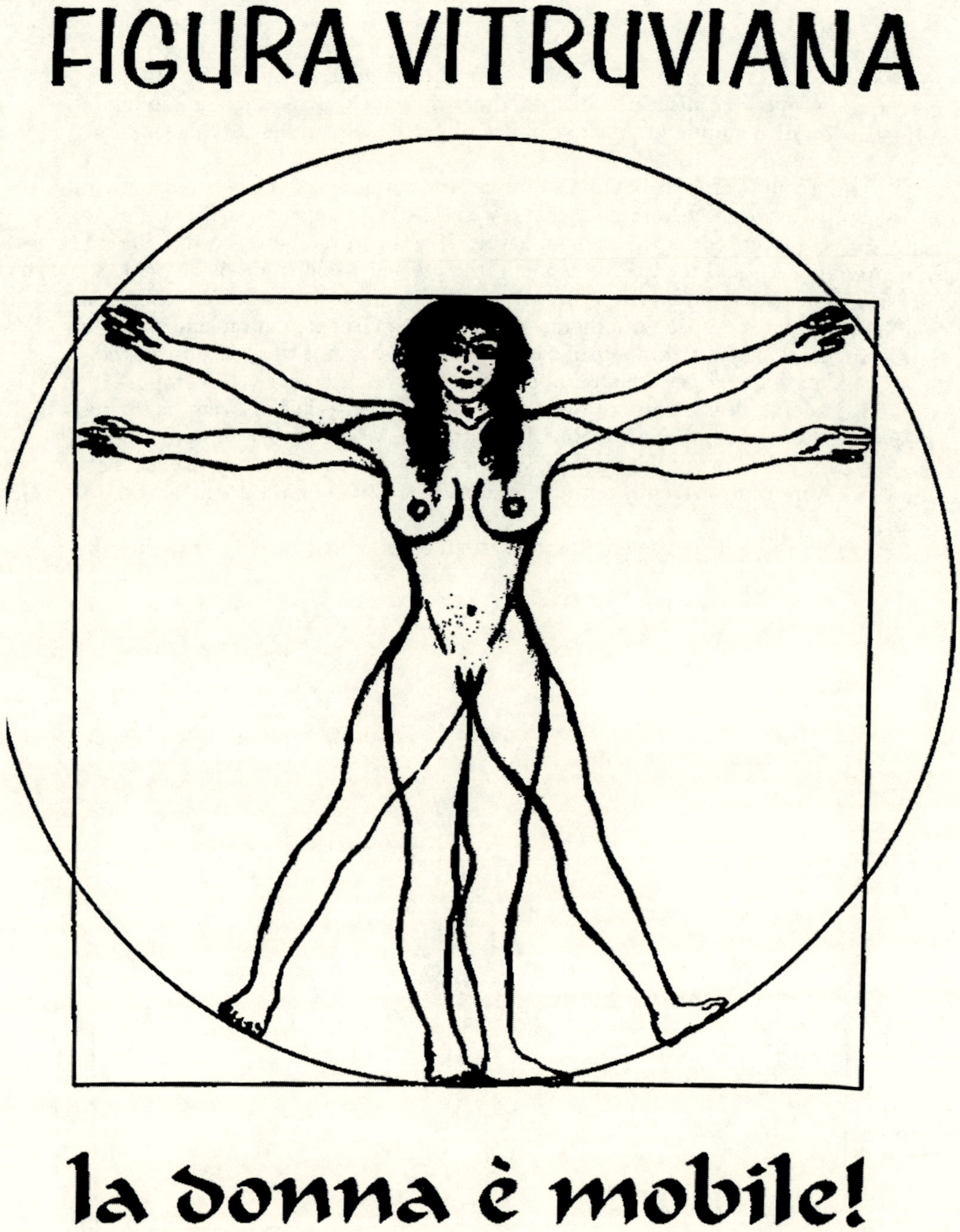

Fig. 85. "Figura Vitruviana." (*Figure néo-vitruvienne hors de serie.*)

envelope. The geometry to which they had recourse did not cause them to discover a new system of construction; instead, its functions were limited to giving novel curves to the arches and to controlling all ornamental design. Geometry afforded to them an intellectual pastime and it engaged the eye through wonderful combinations. [...] Let us enter the Alhambra, for example, one of the latest edifices due to the civilization attributed to the Arabs. And what do we see here? We see a concrete

> structure, like the Roman-Classic ground-plans, with walls of brick and slight porticos, whose delicate marble columns and earthen spandrels support wainscoted ceiling. These masses of brick, lath and plaster are covered with stuccos presenting to the astonished view the most skillful geometrical combinations that can possibly be imagined. [...] In the Arab building, geometry adorns the dress; in the Western medieval edifice, it sustains the body.

To conclude, here is how Viollet-le-Duc sums up the technical value of the same 3–4–5 modular figure, the one which we have repeatedly examined here in a specifically Hispanic — also Islamic — context.

> It cannot be denied that geometry is the starting-point — the very ground-work — of architecture, and that among all geometrical figures the most perfect is the triangle. Of all triangles, those which are best accommodated to the laws of statics and to proportional divisions are the equilateral triangle and [especially] the [Pythagorean] triangle, the one whose base is to the height as 4 to 2.5, and as generated by that other right-angled triangle, that which gives 4, 3, and 5. Consequently, the adoptions of these [inscribed] triangles, also the [inscribed] intersections of their sides with [invisible] vertical lines, supply us with sub-divisions subjected to a single [organizational] principle. The points [of intersection] remind us of the inclinations of the sides, and the proportions necessarily derived from those [inscribed] generating-figures. On the other hand, we now understand how the application of these triangles to architectural designs obliges the designer to make his heights bear a certain proportion to his breadths. Consequently, the more space one part of these heights or breadths occupies, the less he has left for the others. Nonetheless, there is always a [perceived] relationship between these parts, however different they may be.

And thus, thanks to Viollet-le-Duc, we now have in hand a detailed explanation of not only the general principles of architectural proportionality, but also the specific means of application of the most perfect *modulus* of all, the "Pythagorean triangle." As was explained here by Viollet-le-Duc, clearly the 3–4–5 triangular module was a staple of medieval architecture in France. However, as has been argued here by me, this same design module — the 3–4–5 triangular design motif — was similarly an essential element for designing medieval architecture in Islamic Spain, *Al-Andalus*. Finally, and just as this point has been graphically demonstrated here in various ways, this same design module — the Pythagorean-triangle design motif — similarly became a staple of post-medieval, ecclesiastical architecture erected by devout Christian designers in both Spain and, later, throughout Spanish America. As was to be expected, the *conquistadores* had brought their traditional building methods to the New World, and the traditional design techniques recreated there were — as just has been revealed here — what may now be called specifically "neo–Islamic" methods.

Notes

Chapter 1

1. What follows is the considerably expanded version of what was originally a conference-paper, prepared for presentation at the annual meeting of the Society of Architectural Historians in Montreal (April 1989); for a preliminary report on my discovery of the "neo-Islamic" design anomalies at Tepozotlán, see J. F. Moffitt, "¿Tepotzotlán: El Islam latente en America? Observaciones en torno a la portada esculpida hispánica," *Anales del Instituto de Investigaciones Estéticas de la UNAM*, LVIII, 1986, pp. 101–12 (where the one drawing illustrating the leit-motif was prepared, as crudely drawn by me, with pencil, and on tracing paper, and as placed over a xeroxed photo!).

2. For this terminology, see F. Chueca Goitia, *Invariantes Castizos de la Arquitectura Española* (1st ed., 1947); rpt. Madrid: Dossat, 1981. Chueca's book is fundamental to what follows.

3. Although much valuable research has been done for other (non–Spanish) kinds of European architecture, to my knowledge, no such investigation has been previously attempted in the case of routine Hispanic design methods. For a practical scholarly approach to the problem, and serving as my analytical model, see P. Frankl, "The Secret of the Medieval Masons," *Art Bulletin*, XXVII, 1945, pp. 46–50. The "lodge secret" was, in short, the use of "quadrature" and "triangulation," for which, see further F. M. Lund, *Ad Quadratum: A Study of the Geometrical Basis of Classical and Medieval Architecture*, London: Batsford, 1921. Although specifically discussing the method used in designing the Cathedral of Milan, Frankl gave me the essential clue for uncovering what proves to have been the Hispanic practice much later; in short ("Secret," pp. 56, 58), "the Pythagorean triangle is very practical, it being the simplest way to construct the right angle," and the immediate goal "is the enlargement of a small sketch to the desired size of the finished form."

4. One's initial recognition of the general art-historical significance of "proportionality" was due to the pioneering essay by Erwin Panofsky, "Die Entwicklung der Proportionslehre als Abbild der Stilenwicklung," *Monatshefte für Kunstwissenschaft*, XIV, 1921, pp. 188–219 (which, of course, I first read, as a monoglotal undergraduate, in his collection of fundamental essays, *Meaning in the Visual Arts: Papers in and on Art History*, Garden City: Anchor, 1955, pp. 55–107). For the central role of "proportionality" in architectural design, see the fundamental study by Rudolf Wittkower, *Architectural Principles in the Age of Humanism*, New York: Norton, 1971.

5. B. Bevan, *Historia de la Arquitectura Española*, Barcelona: Juventud, 1970, p. 17. For more on this Christian-Islamic cultural fusion, see the classic study by Américo Castro, *España en su historia. Cristianos, moros y judíos*, rpt. Barcelona: Grijalbo, 1996.

6. S. Gauldie, *Architecture*, Oxford Univ. Press, 1969, pp. 37–38; see also P. H. Scholfield, *The Theory of Proportion in Architecture*, Cambridge Univ. Press, 1968.

7. Authorship of the façade of the church at Tepotzotlán has, in fact, been often (probably erroneously) attributed to Lorenzo Rodríguez, an Andalusian architect born and educated in Guadix (Granada). His best known work is the Sagrario of the Cathedral of Mexico City (1749–68); see M. Collier, "New Documents on Lorenzo Rodríguez and His Style," in M. Meiss (ed.), *Latin American Art, and the Baroque Period in Europe*, Princeton Univ. Press, 1963, pp. 203–18; see also G. Kubler and M. Soria, *Art and Architecture in Spain and Portugal and Their American Dominions, 1500–1800*, Baltimore: Pelican, 1959, p. 360, n. 53; R. J. Mullen, *Architecture and Its Sculpture in Viceregal Mexico*, Austin: Texas Univ. Press, 1997, pp. 122, 168, 172 (on Rodríguez); 172–75 (on Tepotzotlán). For historical studies incorporating formal analyses of the specifically Andalusian origins of Latin American architecture, see R. C. Taylor, "Francisco Hurtado and His School," *Art Bulletin*, XXXII,

1950, pp. 25–61; J. F. Moffitt, "Wendel Dietterlin and the Estípite. Observations on Architectural Style in Colonial Mexican Ecclesiastical Facades," *Boletín del Seminario de Estudios de Arte y Arqueología de la Universidad de Valladolid*, L, 1984, pp. 325–49. For the still-vexing question of a lingering Pre-Columbian stylistic component, see two opposing discussions: A. Neumeyer, "The Indian Contribution to Architectural Decoration in Spanish Colonial America," *Art Bulletin*, XXX, 1948, pp. 104–24; G. Kubler, "On the Extinction of the Motifs of Pre-Columbian Art," in *Essays in Pre-Columbian Art and Archaeology*, Harvard Univ. Press, 1961, pp. 14–34. To both arguments, consideration of an extrinsic factor must be now added: most likely, the native master craftsmen had all died off well before 1600. The result was that their orally transmitted, indigenous-traditional skills were irretrievably lost, hence their absence in later Hispanic buildings employing Indian workmen. Due largely to diseases imported by Europeans, Indian populations quickly fell to a tenth (or less) of their pre–Conquest levels. In short, it was *la muerte* that had suppressed all the native architectural traditions; see W. H. McNeill, *Plagues and Peoples*, New York: Mentor, 1976, pp. 176–91; J. F. Moffitt and S. Sebastián, *O Brave New People: The European Invention of the American Indian*, Albuquerque: Univ. of New Mexico Press, 1995, pp. 296–302, 332–36.

8. H. Wöl⁹in, *Principles of Art History*, New York: Dover, 1958.

9. So discussed in Chueca Goitia, *Invariantes Castizos de la Arquitectura Española*.

10. For a comprehensive (but mostly non-architectural) study of these structures in late medieval Spain, see J. Berg Sobré, *Behind the Altar Table: The Development of the Painted Retable in Spain, 1350–1500*, Missouri Univ. Press, 1989. For a detailed discussion of some Mexican examples of the monumentally-scaled Hispanic *retablo*, see J. A. Baird, "The Eighteenth-Century Retables of the Bajío: The Querétaro Style," *Art Bulletin*, XXX, 1953, pp. 195–216.

11. Chueca Goitia, *Invariantes Castizos*, pp. 194–95.

12. For illustrations of both these examples, with a discussion of their relations with the other plastic Hispanic arts, see J. F. Moffitt, *Spanish Painting*, London: Studio Vista, 1973, pp. 45–50; idem., *The Arts in Spain*, London: Thames & Hudson, 1998, pp. 79–82 (*Las Artes en España*, Barcelona: Destino, 1999, pp. 77–82); for another striking Spanish architectural ensemble (with a very different kind of "geometric" basis, namely "anamorphosis"), see idem., "Mystical Meaning and Anamorphic Form in Narciso Tomé's Transparente," *Canadian Art Review*, XIV, 1987, pp. 96–134.

13. For the often "anti-classical" manipulations of Classical elements in Plateresque design and theory, see E. Rosenthal, "The Image of Roman Architecture in Renaissance Spain," *Gazette des Beaux-Arts*, LII, 1958, pp. 329–46. The primary document for all Plateresque concerns is Diego Sagredo's *Medidas del Romano* (1526), now available in facsimile, Madrid: Albatros, 1976. This author often refers to the authority of Vitruvius, whose *De architectura* was evidently well known in Spain before the Spanish translation appeared in 1583 (which is quoted below, in chapter XV).

14. Details of the iconographic program of this structure have been revealed by Santiago Sebastián, *Simbolismo de los Programas Humanísticos de la Universidad de Salamanca*, Salamanca: Universidad, 1973; for its formal characteristics, see Bevan, *Historia de la Arquitectura Española*, pp. 215–18, 244.

15. Chueca, *Invariantes castizos*, p. 136.

16. For a definition of the "Neo-Plateresque" phenomenon in Latin America, with bibliography on the initial Plateresque syndrome, see Moffitt, "Wendel Dietterlin and the Estípite."

17. J. A. Baird, "Mexican Architecture and the Baroque," in M. Meiss (ed.), *Latin American Art, and the Baroque Period in Europe*, Princeton Univ. Press, 1963, pp. 191–202.

Chapter 2

1. On this distinctive Hispanic artistic hybrid, see the now-standard survey by Rafael López Guzmán, *Arquitectura Mudéjar: Del Sincretismo Medieval a las Alternativas Hispanoaméricanos*, Madrid: Cátedra, 2000; see also G. M. Borrás Gualis, *El arte mudéjar*, Teruel: Instituto de Estudios Turolensés, 1990; idem., *El Islam, de Córdoba al Mudéjar*, Madrid: Silex, 1997; J. M. Azcárate, *Arte Gótico en España*, Madrid: Cátedra, 1996, pp. 73–107; P. Lavado Paradinas et al., *El Arte Mudéjar: La estética islámica en el arte cristiano*, Madrid: Electa, 2000 (all five books include specialized bibliography; earlier studies include E. Berteaux, *L'Art Mudéjar. Les Survivances de l'Art Musulman dans l'Art Chrêtien d'Espagne*, Paris, 1912–13; D. Angulo Iñíguez, *Arquitectura Mudéjar Sevillana de los Siglos XIII, XIV y XV*, Seville: Laboratorio del Arte de la Universidad, 1932; F. Pérez Embid, *El Mudejarismo en la Arquitectura Portuguesa de la época Manuelina*, Seville: Marinas, 1944.); for further bibliography (2,013 entries!), see F-A. de Montêquin, *Muslim Architecture of the Iberian Peninsula: Eastern and Western Sources for Hispano-Islamic Building Arts*, West Cornwall, CT: Locust Hill Press, 1987. For the unique social conditions that made this all possible, see V. B. Mann and J. D. Dodds (eds.), *Convivencia: Jews, Muslims, and Christians in Medieval Spain*, New York: Braziller, 1992.

2. For the legal status of these *mudéjares*, see L. P. Harvey, *Islamic Spain, 1250 to 1500*, Chicago Univ. Press, 1990, esp. chapters 4–9. For the pre-Christian artistic context, see Jerrilynn D. Dodds (ed.), *Al-Andalus: The Art of Islamic Spain*, New York: Metropolitan Museum of Art, 1992.

3. There exists a general descriptive study (unfortunately not very analytical) of the Latin American phenomenon: M. Toussaint, *Arte Mudéjar en America*, Mexico DF: Porrúa, 1946; none of the other publications cited here have specifically analyzed in detail the Hispanic design principles exemplified at Tepotzotlán and other buildings discussed (and then cyber-analyzed) herein.

4. O. Hagen, *Patterns and Principles of Spanish Art*, Madison: Wisconsin Univ. Press, 1948.

5. To the contrary, for the scandalous argument that the bust was actually conceived and executed in AD 1897, see J. F. Moffitt, *Art Forgery: The Case of the Lady of Elche*, Gainesville: Florida Univ. Press, 1995; *El caso de la Dama de Elche: Historia de una falsificación*, Barcelona: Destino, 1996.

6. D. Jones, "Surface, Pattern, and Light," in G. Michell (ed.), *Architecture of the Islamic World: Its History and Social Meaning*, London: Thames and Hudson, 1978, pp. 144–75. The grid is still, of course, an essential tool of architectural practice; for its current employment, see (among other discussions) F. D. K. Ching, *Architecture: Form, Space, and Order*, New York: Van Nostrand Reinhold, 1979, pp. 86–87, 238–44.

7. R. Lewcock, "Architects, Craftsmen and Builders: Materials and Techniques," in Michell (ed.), *Architecture of the Islamic World*, pp. 112–43 (including some comments following).

8. E. Panofsky, *Meaning in the Visual Arts: Papers in and on Art History*, Garden City: Anchor, 1955, pp. 60–61, 65. Of particular interest here is his reproduction (Panofsky's plate 18) of a rare working drawing, an Egyptian sculptor's squared-grid, preparatory sketch (ca. 1500 BCE) showing the "inscribed" figure of a sphinx in its ground-plan and elevation.

9. S. Kostof (ed.), *The Architect: Chapters in the History of the Profession*, New York: Oxford Univ. Press, 1977, pp. 7–8, 10. See also A. Badawy, *Ancient Egyptian Architectural Design: A Study of the Harmonic System*, Berkeley: Univ. of California Press, 1965; E. B. Smith, *Egyptian Architecture as Cultural Expression*, New York: American Life Foundation, 1968, esp. Chap. X, "Egyptian Architects and Their Methods."

10. S. Giedion, *The Beginnings of Architecture*, Princeton Univ. Press, 1981, pp. 482–92. For more on the Egyptian cubit, the squared grid, and its initial use to determine proportions in architecture, see B. C. Kielland, *Geometry in Egyptian Art*, London: Tiranti, 1955; E. Iverson, *Canon and Proportions in Egyptian Art*, Warminster: Aris and Philips, 1975.

11. M. C. Ghyka, *The Geometry of Art and Life*, New York: Dover, 1977, pp. 22, 142.

12. The citations following are taken from the standard bi-lingual edition: F. Granger (ed.), *Vitruvius on Architecture*, London: Loeb, 1962.

13. K. J. Conant, "The After-Life of Vitruvius in the Middle Ages," *Journal of the Society of Architectural Historians* 27 (1968): 33–38.

14. Some of the data immediately following is derived from an important unpublished book ms. by Mardith K. Schuetz-Miller, "The Empire Builders. A Socio-Economic History of Architects and Building Artisans from the Neolithic to the Renaissance" (and I thank the author for making her work available to me).

15. Dickie, as quoted in Michell (ed.), *Architecture of the Islamic World*, p. 16.

16. A. G. Walls, *Geometry and Architecture in Islamic Jerusalem: A Study of the Ashrafiyya*, London: Scorpion, 1990; I reproduce here (as **fig. 6**) his Fig. IV/5: "The Madrasa: Anchoring the Fundamental Geometry." Unfortunately, his research did not include the role obviously played in the overall design scheme by the grid.

Chapter 3

1. J. Bourgoin, *Les Eléments de l'art arabe: le trait des entrelacs*, Paris: Eirmin-Didot et Cie, 1879; (facs. rpt.) *Arabic Geometrical Patterns and Designs*, New York: Dover, 1973, Plate 113. Since the Dover publication is readily available, the interested reader can literally duplicate my triangles by cutting out similar templates, then deploying them over the rest of Bourgoin's line drawing. Next, you can do the same with the other, 189 plates—for which task you will need a magnifying glass!

2. Bourgoin, *Arabic Geometrical Patterns and Designs*, Plate 57.

3. A. Donaire, "Lace-Design Methods in Late Medieval Spain," in I. Lavin (ed.), *World Art: Themes of Unity in Diversity*, Penn State Univ. Press, 1989, vol. I, pp. 231–38 (with six diagrams). For the overall context of such decorative details, see O. Grabar, *The Formation of Islamic Art*, Yale Univ. Press, 1973, ch. 7, "Early Islamic Decoration: The Idea of an Arabesque" (unfortunately, like most other scholars, Grabar does not deal with the actual design procedures, evidently involving use of a grid). For the specifically Spanish examples, see the descriptive catalogue raisonné presented by Basilio Pavón Maldonado, *El Arte Hispanomusulmán en su Decoración Geométrica (Una Teoría para un Estilo)*, Madrid: Instituto Hispano-Arabe de Cultura, 1975 (despite its promising title, this lavishly illustrated compendium, with 445 pages of text, also does not provide the reader with a simple explanation for

how these intricate designs were actually executed). For a preliminary report on these findings, see J. F. Moffitt, "Capable of Infinite Extension: From the Arabic 'Interlace' to Alberti's 'Velo' via 'Occam's Razor'," *Source*, XXI/4, 2002, pp. 25–35 (and I thank Prof. Anita Moskowitz [SUNY-Stony Brook] for passing on to me a critique of the terminology used in my earlier effort, since rectified).

4. S. K. Heninger, *Touches of Sweet Harmony: Pythagorean Cosmology and Renaissance Poetics*, San Marino: Huntington Library, 1974, p. 106, showing a geometrical figure taken from Athanasius Kircher's *Arithmologia* (1665) which is duplicated in our **fig. 63**, as taken from the 1583 translation of Vitruvius into Spanish. For another published replication of this standard figure, appearing in 1639, see Fray Laurencio de San Nicolás, *Arte y Uso de Architectura (Madrid, 1639 y 1664)*, Madrid: Albatros, 1989 (facs. ed.), fol. 137v, and as verbally explained on fols. 135v to 137r.

5. *Vitruvius on Architecture*, ed. F. Granger, London: Loeb, 1962, pp. 200–3 (bi-lingual). The rest of this statement is quoted here in Chapter 15.

Chapter 4

1. For what immediately follows, see S. Koslof, (ed.), *The Architect: Chapters in the History of the Profession*, New York: Oxford Univ. Press, 1977, 64–66.

2. Ibn Khaldun, as in G. Michell (ed.), *Architecture of the Islamic World: Its History and Social Meaning*, London: Thames and Hudson, 1978, 132. For the wider context of this observation, see K. A. C. Creswell, *A Short Account of Early Muslim Architecture*, Harmondsworth: Penguin, 1958; the Great Mosque of Córdoba is discussed on pp. 213–28.

3. Isidoro de Sevilla, *Etimologías*, ed. S. Montero Díaz, Madrid: Biblioteca de Autores Cristianos, 1951, pp. 469 (definition), 478 (tools).

4. For a detailed discussion, also defining nomenclature, of the traditional Moslem portal-ensemble, see F-A. de Montêquin, *Compendium of Hispano-Islamic Art and Architecture*, St. Paul, MN: Hamline Univ. Press, 1976, esp. pp. 26, 35; see also M. Toussaint, *Arte Mudéjar en América*, Mexico DF: Porrúa, 1946, pp. 129–31, "Vocabulario de términos mudéjares."

5. F. Chueca Goitia, *Invariantes Castizos de la Arquitectura Española*, Madrid: Dossat, 1981, pp. 75–77; see further E. Camps Cazorla, *Módulo, proporciones y composición en la arquitectura califal cordobesa*, Madrid: CSIC, 1953 (this study is, however, narrowly limited to an analysis of the modular geometry used for designing horseshoe arches).

6. See chapter 5, note 1, for the particulars on this software. Much credit for undertaking the initial graphic research into a wide range of Islamic and Hispanic triangle-schemes goes to my former student, Mr. R. S. Ehlers (B.F.A., M.A.; now resident in Deming, NM). Back in 1986, Mr. Ehlers had initially generated composite images with his Amiga 2000 computer (hardware now gone hopelessly obsolete). The procedure he then followed involved capturing a published photographic image, as a still, in a video camera; once "digitized," the image could be made more or less legible to what computer software was then available. This time-consuming procedure was abandoned once Ehlers discovered the existence of a nearly universal triangular scheme. At that point, he chose simply to inscribe a sheet of plexiglass with a triangle-based grid-pattern. Ehlers then used the transparent template as an overlay, and he put this on top of as many photographs as were made available to him. However, he had not actually hit upon specifically Pythagorean triangles; instead, his modules had highly irregular (and unprecedented) ratios: 5:21:26; 19:26:32; 23:31:36; 27:37:45 (and so forth, and as I have just measured them with a ruler). Nonetheless, my initial working models were originally just photographic montages: Scot's marked, plexiglass plate placed on top of a xeroxed book illustration. In any event, from the consistency of the results produced by Scott's time-consuming experiments, it was made clear to me that a triangular scheme was clearly at work here. The further conclusion was that, once it got itself invented (perhaps a dozen years later), I would then have to master the appropriate software program (for me, with a flattened learning-curve, a heroic feat).

7. On Islamic "centered-thinking," see T. Burckhardt, *La civilización hispano-árabe*, Madrid: Alianza, 1982, p. 241 ff.

Chapter 5

1. See D. Biedny and B. Monroy, *The Official Adobe Photoshop Handbook*, New York: Bantam, 1991; E. Weinmann and P. Lourekas, *Visual Quickstart Guide: Photoshop3 for Windows*, Berkeley: Peachpit, 1996 (admittedly tedious reading, but here cited as essential art-historiographic "documentation."). In any event, these software manuals also show you initially how to prepare your own grid-patterns. Also explained is how to make them in different sizes (and in duplicates); and how to change their size to fit over differently scaled images, also how to move them over your scanned drawings or photographs in order to achieve exact alignments. As mentioned in my text, other than adjustments for parallax problems appearing in my on-site photographs, I have not

in any way altered my graphic material. Once you've mastered the overlay process, you can readily produce dozens of different examples (yours) to corroborate my hypothesis. Other pertinent examples of Hispanic architecture to work with, besides those analyzed here, see Chapter 11, note 12, and Chapter 12, note 3.

2. F. Chueca Goitia, *Invariantes Castizos de la Arquitectura Española*. Madrid: Dossat, 1981, p. 77.

3. See again Erwin Panofsky's fundamental essay, "The History of the Theory of Human Proportions as a Reflection of the History of Styles," as in his *Meaning in the Visual Arts: Papers in and on Art History*, Garden City: Anchor, 1955, pp. 55–107.

4. For the correlation of musical principles to architectural design, see R. Wittkower, *Architectural Principles in the Age of Humanism*, New York: Norton, 1971, Part IV, "The Problem of Harmonic Proportion in Architecture," p. 101 ff.

Chapter 6

1. R. Wittkower, *Architectural Principles in the Age of Humanism*, New York: Norton, 1971, Appendix III, "Bibliographic Notes on the Theory of Proportion," pp. 162–66 (and for specific applications, with illustrations, figs. 74–84, see the Appendix).

2. For an exhaustive examination of the historical evidence, including both graphic and textual materials, supporting this conclusion, see J. S. Ackerman, "'Ars Sine Scientia Nihil Est.' Gothic Theory of Architecture at the Cathedral of Milan," *Art Bulletin*, 31/2, 1949, pp. 84–108.

3. The complete *Cesariano Vitruvius* is easily available for inspection in a facsimile reprint: New York: Benjamin Blom, 1968; the Cathedral of Milan illustration appears on folio xv *recto*.

4. *Cesariano Vitruvius*, folios 52 r (2 diagrams), 52 v, 53 v (2 diagrams), 53 v (grid over an Ionic capital), 63 v (grid over a Corinthian capital), 67 r, 70 r, 75 v, 81 v, 84 r, 98 r, 105 r (with grid), 140 v (grid), 144 r (grids explained as "Platonic" figures), 155 v ("Pythagorean" squares and triangles).

5. F. di Giorgio Martini, *Tratatti di architettura ingegneria e art militare*, ed. C. Maltese, Milan: Polifilo, 1967, pp. 399–401, tavola 233.

6. Ibid., pp. 402–3, tavola 236. It is of interest to note that the diagonal lines intersecting below the figure's knees form two Pythagorean triangles.

7. For this argument, see H. Millon, "The Architectural Theory of Francesco di Giorgio," *Art Bulletin*, XL, 1958, pp. 257–61.

8. F. Colonna, *Hypnerotomachia Poliphili: The Strife of Love in a Dream*, ed. J. Godwin, London: Thames & Hudson, 1999, 42–47; for the significance of this lengthy description of *quadratura* for the history of architecture, see J. Onians, *Bearers of Meaning: The Classical Orders in Antiquity, the Middle Ages, and the Renaissance*, Princeton Univ. Press, 1988, pp. 207–9.

9. S. Serlio, *The First Book of Architecture, made by Sebastian Serly, entreating of Geometry*, London: Robert Peake, 1611; facs. rpt. *The Five Books of Architecture*, New York: Dover, 1982, folio 13 r. The 1611 English translation (taken from an earlier Dutch edition!) is nearly illegible, hence my interpolations; I have also modernized the spelling and grammar.

10. Wittkower, *Architectural Principles*, pp. 126–7.

11. See C. Sambricio, "Un álbum de dibujos poco conocido: el de Silvestre Pérez en la Biblioteca Nacional," *Archivo Español de Arte*, no. 181, 1973, pp. 45–57; P. Navascués, *Historia del Arte Hispánico V: Del Neoclasicismo al Modernismo*, Madrid: Alhambra, 1979, pp. 29–31.

12. F. Chueca Goitia, *Invariantes Castizos de la Arquitectura Española*. Madrid: Dossat, 1981, p. 123.

13. *Ibid.*, p. 134.

Chapter 7

1. R. S. Gustafson (ed.), *Builder 3 & 2*, Washington DC: Navedetra 10646, 1987, Vol. I, pp. 6–12, 13. I owe this bibliographical nugget to Mr. Scott Ehlers, himself a former Seabee (PO-1).

2. For these observations, thanks are due to a civil engineer, Mr. Larry Underwood (Deming, N. M.), who kindly shared with us his experiences of working on various isolated building sites in the state of Chihuahua with Mexican masons and carpenters; it is particularly to him that I owe the contemporary details of the "design-by-string" procedures of traditional Hispanic construction. I also thank Mr. Brian Featheringill (Columbus, Ohio) has also discussed with us similar practices which he observed during the restoration of an eighteenth-century Spanish colonial church in rural Mindanao in the Philippine Islands.

3. On this topic, see S. K. Heninger, *Touches of Sweet Harmony: Pythagorean Cosmology and Renaissance Poetics*, San Marino: Huntington Library, 1974; for the primary documents, see K. S. Guthrie (ed.), *The Pythagorean Sourcebook and Library*, Grand Rapids: Phanes, 1987.

4. For some observations about the specifically Christian, or Spanish Visigothic, contributions to the design of the Great Mosque of Córdoba (with these including the appropriation of earlier Roman *spolia*), see J. F. Moffitt, "La significación histórica de la arquitectura hispanovisigoda," *Goya: Revista de Arte*, no. 186, 1985, pp. 338–47.

5. For a general discussion of what is known about early medieval Arabic "science," including surveying and engineering, see W. Eamon, "Technology, Treatises on," in J. R. Strayer (ed.), *Dictionary of the Middle Ages*, New York: Scribner's, 1988, vol. XII, pp. 640–50. Much of this knowledge, of course, came from earlier techniques worked out by the Greeks and Romans; for these, see L. Sprague de Camp, *The Ancient Engineers*, New York; Ballantine, 1974; Peter James and N. Thorpe, *Ancient Inventions*, New York: Ballantine, 1995. For a modern perspective on the Mohammedan use of geometric design principles in the visual arts, see I. El-Said and A. Paran, *Geometric Concepts in Islamic Art*, London: Tiranti, 1976. Whereas one does not know of any strictly practical Arabic treatise dealing strictly with the craft of building, I am quite familiar with a strictly literary tradition of "poetic" descriptions of Moslem buildings, either real or imaginary; see M. J. Rubiera (ed.), *La Arquitectura en la Literatura Árabe*, Madrid: Nacional, 1981 (with transcriptions of Arabic texts); as the editor observes (p. 25), "Los únicos que no escribieron sobre el tema fueron los arquitectos [y] los alarifes árabes guardaron sus secretos, su concepción de la arquitectura."

6. M-T. Zenner, "Imagining a Building: Latin Euclid and Practical Geometry," in J. J. Contreni (ed.), *Word, Image, Number: Communication in the Middle Ages*, Florence: SISMEL, 2002, 219–46.

7. On this now-famed architectural dilettante, see C. F. Barnes, "Villard de Honnecourt," in *Macmillan Dictionary of Art*, London: Macmillan, 1996, vol. 32, pp. 569–71; M-T. Zenner, "Villard de Honnecourt and Euclidan Geometry," *Nexus Network Journal*, IV/4 (Autumn 2002 @ http:// www. nexusjournal.com /Zenner.html).

Chapter 8

1. Among other publications routinely illustrating this *mudéjar* monument, see J. F. Moffitt, *The Arts in Spain*, London: Thames & Hudson, 1998, fig. 42.

Chapter 9

1. For more on this essential point, see J. F. Moffitt, "Archetypal Micro-Architecture: Prologemena to the 'Custodias Procesionales'," *Konsthistorisk Tidskrift*, XII, 1989, pp. 90–120; for another striking Spanish Host-centered, architectural ensemble (with a very different kind of "geometric" basis), see also idem., "Mystical Meaning and Anamorphic Form in Narciso Tomé's Transparente," *Canadian Art Review*, XIV, 1987, pp. 96–134.

2. J. F. Moffitt, *The Arts in Spain*, London: Thames & Hudson, 1998, pp. 79–81 (there initially giving this novel liturgical interpretation to Siloé's use of an octagonal figure).

Chapter 10

1. At Tepotzotlán, the iconographic programme is, of course, wholly Jesuit in inspiration, celebrating founder-saints (Francis Xavier, Ignatius of Loyola, Francis Borgia, Luis de Gonzaga, Stanislaus Kotska) solemnly standing beneath a representation of the Virgin and Child; see I. Bottineau, *Iberian-American Baroque*, Lausanne: Taschen Verlag, 1970, pp. 52, 134–35 (this book was especially useful to my research due to its high quality photographs and, especially, its accurate line-drawings of certain architectural features). For the written evidence pointing to itinerant journeymen, both builders and sculptors, travelling to different building sites, some even on the farthest frontiers of Nueva España, see Mardith Schuetz, "Professional Artisans in the Hispanic Southwest," *The Americas*, XL/1, July 1983, pp. 17–71 (and to her measured drawings of façades belonging to churches in Texas— her Figs. 1, 2, 7, 9, 15 — I have applied a grid-transparency. So doing, I find that they, too, conform to the 3:4:5 triangulation system).

2. Among others, see F. D. K. Ching, *Architecture: Form, Space and Order*, New York: Van Nostrand Reinhold, 1979, pp. 304–5, "Regulating Lines"; see also the extensive discussion of "inscribed figures" by Viollet-le-Duc, as quoted (and illustrated, figs. 74–84) in the Appendix.

3. E. Harth-Terré, *La Obra de Francisco Becerra en las Catedrales de Lima y Cuzco*, Buenos Aires: Instituto de Arte Americano, 1962. For previous studies analyzing the "*trazados reguladores*" of Latin American architecture, H. Rodríguez Camilloni, "Architectural Principles in the Age of Humanism Applied: The Church of San Francisco, Lima," *Journal of the Society of Architectural Historians*, XXVIII, 1969, pp. 235–53; J. de Mesa and T. Gisbert, "Los Esquemas Armónicos en la Arquitectura del Virreinato Peruano," *Traza y Baza*, V, 1974, pp. 7–29.

4. Rodríguez Camilloni, "Architectural Principles," p. 249.

5. F. Chueca Goitia, *Invariantes Castizos de la Arquitectura Española*. Madrid: Dossat, 1981, p. 134.

Chapter 11

1. For this notable architect's previously unpublished writings, see E. Tundidor de Carrera (ed.), *Discurso del Señor Juan de Herrera, Aposentador Mayor de S. M., "Sobre la Figura Cúbica,"*

Madrid: Nacional, 1976. The architect-author declared (p. 69) that his purpose was "a penetrar la admirable arte lulliana [y es] la figura cúbica raíz y fundamento de la dicha arte lulliana"; see also esp. p. 76 ff., where a "Pythagorean Cube" is discussed and illustrated by the recondite architect. Unfortunately, no scholarly study seems to exist treating specific applications of symbolic arithmetic and geometry during the Siglo de Oro.

2. For the symbolic context and hermetic content of Herrera's major architectural endeavor, see C. von der Osten Sacken, *El Escorial: Estudio Iconológico*, Madrid: Xarait, 1984; see also G. Kubler, *Building the Escorial*, Princeton Univ. Press, 1982 (giving all the pragmatic or technical information).

3. For the contextual material immediately following, see J. F. Moffitt, *The Arts in Spain*, London: Thames & Hudson, 1998, pp. 104–7.

4. My citations are drawn from a modern reprinting: Fray José de Sigüenza, *Fundación del Monasterio de El Escorial*, Madrid: Aguilar, 1963.

5. This curious work is now available in facsimile: J. A. Ramírez (ed.), *Díos, arquitecto. J. F. Villapando y el Templo de Salomón*, Madrid: Siruela, 1991.

6. For other applications in Spanish art of that essentially symbolic focal-point placement, again see J. F. Moffitt, "Archetypal Micro-Architecture: Prologemena to the 'Custodias Procesionales'," *Konsthistorisk Tidskrift*, XII, 1989, pp. 90–120.

7. For more on this archetypal architectural monument, and showing how it was depicted by artists, see J. F. Moffitt, "Anastasis-Templum: 'Subject or Non-Subject' in an Architectural Representation by Jacopo Bellini?" *Paragone*, XXXIII/no, 391, 1982, pp. 3–24.

8. See G. Kubler, *Building the Escorial*, Princeton Univ. Press, 1982, figs. 12, 14.

9. Serlio, *Tercero y Quarto Libro de Arquitectura de Sebastián Serlio*, facsimile edition, Madrid: Albatros, 1977, folio 23 recto.

10. So noted in John Onians, *Bearers of Meaning: The Classical Orders in Antiquity, the Middle Ages, and the Renaissance*, Princeton Univ. Press, 1988, p. 233 (for the larger context, see Moffitt, "Anastasis-Templum").

11. For the history of this palace, see Veronique Gérard, *De Castillo a Palacio: El Alcázar de Madrid en el siglo XVI*, Madrid: Xarait, 1984; for a guided tour (in English) through this Palace, ca. 1650, see J. F. Moffitt, "Velázquez in the Alcázar Palace in 1656: The Meaning of the mise-en-scène of Las Meninas," *Art History*, VI/3, 1983, pp. 272–301; idem., "The Architectural Setting of Velázquez's Las Hilanderas," *Pantheon: Internationale Jahreszeitschrift*, XLIV, 1986, pp. 43–49; idem., "The Prince and the Prime Minister: The Site and the Significance of Velázquez's Equestrian Lesson of Prince Baltasar Carlos," *Studies in Iconography*, XII, 1989, pp. 90–120. On the other hand, for a discussion of a contemporary use in Italy of essentially esoteric, "Pythagorean" architectural layouts, see G. L. Hersey, *Pythagorean Palaces: Magic and Architecture in the Italian Renaissance*, Cornell Univ. Press, 1976.

12. For some more of these (which I have checked with a grid transparency), see the illustrations in I. Bottineau, *Iberian-American Baroque*, Lausanne: Taschen Verlag, 1970, pp. 3 (Santiago, San Francisco), 6 (Umbrete), 8 (Madrid, San Marcos), 14, 21 (Mafra, Palace-Monastery), 42 (Zaragoza, El Pilar), 54–55 (Salamanca, Clerecía), 84 (Madrid, Cárcel de Corte), 89, 95 (Cádiz, Cathedral), etc.

Chapter 12

1. See, among other potential candidates, the examples cited in chapter 11, note 12, also note 3 below.

2. R. J. Mullen, *Architecture and Its Sculpture in Viceregal Mexico*, Austin: Univ. of Texas Press, 1997, pp. 183–4.

3. For a number of other Latin-American examples (some including modern measured drawings of them) that now appear (after being checked with the grid-overlay) to incorporate the 3:4:5 leit-motiv, see S. Sebastián López and J. de Mesa Figueroa, *Arte Iberoamericano, desde la Colonización a la Independencia*, Madrid: Espasa-Calpe, 1985; for façade designs incorporating this module, see Sebastián's Figs. 31, 56, 57, 204 (Acolman), 67 (Actopán), 72 (Coixtlahuaca: a drawing dated 1576), 96 (Angahua), 126, 128, 129 (Huejotzingo), 175 (Cholula), 189 (Tzinzuntzán,), 190 (Teccali), 219 (Ixmiquilpán), 222 (Meztitlán), 225 (Yurihapúndaro), 226–7 (Cuitzeo), 236 (Cathedral, Puebla), 253 (Casa Montejo, Mérida,), 327 (architect's drawing of 1593, Caracas, Venezuela), 335 (Quito, Ecuador), 503, 504, 531 (Cathedral, Mexico City), 516 (Santo Domingo, Oaxaca), 523 (San Bernardo, Mexico City), 524 (Basilica of Guadalupe, Mexico City), etc. For (at least) three retablos designed according to the 3:4:5 format, see Sebastián's Figs. 267–8 (Huejotzingo), 269 (Xochimilco), 529 (Cathedral, Puebla). As is also logical to expect at this point, I have also found that many floor-plans will also fall within this geometric grid-format; see Sebastián's Figs. 29 (Cathedral, Cuzco, Peru), 112 (Cathedral, Santo Domingo, Dominican Republic), 117 (Dominican Fathers Church, Santo Domingo), 155 (Tepeaca, Mexico), 173 (Cholula, Mexico), 220 (Ixmiquilpán, Mexico), 231 (Cathedral, Mexico City), 237 (Cathedral, Guadalajara, Mexico), 372 (La Paz, Bolivia), 380 (Ayacucho, Peru), etc. For more examples (photographs only) of Mexican façades, for which the same overlay procedure was used to find traces of the 3:4:5 triangle design-motif, see

E. W. Weisman, *Art and Time in Mexico, From the Conquest to the Revolution*, New York: Icon, 1985, Figs. 162 (Cathedral, Mexico City), 164 (Cathedral, Puebla), 184 (La Soledad, Oaxaca), 190 (Tlapujahua), 200 (Puebla), 201 (Tepotzotlán), 220 (Ciénega de Mata, Jalisco). Again, these examples do not by any means exhaust the potential corpus of 3:4:5 plans in Latin America and Spain. In any event, because these wholly geometric observations can be objectively duplicated, or even expanded upon by any other interested investigator similarly utilizing a digitized grid-overlay, these findings are, therefore, "scientific."

4. See J. Bourgoin, *Arabic Geometrical Patterns and Designs*, New York: Dover, 1973, Plates 1–42 (hexagons), 42–67 (octagons), 65–105 (dodecagons), 171–90 (pentagons).

5. On this modernist apian-architectural topos, see Juan Antonio Ramírez, *The Beehive Metaphor: From Gaudí to Le Corbusier*, London: Reaktion, 2000.

6. See Bourgoin, *Arabic Geometrical Patterns*, Plates 164–170 ("heptagons"); my fig. 58 is based on his Plate 169, showing one of the occasional heptagonal figures (and, according to my successive overlays over *all* of his 190 drawings, the 3:4:5 triangular-module does not work with either *hexagons, octagons* or *dodecagons*, and certainly not even with *pentagons*).

Chapter 13

1. Leon Battista Alberti, *On the Art of Building in Ten Books*, trans. J. Rykwert et al., Cambridge, Mass.: MIT Press, 1991, p. 7.

2. Fray Laurencio de San Nicolás, *Arte y Uso de Architectura (Madrid, 1639 y 1664)*, Madrid: Albatros, 1989 (facs. ed.), folios 26v — 27r. Since the language is a bit ambiguous, I will need to quote the passage in the original Spanish (and I have modernized the spelling): "Y así el principiante irá acostumbrándose a trazar plantas prolongadas y cuadradas, causando los ángulos con líneas en blanco en el papel do[nde] quiere trazar, y causará los ángulos rectos en la división de la línea, y sacando líneas paralelas, serán los ángulos opuestos también rectos. Y ante todas cosas harás sobre una línea ciertos tamaños, llamados por Vitruvio módulos, y por nosotros comunmente pitipié, [que es el] gobierno que ha de ser de todo el edificio dibujado."

3. E-E. Viollet-le-Duc, *Lectures on Architecture*, London, 1877 (facs. rpt. New York: Dover, 1987), p. 391.

4. Whereas "architectural symbolism," including symbolic valences often attributed to numerical figures, did unquestionably exist, that was, however, employed within a much broader formal context, "typology"; see R. Krautheimer, "Introduction to an 'Iconography' of Medieval Architecture," *Journal of the Warburg and Courtauld Institutes*, V, 1942, pp. 1–33; see also Otto von Simson, *The Gothic Cathedral*, New York: Torchbooks, 1964. For a specific example of well-documented typological symbolism, that traditionally attached to the Holy Sepulcher in Jerusalem, see J. F. Moffitt, "Anastasis-Templum: 'Subject or Non-Subject' in an Architectural Representation by Jacopo Bellini?" *Paragone*, XXXIII/ no, 391, 1982, pp. 3–24.

5. Much modern scholarship had been directed at adducing symbolic or mystical number relationships in medieval architecture. I believe this effort (excepting the publications cited in note 4) to be probably misdirected research because statements about such symbolic numerical-geometrical interpretations as do exist from medieval authors were largely the work of erudite clerics, and these savants were usually writing long after the historical appearance of architectural designs with geometric bases. Surely their arguments were all wholly foreign to the practice of rude builders, either masons or carpenters! Among others touching upon a perhaps inexhaustible subject, see G. Lesser, *Gothic Cathedrals and Sacred Geometry*, London: Tiranti, 1957; E. R. Sunderland, "Symbolic Numbers and Romanesque Church Plans," *Journal of the Society of Architectural Historians*, 18, 1958, pp. 94–103; H. E. Huntley, *The Divine Proportion. A Study in Mathematical Beauty*, New York: Dover, 1970; R. Lawlor, *Sacred Geometry: Philosophy and Practice*, London: Thames and Hudson, 1982; V. F. Hopper, *Medieval Number Symbolism: Its Sources, Meaning, and Influence on Thought and Expression*, rpt. New York: Dover, 2000; N. Hiscock, *The Wise Master Builder: Platonic Geometry in Plans of Medieval Abbeys and Cathedrals*, Oxford: Ashgate, 2000. As it must, "number symbolism" attracts students of the occult, and one example may stand for the unhinged rest of their architectural speculations: Louis Charpentier, *Les Mystères de la Cathedral de Chartres*, Paris: Laffont, 1966; *The Mysteries of Chartres Cathedral*, London: Thorsons, 1972.

6. For the evolution of the professional appellation, see N. Pevsner, "The Term 'Architect' in the Middle Ages," *Speculum*, XVII, 1942, pp. 549–62.

7. A rare exception to this rule of professional nomenclature is presented by José Jaime Ortiz (1656–1707). A Spaniard born in Alicante, and later educated in Valencia, Ortiz came to Quito in 1694, where died thirteen years later. The author of several important architectural projects in Quito, Ortiz is referred to in various documents as either a "Maestro Alarife" or as the "Maestro Arquitecto"; no other known builder of the period carries the second title. Previously unknown to scholars, Ortiz's oeuvre and professional career

have been recently rediscovered thanks to the dedicated archival research of Susan V. Webster, *Arquitectura y Empresa en el Quito Colonial: José Jaime Oritz, Alarife Mayor*, Quito: Ediciones Abya-Yala, 2002.

8. For one of the few in-depth analyses of Hispanic geometric practice, specifically in the work of Rodrigo Gil de Hontañón (ca. 1500–1577) at Segovia and Salamanca, see G. Kubler, "A Late Gothic Computation of Rib Vault Thrusts," *Gazette des Beaux-Arts*, LXXVI, 1944, pp 135–48. I may also mention that if one lays a transparency with the 3:4:5 triangle module (such as the one I have developed) over Kubler's Fig. 6—a section through the nave of the Catedral Nueva in Salamanca, as designed by Gil de Hontañón—it would appear that this provides yet another example of Hispanic design-by-Pythagorean-triangulation.

9. F. Bucher, "Design in Gothic Architecture. A Preliminary Assessment," *Journal of the Society of Architectural Historians*, XXXVII, 1968, pp. 49–71 (esp. pp. 52, 70); see also idem., "Medieval Architectural Design Methods, 800–1560," *Gesta*, XI/2, 1973, pp. 37–51. Particularly germane to this investigation is Bucher's short paper on "The Medieval Architectural Module: A Generating and Ornamental Concept," in I. Lavin (ed.), *World Art: Themes of Unity in Diversity*, Penn State Univ. Press, 1989, vol. I, pp. 223–26.

10. P. Frankl, *The Gothic: Literary Sources and Interpretations through Eight Centuries*, Princeton Univ. Press, 1960, p. 51; emphasis mine. For specific illustrations of such commonplace practices, see the Appendix, and figs. 74–84.

11. Ibid., pp. 51–52; my emphasis. As Frankl also explains (p. 70), "in the Gothic period, one measures according to proportions, and this method did not aim at mysterious calculations of 'beauty,' but rather had the purely practical purpose of measuring—without modern pocket rulers."

12. For the strictly practical nature of Spanish-Renaissance architecture, see the invaluable historiographic survey by A. Bonet Correa, *Figuras, modelos e imágenes en los tratadistas españoles*, Madrid: Alianza, 1993. For the even wider context of "secret" techniques and craft recipes, see W. Eamon, *Science and the Secrets of Nature: Books of Secrets in Medieval and Early Modern Culture*, Princeton Univ. Press, 1994.

13. Cennino d'Andrea Cennini, *The Craftsman's Handbook. The Italian "Il Libro dell'Arte,"* ed. D. V. Thompson, Yale Univ. Press, 1933, pp. 42–43; my emphasis. For the simple geometrical process that was used for finding horizontals and verticals by wielding a compass, so "bisecting an angle," and just as it was discussed by Cennini, see D. Pedoe, *Geometry and the Visual Arts*, New York: Dover, 1976, p. 20 ff.

14. Cennini, *Craftsman's Handbook*, pp. 56–57.

15. Again see E. Panofsky, *Meaning in the Visual Arts: Papers in and on Art History*, Garden City: Anchor, 1955, pp. 60–61, 65; see also S. Giedion, *The Beginnings of Architecture*, Princeton Univ. Press, 1981, pp. 482–92.

16. See E. Panofsky, *The Life and Art of Albrecht Dürer*, Princeton Univ. Press, 1955, p. 247 ff.

17. *Leon Battista Alberti: On Painting*, ed. J. R. Spencer, Yale Univ. Press, 1966, pp. 57, 68–69; idem., *On Painting*, ed. C. Grayson, Harmondsworth: Penguin, 1991, pp. 58, 65–66.

18. Spencer, in ibid., p. 121, n. 28.

19. W. M. Ivins, *Art and Geometry: A Study in Space Intuitions*, New York: Dover, 1964, pp. 72–73; see further, S. Y. Edgerton, *The Renaissance Rediscovery of Linear Perspective*, New York: Icon, 1975; P. Descargues, *Perspective: History, Evolution, Techniques*, New York: Van Nostrand Reinhold, 1982 (with much early graphic materials).

20. For this device, see Peter James and N. Thorpe, *Ancient Inventions*, New York: Ballantine, 1995, pp. 417–18, 534; the diopter illustrated on p. 417. Another kind of portable sighting instrument, and which the Arabs knew well, is the astrolabe; see David A. King, *Islamic Astronomical Instruments*, London: Ashgate, 1987.

21. James and Thorpe, *Ancient Invention*, p. 219; see also R. Temple, *China: Land of Discovery and Invention*, London: Patrick Stephens, 1986, pp. 218–24.

22. Panofsky, *Dürer*, p. 253 (here explicating his fig. 311. Dürer's string-grid is also shown in Panofsky's fig. 310, illustrating Dürer's depiction of a "Draftsman Drawing a Portrait").

23. M. Kemp, *The Science of Art: Optical Themes in Western Art from Brunelleschi to Seurat*, Yale Univ. Press, 1990, pp. 138–9; other net-type devices are discussed pp. 169–73. For more on Pozzo, see I. Sjöström, *Quadratura: Studies in Italian Ceiling Painting*, Stockholm: Almqvist & Wiksell, 1978, pp. 62–67.

24. See R. W. Wycherley, *How the Greeks Built Cities*, Garden City: Anchor, 1969; for an overview of "territorialen Rasters," see André Corboz, *Die Kunst, Stadt und Land zum Sprechen zu bringen*, Basel: Birkhäuser, 2001.

25. See F. Chueca Goitia, *Breve Historia del Urbanismo*, Madrid: Alianza, 1968, esp. p. 126 ff.; A. Bonet Correa, *El Urbanismo en España e Hispanoamérica*, Madrid: Cátedra, 1991, esp. p. 182 ff.; J. M. Morales Folguera, *Arquitectura y Urbanismo Hispanoamericano en Luisiana y Florida Occidental*, Málaga: Universidad, 1987; see further B. E. Mundy, *The Mapping of New Spain: Indigenous Cartography and the Maps of the 'Relaciones Geográficas'*, Univ. of Chicago Press, 1996; J. F. Moffitt, "Ptolemy's Chorographia in the New World: Revelations from the *Relaciones Geográficas de la Nueva España* of 1571–1581," *Art History*, XXI/3, Sept. 1998, pp. 367–92; R. Kagan and

F. Marías, *Urban Images of the Hispanic World, 1493–1793*, Yale Univ. Press, 1999.

26. Toribio de Benavente ("Motolinía"), *Historia de los indios de la Nueva España*, Madrid: Alianza, 1988, pp. 296–97 (Bk. III, ch. 17).

Chapter 14

1. This has now been published, with an English translation and a transcription of the original Spanish text; see Mardith K. Schuetz (ed.), *Architectural Practice in Mexico City: A Manual for Journeyman Architects of the Eighteenth Century*, Tucson: Univ. of Arizona Press, 1987.

2. Ibid., pp. 20, 21, 26, 27, 43–44, 51–53, 70; I am, however, quoting from the corresponding Spanish text (which I have modernized), as given on pp. 82, 86, 87, 88, 98–99, 102–3, 116 (and my translation differs from Schuetz's in various particulars).

Chapter 15

1. The citations following (giving book, chapter, and sections numbers) are again taken from the standard bi-lingual edition: F. Granger (ed.), *Vitruvius on Architecture*, London: Loeb, 1962.

2. For the essentially rhetorical issues in the Renaissance propelling such "theory" (as opposed to practical workshop practice and procedures), see V. Hart and P. Hicks (eds.), *Paper Palaces: The Rise of the Renaissance Architectural Treatise*, Yale Univ. Press, 1998; Alina A. Payne, *The Architectural Treatise in the Italian Renaissance: Architectural Invention, Ornament, and Literary Culture*, Cambridge Univ. Press, 1999.

3. Leon Battista Alberti, *On the Art of Building in Ten Books*, trans. J. Rykwert et al., Cambridge, Mass.: MIT Press, 1991, pp. 302–3.

4. See K. Conant, "The After-Life of Vitruvius in the Middle Ages," *Journal of the Society of Architectural Historians*, XXXVII, 1968, pp. 33–38. Particularly this author emphasizes the survival of Vitruvian proportionate systems ("*eurhythmia*" and "*symmetria*"), which he sees commonly expressed in the medieval "*diagon*" and "*auron*" triangular figures. However, Paul Frankl (*The Gothic: Literary Sources and Interpretations through Eight Centuries*, Princeton Univ. Press, 1960, p. 102) stated the case more succinctly: "The secret of the Gothic [builders'] lodges was derived from the *ordinatio* and the *symmetria* of Vitruvius."

5. These, and other, authority-figures are cited in S. K. Heninger, *Touchs of Sweet Harmony: Pythagorean Cosmology and Renaissance Poetics*, San Marino: Huntington Library, 1974, p. 43, n. 55.

6. R. Wittkower, *Gothic vs. Classic: Architectural Projects in Seventeenth-Century Italy*, New York: Braziller, 1974, p. 25. The Cathedral of Milan illustration appears on folio xv *verso* of the *Cesariano Vitruvius*.

7. Frankl, *The Gothic*, p. 67.

8. Miguel de Urrea (trans.), *M. Vitrvvio Pollión De Architectura* (Alcalá de Henares: Juan Gracián, 1583); facsimile rpt., Madrid: Albatros Ediciones, 1978, pp. 112–13 (I have modernized the Spanish orthography).

9. The woodcut by Urrea discussed here (my fig. 63) was copied from the *Cesariano Vitruvius* (1521), where the "*Normae Pythagorae*" are shown on folio 145 v. (but Urrea had independently added the notched-ruler motif to his woodcut).

10. In Cesariano's print (fol. 145 v), which was unquestionably Urrea's graphic model, this motif is labeled "*Duplae Scalae Iconographia.*"

11. Fray Laurencio de San Nicolás, *Arte y Uso de Architectura (Madrid, 1639 y 1664)*, Madrid: Albatros, 1989 (facs. ed.), fols. 79 verso to 81 recto. Generation of the other *cartabones* is graphically illustrated on fol. 80r; this woodcut looks very much like our fig. 67 (center, and as taken from a ca. 1640 Mexican manuscript). Another reference to the Pythagorean rectangle, serving to demonstrate the square root, is found on fol. 136v.

Chapter 16

1. Diego López de Arenas, *Breve Compendio de la Carpintería de lo Blanco y Tratado de Alarifes*, ed. M. A. Toajas Roger, Madrid: Visor, 1997, with extensive bibliography. For the Spanish-Renaissance context of this wholly practical *tratado*, see A. Bonet Correa, *Figuras, modelos e imágenes en los tratadistas españoles*, Madrid: Alianza, 1993; for the even wider context of López de Arenas' manual, full of "secret" techniques and craft recipes, again see W. Eamon, *Science and the Secrets of Nature: Books of Secrets in Medieval and Early Modern Culture*, Princeton Univ. Press, 1994; see also David Hockney, *Secret Knowledge: Rediscovering the Lost Techniques of the Old Masters*, London: Thames & Hudson, 2001.

2. My great thanks go to my Iberian colleague, Dr. Pedro Lavado Paradinas (Madrid, Ministerio de Cultura), who kindly sent me a facsimile of the holograph edition (1618) of this standard how-to-do-it manual: M. Gómez-Moreno (ed.), *Primera y segunda parte de las Reglas de la Carpintería, hecho por Diego López de Arenas en este año de MDCXVIII*, Madrid: Instituto Valencia de Don Juan, 1966 (this drawing is found on p. 15).

Chapter 17

1. M. A. Toajas (ed.), in Diego López de Arenas, *Breve Compendio de la Carpintería de lo*

Blanco y Tratado de Alarifes, Madrid: Visor, 1997, p. 53.

2. Again my warmest thanks are extended to Pedro Lavado for initially sending me a partial copy of the ms. by Fray Andrés de San Miguel, so spurring my interest; see E. Báez Macías (ed.), *Obras de Fray Andrés de San Miguel*, Cd. México: Universidad Autónoma, 1969 (ms. transcription on pp. 91–246); see also Manuel Toussaint, "Fray Andrés de San Miguel, arquitecto de la Nueva España," *Anales del Instituto de Investigaciones Estéticas*, XIII, 1945, pp. 5–14; S. Sebastián and J. de Mesa Figueroa, *Arte Iberoamericano, desde la Colonización a la Independencia*, Madrid: Espasa-Calpe, 1985, pp. 44, 45, 71, 84, 85, 116, 493–97.

3. Báez Macías (ed.), *Obras de Fray Andrés de San Miguel*, pp. 91–98.

4. Ibid., plate 49; as described on p. 166.

5. Ibid., p. 166.

6. Ibid., plate 51; as described on p. 167.

7. Ibid., pp. 163–74, "De la carpintería de lo blanco." San Miguel's verbal descriptions, as clarified by his meticulous ink drawings (plates 47–63), are clearer than those by López de Arenas.

8. Ibid., p. 167.

9. For other plans of this sort, see ibid., plates 73–78. Application of a grid-transparency with 3:4:5 triangles motifs shows me that these *plantas* also to conform to the same scheme.

10. "Regla," in ibid., p. 35.

11. San Miguel, in *ibid.*, pp. 103–4.

12. Ibid., p. 110.

13. Ibid., p. 114; plate 2, fig. 18; the further description of "*el triángulo perfecto*" is found on pp. 116–17; plate 2, figs. 29, 32.

14. Ibid., plate 69 (not otherwise described in San Miguel's text).

Chapter 18

1. Mendieta, as quoted in John McAndrew, *The Open-Air Churches of Sixteenth-Century Mexico*, Harvard Univ. Press, 1969, p. 194.

2. So observed in ibid., pp. 128, 333, 587.

3. Ibid., p. 183.

4. Ibid., pp. 199–200 (emphasis mine).

5. Ibid., pp. 196–97, 316.

6. Puga, as in ibid., p. 121.

7. Mendieta, as in ibid., pp. 202–3. For an extensive discussion of various examples of open-patio churches, see ibid., pp. 202–34.

8. Ibid., p. 202.

9. Ibid., p, 235.

10. Ibid., p. 236.

11. For what is actually known of the early history (and eventual fate) of San José, see ibid., pp. 374–88.

12. Zumárraga, as in ibid., pp. 375, 376.

13. Cervantes, as in ibid., p. 381.

14. Gante, as in ibid., pp. 378, 379.

15. Cervantes, as in ibid., pp. 381, 384.

16. See ibid., pp. 388–93.

17. Ibn Jubayr, as in K. A. C. Creswell, *A Short Account of Early Muslim Architecture*, Harmondsworth: Penguin, 1958, p. 13.

18. Edict, as fully transcribed in M. Toussaint, *Arte Mudéjar en America*, Mexico DF: Porrúa, 1946, pp. 119–20; the Arabic names of the unwelcome settlers are given on p. 9, the citation from Dr. Cervantes appears on p. 8.

19. Decrees, as cited in E. W. Palm, *Los Monumentos Arquitectónicos de La Española*, Santo Domingo: Editora de Santo Domingo, 1984, p. 89.

20. See Toussaint, *Arte Mudéjar en America*, pp. 11, 15, and illustrated in fig. 1.

21. For an extensive discussion of the Christian sources of the horseshoe arch (and citing the specialized literature), see Jerrilynn D. Dodds, *Architecture and Ideology in Early Medieval Spain*, Pennsylvania State Univ. Press, 1990.

22. Ibid., pp. 98, 105; for the broader, mostly political, context of this quite different architectural arrangement, see ibid., pp. 83–109.

23. John D. Hoag, *Western Islamic Architecture*, New York: Braziller, 1967, p. 9. For the many Koranic descriptions of the "paradise" promised to the faithful, see the index (s.v. "paraíso") to *El Corán*, trans. Julio Cortés, Madrid: Editorial Nacional, 1980. For the Christian convent-garden as a "*paradisus claustri*," see (among others) Santiago Sebastián López, *Iconografía Medieval*, San Sebastián: Etor, 1988, pp. 165–70.

24. For the former mosque now called El Salvador, see G. M. Borrás Gualis, *El Islam, de Córdoba al Mudéjar*, Madrid: Silex, 1997, p. 126. For a plan of the Almohad *masjid alhama* of Seville (now become the Cathedral of Seville), see Jerrilynn D. Dodds (ed.), *Al-Andalus: The Art of Islamic Spain*, New York: Metropolitan Museum of Art, 1992, p. 78; its geographical relationship to El Salvador is shown on p. 168.

25. For what is actually known of the early history of the Capilla Real at Cholula, see McAndrew, *The Open-Air Churches of Sixteenth-Century Mexico*, pp. 400–11. Manuel Toussaint (*Arte Mudéjar en America*, p. 26) cited the Cholula monument as the best surviving example of the Mexican "capillas abiertas en forma de mezquitas musulmanas." Other authors supporting the mosque derivation of the Capilla Real include Santiago Sebastián López and José de Mesa Figueroa, *Arte Iberoamericano, desde la Colonización a la Independencia*, Madrid: Espasa-Calpe, 1985, pp. 155–59; S. Y. Edgerton, *Theaters of Conversion: Religious Architecture and Indian Artisans in Colonial Mexico*, Albuquerque: Univ. of New Mexico Press, 2001, pp. 58–60.

26. *Relación de Cholula*, as in McAndrew, *Open-Air Churches*, p. 400; for the identification

of Luis de Arciniega as the most likely architect of the Capilla Real, see p. 406; for a graphic reconstruction of the original, Córdoban façade of the Capilla Real, see ibid., p. 408. However, our isometric reconstruction of the entire building (fig. 71), likewise its measured floor-plan (fig. 72), have been taken from Santiago Sebastián López and J. de Mesa Figueroa, *Arte Iberoamericano, desde la Colonización a la Independencia*, Madrid: Espasa-Calpe, 1985, figs. 172, 173.

27. Cobo, as in McAndrew, ibid. p. 410.

28. For other patio-churches built in the implicitly mosque-like manner of San José de los Naturales, namely Jilotepec, Etzatlán and Toluca, see ibid., pp. 411–17.

29. Charles V, as in Creswell, *A Short Account of Early Muslim Architecture*, p. 216.

30. Ibn Idari, as in M. J. Rubiera (ed.), *La Arquitectura en la Literatura Árabe*, Madrid: Nacional, 1981, p. 111. For the Visigothic sources of the Great Mosque of Córdoba and other Hispano-Islamic buildings, see Dodds, *Architecture and Ideology in Early Medieval Spain*.

Chapter 19

1. I have tried to document the use of the "design-by-string" methodology by native builders in New Mexico. Alas, to date, I have not been able to photograph this process, such as it is (probably still, I think) being done today. Although modern technology is steadily eroding traditional practice, all the local builders are, of course, still quite familiar with the ancient practice of initially "drawing lines" across façades by string in order to lay out rough designs.

2. Plutarch, as quoted in E-E. Viollet-le-Duc, *Lectures on Architecture*, (rpt.) New York: Dover, 1987, p. 391 (see also the Appendix). For more on this citation, see P. Frankl, *The Gothic: Literary Sources and Interpretations through Eight Centuries*, Princeton Univ. Press, 1960, p. 69.

3. A few colonial-era drawings of Latin-American buildings have been published; see Diego Angulo Iñiguez, *Planos de monumentos arquitectónicos de América y Filipinas existentes en el Archivo de Indias, Sevilla*, Seville: Laboratorio del Arte de la Universidad, 1933–40 (5 vols.). The problem with these *planos* is that they only include constructions directly funded by the Crown, that is, government buildings. Consequently, the total number of buildings represented here is less than two dozen, and none show a measured drawing of a *fachada-retablo*. Among these, there are (of course) no working sketches, and certainly none with inscribed figures.

4. For the most likely candidates, again see (among others) the buildings listed above, in chapter 11, note 12, and chapter 12, note 3.

5. E. Camps Cazorla, *Módulo, proporciones y composición en la arquitectura califal cordobesa*, Madrid: CSIC, 1953, pp. 13, 93–96; figs. 58–61.

6. Whereas my "inscribed grid" proposal is simple, even rudimentary ("Neo-Mudéjar Design for Dummies"), for a much more elaborate explanation of the unquestionably "geometric" character of Latin-American churches, see Mardith Schuetz, "Proportionate System and Ancient Geometry," *Bulletin of the Southwestern Mission Research Center*, 1980, pp. 1–7. Attractive as are her eleven drawings—and her "geometric" thesis is certainly correct in the broadest sense — nonetheless she provides neither written nor graphic evidence of contemporary workshop procedures to back up the raison d'être for her complicated diagrams, with immense circles and equilateral triangles seemingly extending into outer space. In any event, the actual (or "workable") geometric basis of each of the two façades she illustrated — San Francisco Borja, Baja California, and La Concepción in San Antonio, Texas—could be explained far more simply with applications (as transparencies) of the 3:4:5 grid-pattern overlay I have used. The principle of "Occam's razor" works here too: *the simplest, or most mundane, explanation is always preferable* (for equivalent examples, as "secretly" done by post-medieval painters, see D. Hockney, *Secret Knowledge: Rediscovering the Lost Techniques of the Old Masters*, London: Thames & Hudson, 2001.

7. See Carmen C. Bambach, *Drawing and Painting in the Italian Renaissance Workshop: Theory and Practice, 1300–1600*, Cambridge Univ. Press, 2000.

8. J. Ackerman, "Architectural Practice in the Italian Renaissance," *Journal of the Society of Architectural Historians*, XIII, 1954, pp. 3–11.

9. For the most likely candidates, once again see (among many others) the buildings listed above, in chapter 11, note 12, and chapter 12, note 3.

Appendix

1. The citations (and illustrations) following are drawn from the standard English translation: Eugène-Emmanuel Viollet-le-Duc, *Lectures on Architecture*, New York: Dover, 1987, vol. I, p. 388 ff. (in "Lecture IX").

Bibliography

An [*] indicates a primary source or historical document.

Ackerman, J. S. "Architectural Practice in the Italian Renaissance," *Journal of the Society of Architectural Historians*, XIII, 1954, pp. 3–11.

_____. "'Ars Sine Scientia Nihil Est': Gothic Theory of Architecture at the Cathedral of Milan," *Art Bulletin*, 31/2, 1949, pp. 84–108.

[*] Alberti, Leon Battista. *Leon Battista Alberti: On Painting*, ed. J. R. Spencer, Yale Univ. Press, 1966.

[*] _____. *On Painting*, ed. C. Grayson, Harmondsworth: Penguin, 1991.

[*] _____. *On the Art of Building in Ten Books*, trans. J. Rykwert et al., Cambridge, Mass.: MIT Press, 1991.

Angulo Iñíguez, Diego. *Arquitectura Mudéjar Sevillana de los Siglos XIII, XIV y XV*. Seville: Laboratorio del Arte de la Universidad, 1932.

[*] _____. *Planos de monumentos arquitectónicos de América y Filipinas existentes en el Archivo de Indias, Sevilla*. Seville: Laboratorio del Arte de la Universidad, 1933–40 (5 vols.).

Azcárate, J. M. *Arte Gótico en España*. Madrid: Cátedra, 1996.

Badawy, A. *Ancient Egyptian Architectural Design: A Study of the Harmonic System*. Berkeley: Univ. of California Press, 1965.

Baird, J. A. "The Eighteenth-Century Retables of the Bajío: The Querétaro Style," *Art Bulletin*, XXX, 1953, pp. 195–216.

_____. "Mexican Architecture and the Baroque," in M. Meiss (ed.), *Latin American Art, and the Baroque Period in Europe*. Princeton Univ. Press, 1963, pp. 191–202.

Bambach, Carmen C. *Drawing and Painting in the Italian Renaissance Workshop: Theory and Practice, 1300–1600*. Cambridge Univ. Press, 2000.

Barnes, C. F. "Villard de Honnecourt," in *Macmillan Dictionary of Art*. London: Macmillan, 1996, vol. 32, 569–71.

Berg-Sobré, Judith. *Behind the Altar Table: The Development of the Painted Retable in Spain, 1350–1500*. Missouri Univ. Press, 1989.

Berteaux, E. *L'Art Mudéjar. Les Survivances de l'Art Musulman dans l'Art Chrêtien d'Espagne*. Paris, 1912–13.

Bevan, Bernard. *Historia de la Arquitectura Española*. Barcelona: Juventud, 1970.

Biedny, D., and B. Monroy. *The Official Adobe Photoshop Handbook*. New York: Bantam, 1991.

Blair, Sheila S. *Islamic Inscriptions*. New York: NYU Press, 1998.

Bonet Correa, Antonio. *El Urbanismo en España e Hispanoamérica*. Madrid: Cátedra, 1991.

_____. *Figuras, modelos e imágenes en los tratadistas españoles*. Madrid: Alianza, 1993.

Borrás Gualis, G. M. *El arte mudéjar*. Teruel: Instituto de Estudios Turolenses, 1990.

_____. *El Islam, de Córdoba al Mudéjar*. Madrid: Silex, 1997.

Bottineau, Ives. *Iberian-American Baroque*. Lausanne: Taschen Verlag, 1970.

[*] Bourgoin, Jean. *Les Eléments de l'art arabe: le trait des entrelacs*. Paris: Eirmin-Didot et Cie, 1879; (facs. rpt.) *Arabic Geometrical Patterns and Designs*. New York: Dover, 1973.

Bucher, François. "Design in Gothic Architecture. A Preliminary Assessment," *Journal of the Society of Architectural Historians*, XXXVII, 1968, pp. 49–71.

_____. "Medieval Architectural Design Methods, 800–1560," *Gesta*, XI/2, 1973, pp. 37–51.

_____. "The Medieval Architectural Module: A Generating and Ornamental Concept," in I. Lavin (ed.), *World Art: Themes of Unity in Diversity*. Penn State Univ. Press, 1989, vol. I, pp. 223–26.

Burckhardt, Titus. *La civilización hispano-árabe*. Madrid: Alianza, 1982.

Camps Cazorla, Emilio. *Módulo, proporciones y composición en la arquitectura califal cordobesa*. Madrid: CSIC, 1953.

Castro, Américo. *España en su historia. Cristianos, moros y judíos*. Barcelona: Grijalbo, 1996.

[*] Cennini, Cennino d'Andrea. *The Craftsman's Handbook. The Italian "Il Libro dell'Arte,"* ed. D. V. Thompson. Yale Univ. Press, 1933.

[*] Cesariano, Ceasre. *Cesariano Vitruvius*. facs. rpt: New York: Benjamin Blom, 1968.

Charpentier, Louis. *Les Mystères de la Cathedral de Chartres*. Paris: Laffont, 1966; *The Mysteries of Chartres Cathedral*. London: Thorsons, 1972.

Ching, F. D. K. *Architecture: Form, Space, and Order*. New York: Van Nostrand Reinhold, 1979.

Chueca Goitia, Fernando. *Breve Historia del Urbanismo*. Madrid: Alianza, 1968.

_____. *Invariantes Castizos de la Arquitectura Española*. Madrid: Dossat, 1981.

Collier, Margaret. "New Documents on Lorenzo Rodríguez and His Style," in M. Meiss (ed.), *Latin American Art, and the Baroque Period in Europe*. Princeton Univ. Press, 1963, pp. 203–18.

[*] Colonna, Francesco. *Hypnerotomachia Poliphili: The Strife of Love in a Dream*, ed. J. Godwin. London: Thames & Hudson, 1999.

Conant, Kenneth. "The After-Life of Vitruvius in the Middle Ages," *Journal of the Society of Architectural Historians*, XXVII, 1968, pp. 33–38.

Corboz, André. *Die Kunst, Stadt und Land zum Sprechen zu bringen*. Basel: Birkhäuser, 2001.

Creswell, K. A. C. *A Short Account of Early Muslim Architecture*. Harmondsworth: Penguin, 1958.

Curtius, Ernst Robert. *European Literature and the Latin Middle Ages*. Princeton Univ. Press, 1990.

Descargues, Pierre. *Perspective: History, Evolution, Techniques*. New York: Van Nostrand Reinhold, 1982.

Dodds, Jerrilynn D. *Architecture and Ideology in Early Medieval Spain*. Pennsylvania State Univ. Press, 1990.

Dodds, Jerrilynn D. (ed.). *Al-Andalus: The Art of Islamic Spain*. New York: Metropolitan Museum of Art, 1992.

Donaire, Alberto. "Lace-Design Methods in Late Medieval Spain," in I. Lavin (ed.), *World Art: Themes of Unity in Diversity*. Penn State Univ. Press, 1989, vol. I, pp. 231–38.

Eamon, William. "Technology, Treatises on," in J. R. Strayer (ed.), *Dictionary of the Middle Ages*. New York: Scribner's, 1988, vol. XII, pp. 640–50.

Eamon, William. *Science and the Secrets of Nature: Books of Secrets in Medieval and Early Modern Culture*. Princeton Univ. Press, 1994.

Edgerton, S. Y. *The Renaissance Rediscovery of Linear Perspective*. New York: Icon, 1975.

_____. *Theaters of Conversion: Religious Architecture and Indian Artisans in Colonial Mexico*. Albuquerque: Univ. of New Mexico Press, 2001.

El-Said, I., and A. Paran. *Geometric Concepts in Islamic Art*. London: Tiranti, 1976.

Frankl, Paul. "The Secret of the Medieval Masons," *Art Bulletin*, XXVII, 1945, pp. 46–50.

_____. *The Gothic: Literary Sources and Interpretations through Eight Centuries*. Princeton Univ. Press, 1960.

Gauldie, S. *Architecture*. Oxford Univ. Press, 1969.

Gérard, Veronique. *De Castillo a Palacio: El Alcázar de Madrid en el siglo XVI*. Madrid: Xarait, 1984.

Ghyka, M. C. *The Geometry of Art and Life*. New York: Dover, 1977.

Giedion, Siegfried. *The Beginnings of Architecture*. Princeton Univ. Press, 1981.

[*] Giorgio Martini, Francesco di. *Tratatti di architettura ingegneria e art militare*, ed. C. Maltese. Milan: Polifilo, 1967.
Grabar, Oleg. *The Formation of Islamic Art*. Yale Univ. Press, 1973.
Gustafson, R. S. (ed.). *Builder 3 & 2*. Washington DC: Navedetra 10646, 1987.
[*] Guthrie, K. S. (ed.). *The Pythagorean Sourcebook and Library*. Grand Rapids: Phanes, 1987.
Hagen, Oscar. *Patterns and Principles of Spanish Art*. Madison: Wisconsin Univ. Press, 1948.
Hart, V., and P. Hicks (eds.). *Paper Palaces: The Rise of the Renaissance Architectural Treatise*. Yale Univ. Press, 1998.
Harth-Terré, Emilio. *La Obra de Francisco Becerra en las Catedrales de Lima y Cuzco*. Buenos Aires: Instituto de Arte Americano, 1962.
Harvey, L. P. *Islamic Spain, 1250 to 1500*. Chicago Univ. Press, 1990.
Heninger, S. K. *Touches of Sweet Harmony: Pythagorean Cosmology and Renaissance Poetics*. San Marino: Huntington Library, 1974.
[*] Herrera, Juan de. *Discurso del Señor Juan de Herrera, Aposentador Mayor de S. M., "Sobre la Figura Cúbica,"* ed. E. Tundidor de Carrera. Madrid: Nacional, 1976.
Hersey, G. L. *Pythagorean Palaces: Magic and Architecture in the Italian Renaissance*. Cornell Univ. Press, 1976.
Hiscock, N. *The Wise Master Builder: Platonic Gerometry in Plans of Medieval Abbeys and Cathedrals*. Oxford: Ashgate, 2000.
Hoag, John D. *Western Islamic Architecture*. New York: Braziller, 1967.
Hockney, David. *Secret Knowledge: Rediscovering the Lost Techniques of the Old Masters*. London: Thames & Hudson, 2001.
Hopper, V. F. *Medieval Number Symbolism: Its Sources, Meaning, and Influence on Thought and Expression*. New York: Dover, 2000.
Huntley, H. E. *The Divine Proportion. A Study in Mathematical Beauty*. New York: Dover, 1970; Lawlor, R. *Sacred Geometry: Philosophy and Practice*. London: Thames and Hudson, 1982.
[*] Isidoro de Sevilla. *Etimologías*, ed. S. Montero Díaz. Madrid: Biblioteca de Autores Cristianos, 1951.
Iverson, E. *Canon and Proportions in Egyptian Art*. Warminster: Aris and Philips, 1975.
Ivins, W. M. *Art and Geometry: A Study in Space Intuitions*. New York: Dover, 1964.
James, Peter, and N. Thorpe. *Ancient Inventions*. New York: Ballantine, 1995.
Jones, Dalu. "Surface, Pattern, and Light," in G. Michell (ed.), *Architecture of the Islamic World: Its History and Social Meaning*. London: Thames and Hudson, 1978, pp. 144–75.
Kagan, R., and F. Marías. *Urban Images of the Hispanic World, 1493–1793*. Yale Univ. Press, 1999.
Kemp, Martin. *The Science of Art: Optical Themes in Western Art from Brunelleschi to Seurat*. Yale Univ. Press, 1990.
Kielland, B. C. *Geometry in Egyptian Art*. London: Tiranti, 1955.
King, David A. *Islamic Astronomical Instruments*. London: Ashgate, 1987.
Koslof, Spiro (ed.). *The Architect: Chapters in the History of the Profession*. New York: Oxford Univ. Press, 1977.
_____. *A History of Architecture: Settings and Ritual*. New York: Oxford Univ. Press, 1995.
Krautheimer, Rudolf. "Introduction to an 'Iconography' of Medieval Architecture," *Journal of the Warburg and Courtauld Institutes*, V, 1942, pp. 1–33.
Kubler, George, and M. Soria. *Art and Architecture in Spain and Portugal and Their American Dominions, 1500–1800*. Baltimore: Pelican, 1959.
_____. *Building the Escorial*. Princeton Univ. Press, 1982.
_____. "A Late Gothic Computation of Rib Vault Thrusts," *Gazette des Beaux-Arts*, LXXVI, 1944, pp 135–48.
_____. "On the Extinction of the Motifs of Pre-Columbian Art," in *Essays in Pre-Columbian Art and Archaeology*. Harvard Univ. Press, 1961.
Lavado Paradinas, Pedro, et al. *El Arte Mudéjar: La estética islámica en el arte cristiano*. Madrid: Electa, 2000.
Lavin, Irving (ed.). *World Art: Themes of Unity in Diversity*. Penn State Univ. Press, 1989.
Lesser, G. *Gothic Cathedrals and Sacred Geometry*. London: Tiranti, 1957.

Lewcock, Ronald. "Architects, Craftsmen and Builders: Materials and Techniques," in G. Michell, (ed.), *Architecture of the Islamic World: Its History and Social Meaning*. London: Thames & Hudson, 1978, pp. 112–43.

[*] López de Arenas, Diego. *Breve Compendio de la Carpintería de lo Blanco y Tratado de Alarifes*, ed. M. A. Toajas Roger. Madrid: Visor, 1997.

[*] _____. *Primera y segunda parte de las Reglas de la Carpintería, hecho por Diego López de Arenas en este año de MDCXVIII*, ed. M. Gómez-Moreno. Madrid: Instituto Valencia de Don Juan, 1966.

López Guzmán, R. *Arquitectura Mudéjar: Del Sincretismo Medieval a las Alternativas Hispanoaméricanas*. Madrid: Cátedra, 2000.

Lund, F. M. *Ad Quadratum: A Study of the Geometrical Basis of Classical and Medieval Architecture*. London: Batsford, 1921.

Mann, V. B. and J. D. Dodds (eds.). *Convivencia: Jews, Muslims, and Christians in Medieval Spain*. New York: Braziller, 1992.

McAndrew, John. *The Open-Air Churches of Sixteenth-Century Mexico*. Harvard Univ. Press, 1969.

McNeill, W. H. *Plagues and Peoples*. New York: Mentor, 1976.

Meinecke, Michael. *Patterns of Stylistic Change in Islamic Architecture: Local Traditions versus Migrating Artists*. New York: NYU Press, 1996.

Mesa, J. de, and T. Gisbert. "Los Esquemas Armónicos en la Arquitectura del Virreinato Peruano," *Traza y Baza*, V, 1974, pp. 7–29.

Michell, George (ed.). *Architecture of the Islamic World: Its History and Social Meaning*. London: Thames and Hudson, 1978.

Millon, Henry. "The Architectural Theory of Francesco di Giorgio," *Art Bulletin*, XL, 1958, pp. 257–61.

Moffitt, J. F. "Anastasis-Templum: 'Subject or Non-Subject' in an Architectural Representation by Jacopo Bellini?" *Paragone*, XXXIII/ no, 391, 1982, pp. 3–24.

Moffitt, J. F. "Archetypal Micro-Architecture: Prologemena to the 'Custodias Procesionales'," *Konsthistorisk Tidskrift*, XII, 1989, pp. 90–120.

_____. "The Architectural Setting of Velázquez's Las Hilanderas," *Pantheon: Internationale Jahreszeitschrift*, XLIV, 1986, pp. 43–49.

Moffitt, J. F. *Art Forgery: The Case of the Lady of Elche*. Gainesville: Florida Univ. Press, 1995 (*El caso de la Dama de Elche: Historia de una falsificación*. Barcelona: Destino, 1996).

_____. *The Arts in Spain*. London: Thames & Hudson, 1998 (*Las Artes en España*, Barcelona: Destino, 1999).

_____. "Capable of Infinite Extension: From the Arabic 'Interlace' to Alberti's 'Velo' via 'Occam's Razor'." *Source*, XXI/4, 2002, pp. 25–35.

_____. "Mystical Meaning and Anamorphic Form in Narciso Tomé's Transparente," *Canadian Art Review*, XIV, 1987, pp. 96–134.

_____. "The Prince and the Prime Minister: The Site and the Significance of Velázquez's Equestrian Lesson of Prince Baltasar Carlos," *Studies in Iconography*, XII, 1989, pp. 90–120.

_____. "Ptolemy's Chorographia in the New World: Revelations from the *Relaciones Geográficas de la Nueva España* of 1571–1581," *Art History*, XXI/3, Sept. 1998, pp. 367–92.

_____. "La significación histórica de la arquitectura hispano-visigoda," *Goya: Revista de Arte*, no. 186, 1985, pp. 338–47.

_____. *Spanish Painting*. London: Studio Vista, 1973.

_____. "¿Tepotzotlán: El Islam latente en America? Observaciones en torno a la portada esculpida hispánica," *Anales del Instituto de Investigaciones Estéticas de la UNAM*, LVIII, 1986, pp. 101–12.

_____. "Velázquez in the Alcázar Palace in 1656: The Meaning of the mise-en-scène of Las Meninas," *Art History*, VI/3, 1983, pp. 272–301.

_____. "Wendel Dietterlin and the Estípite. Observations on Architectural Style in Colonial Mexican Ecclesiastical Facades," *Boletín del Seminario de Estudios de Arte y Arqueología de la Universidad de Valladolid*, L, 1984, pp. 325–49.

_____, and S. Sebastián. *O Brave New People: The European Invention of the American Indian*. Albuquerque: UNM Press, 1996.

Montêquin, F-A. de (ed.). *Compendium of Hispano-Islamic Art and Architecture.* St. Paul, MN: Hamline Univ. Press, 1976.

_____. de *Muslim Architecture of the Iberian Peninsula: Eastern and Western Sources for Hispano-Islamic Building Arts.* West Cornwall, CT: Locust Hill Press, 1987.

Morales Folguera, J. M. *Arquitectura y Urbanismo Hispanoamericano en Luisiana y Florida Occidental.* Málaga: Universidad, 1987

[*] Motolinía (Fray Toribio de Benavente). *Historia de los indios de la Nueva España.* Madrid: Alianza, 1988.

Mullen, R. J. *Architecture and Its Sculpture in Viceregal Mexico.* Austin: Texas Univ. Press, 1997.

Mundy, B. E. *The Mapping of New Spain: Indigenous Cartography and the Maps of the 'Relaciones Geográficas'.* Chicago Univ. Press, 1996.

Navascués, P. *Historia del Arte Hispánico V: Del Neoclasicismo al Modernismo.* Madrid: Alhambra, 1979.

Neumeyer, Alfred. "The Indian Contribution to Architectural Decoration in Spanish Colonial America," *Art Bulletin,* XXX, 1948, pp. 104–24.

Onians, John. *Bearers of Meaning: The Classical Orders in Antiquity, the Middle Ages, and the Renaissance.* Princeton Univ. Press, 1988.

Osten Sacken, Cornelia von der. *El Escorial: Estudio Iconológico.* Madrid: Xarait, 1984.

Palm, E. W. *Los Monumentos Arquitectónicos de La Española.* Santo Domingo: Editora de Santo Domingo, 1984.

Panofsky, Erwin. "Die Entwicklung der Proportionslehre als Abbild der Stilenwicklung," *Monatshefte für Kunstwissenchaft,* XIV, 1921, pp. 188–219.

_____. *The Life and Art of Albrecht Dürer.* Princeton Univ. Press, 1955.

_____. *Meaning in the Visual Arts: Papers in and on Art History.* Garden City: Anchor, 1955.

Pavón Maldonado, Basilio. *El Arte Hispanomusulmán en su Decoración Geométrica (Una Teoría para un Estilo).* Madrid: Instituto Hispano-Arabe de Cultura, 1975.

Payne, Alina A. *The Architectural Treatise in the Italian Renaissance: Architectural Invention, Ornament, and Literary Culture.* Cambridge Univ. Press, 1999.

Pedoe, Daniel. *Geometry and the Visual Arts.* New York: Dover, 1976.

Pérez Embid, F. *El Mudejarismo en la Arquitectura Portuguesa de la época manuelina.* Seville: Marinas, 1944.

Pevsner, N. "The Term 'Architect' in the Middle Ages," *Speculum,* XVII, 1942, pp. 549–62.

[*] *Qu'ran: El Corán,* trans. Julio Cortes. Madrid: Editorial Nacional, 1980.

Ramírez, Juan Antonio (ed.). *Díos, arquitecto. J. F. Villapando y el Templo de Salomón.* Madrid: Siruela, 1991.

_____. *The Beehive Metaphor: From Gaudí to Le Corbusier.* London: Reaktion, 2000.

Rodríguez Camilloni, H. "Architectural Principles in the Age of Humanism Applied: The Church of San Francisco, Lima," *Journal of the Society of Architectural Historians,* XXVIII, 1969, pp. 235–53.

Rosenthal, Earl. "The Image of Roman Architecture in Renaissance Spain," *Gazette des Beaux-Arts,* LII, 1958, pp. 329–46.

[*] Rubiera, M. J. (ed.). *La Arquitectura en la Literatura Árabe.* Madrid: Editorial Nacional, 1981.

Rudofsky, Bernard. *Architecture Without Architects: A Short Introduction to Non-Pedigreed Architecture.* New York: Doubleday, 1964.

[*] Sagredo, Diego. *Medidas del Romano* (1526); facs. ed. Madrid: Albatros, 1976.

Sambricio, C. "Un álbum de dibujos poco conocido: el de Silvestre Pérez en la Biblioteca Nacional," *Archivo Español de Arte,* no. 181, 1973, pp. 45–57.

[*] San Miguel, Fray Andrés de. *Obras de Fray Andrés de San Miguel,* ed. E. Báez Macías. Cd. México: Universidad Autónoma, 1969.

[*] San Nicolás, Fray Laurencio de. *Arte y Uso de Architectura (Madrid, 1639 y 1664),* facs. ed. Madrid: Albatros, 1989.

Scholfield, P. H. *The Theory of Proportion in Architecture.* Cambridge Univ. Press, 1968.

Schuetz, Mardith. "Professional Artisans in the Hispanic Southwest," *The Americas,* XL/1, July 1983, pp. 17–71.

_____. "Proportionate System and Ancient Geometry," *Bulletin of the Southwestern Mission Research Center*, 1980, pp. 1–7.
Schuetz, Mardith K. (ed.). *Architectural Practice in Mexico City: A Manual for Journeyman Architects of the Eighteenth Century*. Tucson: Arizona Univ. Press, 1987.
Schuetz-Miller, Mardith K. "The Empire Builders. A Socio-Economic History of Architects and Building Artisans from the Neolithic to the Renaissance" (unpubl. book ms., 2001).
Sebastián López, Santiago. *Iconografía Medieval*. San Sebastián: Etor, 1988.
_____. *Simbolismo de los Programas Humanísticos de la Universidad de Salamanca*. Salamanca: Universidad, 1973.
_____, and J. de Mesa Figueroa. *Arte Iberoamericano, desde la Colonización a la Independencia*. Madrid: Espasa-Calpe, 1985.
[*] Serlio, Sebastiano. *The First Book of Architecture, made by Sebastian Serly, entreating of Geometry*. London: Robert Peake, 1611 (facs. rpt., New York: Dover, 1982).
[*] _____. *The Five Books of Architecture*. New York: Dover, 1982.
[*] _____. *Tercero y Quarto Libro de Arquitectura de Sebastián Serlio*, fasc. ed. Madrid: Albatros, 1977.
[*] Sigüenza, Fray José de. *Fundación del Monasterio de El Escorial*. Madrid: Aguilar, 1963.
Simson, Otto von. *The Gothic Cathedral*. New York: Torchbooks, 1964.
Sjöström, I. *Quadratura: Studies in Italian Ceiling Painting*. Stockholm: Almqvist & Wiksell, 1978,.
Smith, E. B. *Egyptian Architecture as Cultural Expression*, rpt. New York: American Life Foundation, 1968.
Sprague de Camp, L. *The Ancient Engineers*. New York; Ballantine, 1974.
Strayer, J. R. (ed.). *Dictionary of the Middle Ages*. New York: Scribner's, 1988.
Sunderland, E. R. "Symbolic Numbers and Romanesque Church Plans," *Journal of the Society of Architectural Historians*, 18, 1958, pp. 94–103.
Taylor, R. C. "Francisco Hurtado and His School," *Art Bulletin*, XXXII, 1950, pp. 25–61.
Temple, R. *China: Land of Discovery and Invention*. London: Patrick Stephens, 1986.
Toussaint, Manuel. *Arte Mudéjar en América*. Mexico DF: Porrúa, 1946.
_____. "Fray Andrés de San Miguel, arquitecto de la Nueva España," *Anales del Instituto de Investigaciones Estéticas*, XIII, 1945, pp. 5–14.
[*] Urrea, Miguel de (trans.). *M. Vitrvvio Pollión De Architectura* (Alcalá de Henares: Juan Gracián, 1583); facsimile rpt. Madrid: Albatros Ediciones, 1978 (see also Vitruvius).
[*] Viollet-le-Duc, E-E. *Lectures on Architecture* (rpt.). New York: Dover, 1987.
[*] Vitruvius. *Vitruvius on Architecture*, ed. F. Granger. London: Loeb, 1962 (see also Urrea).
Walls, A. G. *Geometry and Architecture in Islamic Jerusalem: A Study of the Ashrafiyya*. London: Scorpion, 1990.
Webster, Susan V. *Arquitectura y Empresa en el Quito Colonial: José Jaime Oritz, Alarife Mayor*. Quito: Ediciones Abya-Yala, 2002.
Weinmann. E., and P. Lourekas. *Visual Quickstart Guide: Photoshop3 for Windows*. Berkeley: Peachpit, 1996.
Weisman, E. W. *Art and Time in Mexico, From the Conquest to the Revolution*. New York: Icon, 1985.
Wittkower, Rudolf. *Architectural Principles in the Age of Humanism*. New York: Norton, 1971.
_____. *Gothic vs. Classic: Architectural Projects in Seventeenth-Century Italy*. New York: Braziller, 1974.
Wöl°in, Heinrich. *Principles of Art History*. New York: Dover, 1958.
Wycherley, R. W. *How the Greeks Built Cities*. Garden City: Anchor, 1969.
Zenner, M-T. "Imagining a Building: Latin Euclid and Practical Geometry," in J. J. Contreni (ed.), *Word, Image, Number: Communication in the Middle Ages*. Florence: SISMEL, 2002, 219–46.
_____. "Villard de Honnecourt and Euclidan Geometry," *Nexus Network Journal*, IV/4 (Autumn 2002 @ http://www.nexusjournal.com /Zenner.html).

Index

Numbers in **bold** represent illustrations